Microelectronics, Microsystems and Nanotechnology

Microelectronics, Microsystems and Nanotechnology

Papers presented at MMN 2000

Athens, Greece   20–22 November 2000

Editors

Androula G. Nassiopoulou
Xanthi Zianni

Institute of Microelectronics/National Center for
Scientific Research "Demokritos"

World Scientific
New Jersey • London • Singapore • Hong Kong

Published by
World Scientific Publishing Co. Pte. Ltd.
P O Box 128, Farrer Road, Singapore 912805
USA office: Suite 1B, 1060 Main Street, River Edge, NJ 07661
UK office: 57 Shelton Street, Covent Garden, London WC2H 9HE

British Library Cataloguing-in-Publication Data
A catalogue record for this book is available from the British Library.

MICROELECTRONICS, MICROSYSTEMS AND NANOTECHNOLOGY (MMN 2000)
Copyright © 2001 by World Scientific Publishing Co. Pte. Ltd.
All rights reserved. This book, or parts thereof, may not be reproduced in any form or by any means, electronic or mechanical, including photocopying, recording or any information storage and retrieval system now known or to be invented, without written permission from the Publisher.

For photocopying of material in this volume, please pay a copying fee through the Copyright Clearance Center, Inc., 222 Rosewood Drive, Danvers, MA 01923, USA. In this case permission to photocopy is not required from the publisher.

ISBN 981-02-4769-9

Printed in Singapore by Uto-Print

ORGANIZERS

Institute of Microelectronics, NCSR "Demokritos" (Athens, Greece)
Greek Network on Microelectronics, Microsystems and Nanotechnology

*Under the auspices of the General Secretariat of Research and Technology
With the support of IEEE, Greece Section*

SPONSORS

ATMEL
IMEL
INTRASOFT
PANAFON
Innovation Relay Centre HELP–FORWARD
MINATECH
THETA Microelectronics
VLSI Design Laboratory–University of Patras

COMMITTEES

Organizing Committee
Chairperson: A.G. Nassiopoulou, IMEL/NCSR "Demokritos", Greece
Members: G. Constantinidis, FORTH/Crete, Greece
C. Tsamis, IMEL/NCSR "Demokritos", Greece
X. Zianni, IMEL/NCSR "Demokritos", Greece

Programme Committee
A.G. Nassiopoulou, IMEL/NCSR "Demokritos", Greece
A. Arapogianni, University of Athens, Greece
I. Avaritsiotis, Electrical and Electronics Engineers Dept., NTUA, Greece
A. Birbas, University of Patras, Greece
S. Blionas, Intracom, Greece
Ch. Dimitriadis, University of Thessaloniki, Greece
A. Dimoulas, IMS/NCSR "Demokritos", Greece
A. Georgakilas, University of Crete and FORTH, Crete, Greece
C. Goutis, University of Patras, Greece
S. Katsafouros, IMEL/NCSR "Demokritos", Greece
O. Koufopavlou, University of Patras, Greece
K. Misiakos, IMEL/NCSR "Demokritos", Greece

N. Ouzounoglou, National Technical University of Athens, Greece
A. Thanailakis, University of Thrace, Greece
D. Tsoukalas, IMEL/NCSR "Demokritos", Greece
I. Xanthakis, National Technical University of Athens, Greece

Advisory Committee
D. Antoniadis, MIT, Boston, USA
C. Claeys, IMEC, Belgium
M. Hatzakis, IMEL/NCSR "Demokritos", Greece
G. Kamarinos, LPCS/Institute National Polytechnique de Grenoble, France
Ing. Lundstrom, University of Linkoping, Sweden
S. Pantelides, Vanderbilt University, USA
I. Stoemenos, University of Thessaloniki, Greece

FOREWORD

This volume contains the papers presented at the First Conference on "Microelectronics, Microsystems and Nanotechnology", held at NCSR "Demokritos" in Athens-Greece in November 20-22, 2000. The Conference was organized by the Institute of Microelectronics within the framework of the Greek Network on Microelectronics, Microsystems and Nanotechnology with the aim to gather together in an interactive forum all scientists and engineers working in Greece in the above fields, as well as internationally recognized specialists who were invited to present the latest achievements worldwide.

This first Conference of the year 2000 has been attended by 140 scientists from research centers, universities and Greek companies. There have been around 40 oral contributions from which 6 were invited papers, while around 60 papers were presented in the form of posters in the corresponding poster sessions.

The main subjects covered by the Conference were related to Nanotechnology and Quantum Devices, Sensors and Microsystems, C-MOS Devices and Devices based on Compound Semiconductors, Processing, Silicon Integrated Technology and Integrated Circuit Design. An industry exhibition took also place at the Conference area, where the main Greek companies involved in the field presented their products and R&D activities.

Nanoscience and Nanotechnology as well as Microsystems and Sensors are among the most challenging research areas for the next years, involving multidisciplinary research, covering from Physics, Chemistry and Microelectronics to Materials Science, Biology, Medicine and many others. In Nanotechnology the presented papers covered from "Molecular Electronics" and "Strategies of DNA chip technology" to fundamental studies of Semiconductor Nanocrystals, Silicon Nanocrystal Memories, Nanotechnology Processes and Materials for novel Magnetic Memories and other devices. In C-MOS technology, the latest new and future challenges were presented. There were also papers on the state-of-the art in materials and devices based on III-V semiconductors, including novel approaches of wafer bonding of GaAs on Si for optoelectronic interconnect applications. Integrated Circuit Design for high speed-low power applications involved also interesting papers, and it was combined with micromachining techniques and sensors for novel microsystems.

In conclusion, the MMN 2000 covered nicely different interesting fields of Micro/Nanotechnology, Microelectronics and Microsystems and gathered together scientists from research organizations and universities with scientists and engineers from the industry, in a multidisciplinary forum, in an effort to strengthen linkages between research and applications.

The organizers hope that the Conference will be held in Greece on a biannual basis and that it will attract in the future a great number of scientists from the International Scientific Community.

Dr A.G.Nassiopoulou Athens, July 2001
Chairperson of MNN 2000

CONTENTS

Foreword vii

Nanotechnology and Quantum Devices

A New Strategy for *In Situ* Synthesis of Oligonucleotides Arrays for
DNA Chip Technology 3
    F. Vinet, A. Hoang, F. Mittler and C. Rosilio (invited)

Magnetotransport Properties of La-Ca-Mn-O Multilayers 13
    C. Christides

A Novel Method for the Calculation of the Local Electric Field at the
Emitting Surface of a Carbon Single-Wall nanotube 17
    G.C. Kokkorakis, A. Modinos and J.P. Xanthakis

Study of Photoluminescence and Micro-Photoluminesence of
V-Shaped Quantum Wires 21
    M. Tsetseri, G.P. Triberis, V. Voliotis and R. Grousson

Catalytic Action of Ni Atoms in the Formation of Carbon Nanotubes:
A Combined *Ab-Initio* and Molecular Dynamics study 25
    A.N. Andriotis, M. Menon and G. Froudakis

Si Nanocrystal MOS Memory Obtained by Low-Energy Ion
Beam Synthesis 29
    E. Kapetanakis, P. Normand, D. Tsoukalas, K. Beltsios,
    S. Zhang, J. van den Berg and J. Stoemenos

Charge Effects and Related Transport Phenomena in Nanosize
Silicon/Insulator Structures 33
    J.A. Berashevich, A.L. Danilyuk, A.N. Kholod, F. Arnaud,
    D'Avitaya and V.E. Borisenko

Radiative Recombination from Silicon Quantum Dots in Si/SiO_2
Superlattices 37
    P. Photopoulos, T. Quisse, D.N. Kouvatsos and
    A.G. Nassiopoulou

Avalanche Porous Silicon Light Emitting Diodes for Optical
Intra-Chip Interconnects 41
    S.K. Lazarouk, P.V. Jaguiro, A.A. Leshok and V.E. Borisenko

Infrared Absorption in Strained Si/Si$_{1-x}$Ge$_x$/Si Quantum Wells 45
    G. Hionis and G.P. Triberis

Thermopower Calculations at Filling Factor 3/2 and 1/2 for
Two-Dimensional Systems 49
    V.C. Karavolas and G.P. Triberis

Thermoelectric Properties of Composite Fermions 53
    M. Tsaousidou and G.P. Triberis

Design and Fabrication of Supported-Metal Catalysts Through
Nanotechnology 57
    I. Zuburtikudis

Calculated Spontaneous Emission Rates in Silicon Quantum Wires
Grown in {100} Plane 61
    X. Zianni and A.G. Nassiopoulou

Electrical Modeling and Characterization of Si/SiO$_2$ Superlattices 65
    T. Ouisse, A.G. Nassiopoulou and D.N. Kouvatsos

Ge/SiO$_2$ Thin Layers Through Low-Energy Ge$^+$ Implantation and
Annealing: Nanostructure Evolution and Electrical Characteristics 69
    K. Beltsios, P. Normand, E. Kapetanakis, D. Tsoukalas,
    A. Travlos, J. Gautier, F. Jourdan and P. Holliger

Vertical Transport Mechanisms in nc-Si / CaF$_2$ Multi-layer 73
    V. Ioannou-Sougleridis, A.G. Nassiopoulou, F. Bassani and
    F. Arnaud d'Avitaya

Photo- and Electroluminescence from nc - Si / CaF$_2$ Superlattices 77
    V. Ioannou-Sougleridis, T. Ouisse, A.G. Nassiopoulou,
    F. Bassani and F. Arnaud d'Avitaya

Ab Initio Calculation of the Optical Gap in Small Silicon Nanoparticles 81
    C.S. Garoufalis and A.D. Zdetsis

Ground State Electronic Structure of Small Si Quantum Dots 85
    C.S. Garoufalis, A.D. Zdetsis and J.P. Xanthakis

Processing

Technology Roadmap Challenges for Deep Submicron CMOS 91
    C.L. Claeys and H.E. Maes (invited)

Photolithographic Materials for Novel Biocompatible Lift Off Processes 103
    A. Douvas, C.D. Diakoumakos, P. Argitis, K. Misiakos,
    D. Dimotikali, C. Mastihiadis and S. Kakabakos

Polycrystalline Silicon Thin Film Transistors Having Gate Oxides
Deposited Using TEOS 107
    V.Em. Vamvakas, D.N. Kouvatsos and D. Davazoglou

Solid Interface Studies with Applications in Microelectronics 111
    S. Kennou, S. Ladas, A. Siokou, I. Dontas and V. Papaefthimiou

A Comparison Between Point Defect Injecting Processes in Silicon Using
Extended Defects and Dopant Marker Layers as Point Defect Detectors 115
    D. Skarlatos, D. Tsoukalas, C. Tsamis, M. Omri, L.F. Giles,
    A. Claverie and J. Stoemenos

Rapid Thermal Annealing of Arsenic Implanted Silicon for the
Formation of Ultra Shallow n^+p Junctions 119
    N. Georgoulas, D. Girginoudi, A. Mitsinakis, M. Kotsani
    and A. Thanailakis

Simulation of the Formation and Characterization of Roughness in
Photoresists 123
    G.P. Patsis, V. Constantoudis and E. Gogolides

F_2 laser (157 nm) Lithography: Materials and Processes 127
    E. Tegou, E. Gogolides, P. Argitis, C.D. Diakoumakos,
    A. Tserepi, A.C. Cefalas, E. Sarantopoulou, J. Cashmore,
    and P. Grunewald

Fabrication of Fine Copper Lines on Silicon Substrates Patterned with
AZ 5214™ Photoresist via Selective Chemical Vapor Deposition 131
    D. Davazoglou, S. Vidal and A. Gleizes

Investigation of the Nitridation of Al_2O_3 (0001) Substrates by a
Nitrogen Radio Frequency Plasma Source 135
    S. Mikroulis, V. Cimalla, A. Kostopoulos, G. Constandinidis,
    G. Drakakis, M. Zervos, M. Cengher and A. Georgakilas

Simulation of Si and SiO$_2$ Feature Etching in Fluorocarbon Plasmas 139
    G. Kokkoris, E. Gogolides and A.G. Boudouvis

Epitaxial ErSi$_2$ on Strained and Relaxed Si$_{1-x}$Ge$_x$ 143
    *G. Apostolopoulos, N. Boukos, P. Papandreopoulos
and A. Travlos*

Development of a New Low Energy Electron Beam Lithography
Simulation Tool 147
    D. Velessiotis, X. Zianni, N. Glezos and K.N. Trohidou

CMOS Devices and Devices Based on Compound Semiconductors

Advanced SOI Device Architectures for CMOS ULSI 153
    F. Balestra (invited)

Recent Developments and Reliability of Polycrystalline Silicon Thin
Film Transistors 167
    *C.A. Dimitriadis, J. Stoemenos, F.V. Farmakis, J. Brini
and G. Kamarinos*

Different Types of Single Crystalline Gallium Nitride Thin Films Grown
Directly on Vicinal (100) Gallium Arsenide Substrates 171
    *A. Georgakilas, K. Amimer, M. Androulidaki, K. Tsagaraki,
B. Pecz, L. Toth and M. Calamiotou*

Epitaxial Y$_2$O$_3$ on Si (001) by MBE for High-k Gate Dielectric
Applications 175
    G. Vellianitis, A. Dimoulas and A. Travlos

Performance of GaAs/AlGaAs Laser Diodes Fabricated by Epitaxial
Material with Significantly Different Numbers of Quantum Wells 179
    *D. Cengher, G. Deligeorgis, E. Aperathitis, M. Sfendourakis,
G. Halkias, Z. Hatzopoulos and A. Georgakilas*

Microhardness Characterization of Epitaxially Grown GaN Films.
Effect of Light Ion Implantation 183
    *P. Kavouras, M. Katsikini, Ph. Komninou, E.C. Paloura,
J.G. Antonopoulos and Th. Karakostas*

Multiple Quantum Well Solar Cells Under AM1 and Concentrated
Sunlight 187
        E. Aperathitis, Z. Hatzopoulos, M. Kayambaki, V. Foukaraki,
        M. Ružinský, V. Šály, P. Sirotný and P. Panayotatos

Processing with In Situ Diagnostic Techniques for the Integration of
GaAs-based Opto-Electronic Devices Bonded on Si CMOS Wafers 193
        G. Deligeorgis, E. Aperathitis, D. Cengher, Z. Hatzopoulos
        and A. Georgakilas

Investigation of Different Si (111) Substrate Preparation Methods
for the Growth of GaN by RF Plasma-Assisted Molecular
Beam Epitaxy 197
        M. Androulidaki, K. Amimer, K. Tsagaraki, M. Zervos,
        G. Constantinidis, Z. Hatzopoulos, A. Georgakilas, F. Peiro
        and A. Cornet

Material Properties of GaN Films with Ga- or N-Face Polarity
Grown by MBE on Al_2O_3 (0001) Substrates Under Different
Growth Conditions.. 201
        A. Kostopoulos, S. Mikroulis, E. Dimakis, E.-M. Pavelescu,
        M. Androulidaki, K. Tsagaraki, G. Constantinidis, A. Georgakilas,
        Ph. Komninou, Th. Kehagias and Th. Karakostas

The Influence of Silicon Interstitial Clustering on the Reverse Short
Channel Effect... 205
        C. Tsamis and D. Tsoukalas

Noise Modeling of Interdigitated Gate CMOS Devices................................... 209
        E.F. Tsakas and A.N. Birbas

Sub-Threshold Characteristics of 0.15 μm SOI-MOSFETs After
Hot-Carrier Stress 213
        P. Dimitrakis, G.J. Papaioannou, J. Jomaah and F. Balestra

A Low Voltage Bias Technique to Increase Sensitivity of MOSFETs
Dosimeters 217
        G. Fikos, S. Siskos, A. Chatzigiannaki and G. Sarrabayrouse

High Precision CMOS Euclidean Distance Computing Circuit....................... 221
        G. Fikos and S. Siskos

Microsystems

Pressure Sensors Based on 3C-SiC on Si-on-Insulator for High
Temperature Applications 227
    S. Zappe, M. Eickhoff and J. Stoemenos

Micromachined L.T.GaAs/AlGaAs Membranes as Support for
38 GHz and 77 GHz Filters 234
    G. Deligeorgis, M. Lagadas, G. Konstantinidis, N. Kornilios,
    A. Müller, S. Iordanescu, I. Petrini, D. Vasilache and P. Blondy

Alternative Signal Extraction Technique for Miniature Fluxgates 238
    P.D. Dimitropoulos and J.N. Avaritsiotis

Integrated Gas Flow Sensor Fabricated by Porous Silicon Technology 242
    G. Kaltsas and A.G. Nassiopoulou

Linear Arrays of Poly Si-Ge Uncooled Microbolometers with CMOS
Readout as Long Wavelength Infrared Sensors 246
    S. Kavadias, P. De Moor, M. Gastal and C. van Hoof

Silicon Capacitive Pressure Sensors and Pressure Switches Fabricated
Using Silicon Fusion Bonding 250
    S. Koliopoulou, D. Goustouridis, S. Chatzandroulis and
    D. Tsoukalas

Low Power Silicon Microheaters on a Thin Dielectric Membrane with
Thick Film Sensing Layer for Gas Sensor Applications 254
    V. Guarnieri, S. Brida, B. Margesin, F. Giacomozzi, M. Zen,
    A.A. Vasiliev, A.V. Pisliakov, G. Soncini, G. Pignatel,
    D. Vincenzi, M.A. Butturi, M. Stefancich, M.C. Carotta and
    G. Martinelli

Microsystems for Acoustical Signal Detection Applications 259
    D.K. Fragoulis and J.N. Avaritsiotis

Capillary Format Bioanalytical Microsystems 263
    K. Misiakos, C. Mastichiadis and S.E. Kakabakos

Ultrathin Nanoporous and Microporous SiO_2 Coatings for Gas/Vapor
Separation and Sensor Applications 267
    K. Beltsios, N. Kanellopoulos, E. Soterakou and G. Tsangaris

Structure Control of Thin Microporous Carbon Coatings for Gas/Vapor
Separation and Sensor Applications 271
    K. Beltsios, G. Pilatos, F. Katsaros, N. Kanellopoulos and
    A. Andreopoulos

2-D Simulation of On-Chip BAW Resonators 275
    E.D. Tsamis and J.N. Avaritsiotis

Theoretical Calculation of the X-Ray Photon Detector Response
Fabricated on Si Gallium Arsenide 279
    V. Theonas, P. Dimitrakis and G. Papaioannou

Effectiveness of Local Thermal Isolation by Porous Silicon in a Silicon
Thermal Sensor 283
    D. Pagonis, C. Tsamis, G. Kaltsas and A.G. Nassiopoulou

Thermal Conductivity of Porous Silicon Layers Probed by Micro-
Raman Spectroscopy 287
    D. Papadimitriou, P. Tassis, L. Tsoura, C. Tsamis and
    A.G. Nassiopoulou

Silicon Integrated Technology and Integrated Circuit Design

A MCM-L Board for the Baseband Processor of a Dual Mode
Wireless Terminal 293
    C. Drosos, C. Dre, K. Potamianos and S. Blionas

A CAD Tool for Automatic Generation of RNS & QRNS Converters 297
    M.M. Dasigenis, D.J. Soudris, S.K. Vasilopoulou and
    A.T. Thanailakis

MOSFET Model Benchmarking Using a Novel CAD Tool 301
    N.A. Nastos and Y. Papananos

An HBT-BiCMOS Laser Driver with Independently Adjustable DC
and Modulation Currents for High Speed Optical Interconnections 305
    P. Robogiannakis, S.G. Katsafouros, E. Kyriakis-Bitzaros,
    N. Haralabidis and G. Halkias

On the Design of a Low Power Modulator/Demodulator for
DECT/GSM 309
    C. Drosos, C. Dre, S. Blionas, D. Soudris and G. Kalivas

Bluetooth Encryption / Decryption Algorithm Architecture and
Implementation 313
    P. Kitsos and O. Koufopavlou

Passive Element Design Issues for Fully Integrated RF VCOs 317
    A. Kyranas and Y. Papananos

Power Amplifier Linearisation Techniques: An Overview 321
    N. Naskas and Y. Papananos

Low-Power Implementation of an Encryption /Decryption System
with Asychronous Techniques 325
    N. Sklavos and O. Koufopavlou

A 0.25 CMOS Fast Current Amplifier with Leakage Current
Compensation for Solid State Detector Applications 329
    E. Zervakis and N. Haralabidis

Design and Simulation of On-Chip Bandpass Filters 333
    A.T. Kollias, E.D. Tsamis and J.N. Avaritsiotis

The Design of a Ripple Carry Adiabatic Adder 337
    V. Pavlidis, D. Soudris and A. Thanailakis

Maximum Power Estimation in CMOS VLSI Circuits 341
    N.E. Evmorfopoulos, J.N. Avaritsiotis and G.I. Stamoulis

Designing a Microwave VCO 945MHz with Computer Aided Design
Simulation 345
    S. Panagiotopoulos and A. Kouphoyannidis

Power Dissipation Considerations in Low-Voltage CMOS Circuits ... 349
    A.A. Hatzopoulos

Microelectronics Networks / Technology Transfer and Exploitation

EURACCESS: A European Platform for Access to CMOS Processing 355
    C.L. Claeys

MMN: Greek Network on Microelectronics, Microsystems and
Nanotechnology 359
    A.G. Nassiopoulou

Technology Transfer in Action — HELP-FORWARD's Support for
Developing and Exploiting Technologies 360
    V. Tsakalos

Supporting European Research Alliances in Microelectronics — The
MINATECH project 375
    D. Tsamtsakis

Simulations of Molecular Electronics 380
    S.T. Pantelides, M. Di Ventra and N.D. Lang

Author Index 387

Nanotechnology and Quantum Devices

A NEW STRATEGY FOR IN SITU SYNTHESIS OF OLIGONUCLEOTIDES ARRAYS FOR DNA CHIP TECHNOLOGY

F. VINET, A. HOANG AND F. MITTLER
CEA/LETI- 17 Rue des Martyrs
38054- Grenoble CEDEX 9
France

C. ROSILIO (invited)
CEA /LETI- Saclay
91191- Gif – sur –Yvette
France

We show in this paper the principle and the application of a new approach to the fabrication of oligonucleotide probe arrays on a silicon substrate structured with microwells . This new strategy uses a principle of selective protection by a polymer for in situ oligonuceotides synthesis. The polymer is deposited on chosen microwells before one of the four steps of the DNA synthesis based on phosphoramidite chemistry.

The process combines an automatic synthesizer with a micro drop dispenser for spatially localized deposition of the protective polymer. A great choice of polymer/solvent couples allows the integration of the process in the synthesis cycle of an automated synthesizer.

The preliminary results for the growing of oligonucleotides (20 nucleotides in length) in a multi step fashion are described and discussed. Coupling efficiencies, synthesis yield and obtained sequence purity, show the possibility of protective polymer strategy for the in situ synthesis DNA chip fabrication.

1 Introduction

New and powerful molecular biology tools for biological analysis have been well developed for investigations of diagnostic and research on gene expression, mutations and polymorphism [1]. These analytical methods are based on detection the of nucleic acids hybridization between the oligonucleotide probes of known sequences immobilized on the chip and fluorescent labelled oligonucleotide targets extracted from biological sample. The fabrication of these DNA chips relies on techniques of oligonucleotide immobilization onto a solid substrate.

Different technical methods for the realization of dense array of oligonucleotides on a chip for DNA biosensor have been investigated by the main industrial actors in the field of biological analysis [2]. Three principal approaches have been proposed [3]:

    3. pre- synthesized oligonucleotide immobilization, by covalent attachment on a derivatized substrate [4] (silanization and linker or bi-functional crosslinking reagents), or on functionalised polymers

[5] (polyacrylamide – hydrazine gel) or by grafting on a electropolymerised conductive polymer [6]
4. microrobotic deposition, or ink jet printing of c-DNA or PCR products on an active solid support (polylysine on glass) [7]
5. in situ synthesis of oligonucleotide probes.

The last method follows the work of Southern [8] considered as pioneering in this field, which is based on solid phase synthesis of oligonucleotides on a substrate. The delimitation of the surface area after each nucleotide unit addition is obtained by using a mobile mechanical mask.

A technique of in situ synthesis developed by Affymetrix utilizes light-directed combinatorial synthesis [9]. Another method, proposed by Protogene, is based on a mechanically addressable spotting of the four activated nucleotides on a substrate at discrete regions of the surface [10,11]. In situ synthesis can be achieved on a polypyrole-conducting polymer on top of gold electrodes, by local electro deprotection of a 5' electro labile protecting group [12].

The principle of the process of in situ synthesis presented in this paper is based on the strategy of selective protective polymer deposition by a micro robotic dispenser .The polymer film formed after drying constitutes a protective barrier during the synthesis cycle. After each cycle of nucleotide unit coupling, polymer will be stripped in a rinse step with a suitable solvent. In the next cycle, a subsequent deposition of polymer droplets on other selected sites is performed, and so on up to the required sequences.

An automatic synthesizer and a polymer micro-dispensing robotic device have been combined to achieve synthesis directed by a polymer protection. Various polymers were proposed and adapted to conventional protocols of oligonucleotide synthesis, for fabricating DNA microarray of sequence with desired compositions [13].

Hybridization experiments of the synthesized probes with complementary labelled targets were investigated for evaluating the feasibility and the potentiality of such a process.

2 Selective polymer masking for oligonucleotides arrays synthesis

All the in situ synthesis methods outlined previously, involve the conventional phosphoramidite chemistry [14] for successive coupling of nucleotide units protected in 5' position by a well-suited function (photolabile, electrolabile or acido labile). The synthesis cycle consists of four successive steps i.e. deprotection (or detrytilation), coupling, capping and oxidation, which allows the growing of the oligonucleotide from an activated substrate surface.

2.1 Process description

The aim of the proposed process is the use of a selective polymer deposition on a silicon substrate, structured by "functionalised" microwells in combination with the phosphoramidite chemistry. The polymer will protect the selected microwells during the step of coupling with the next nucleotide unit as described in Fig.1.

A mechanically addressable spotting realizes like as a cover of solid polymer and must provide a tight and close fitting cap over wells. In such a way, we can determine "activated" areas towards chemical reagents. The polymer will be eliminated by dissolution in a solvent after each cycle, or only just after the detritylation step in the synthesis cycle. The polymer removal is included in the classical cycle by one of the rinse steps or by an additional dissolution step in a compatible solvent with the synthesis protocol such as dichloromethane, acetonitrile or tetrahydrofurane

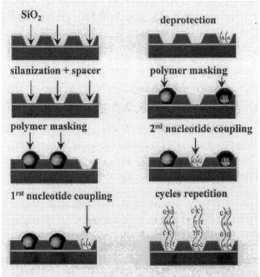

Fig. 1. Process steps description of in situ synthesis of oligonucleotides using polymer masking

The process offers the possibility of a great deal of protective polymers with suitable properties of solubility, tightness and compatibility with either one or all the steps of the synthesis cycle. A microdrop dispenser associated with an automatic synthesizer can realize the different process steps.

2.1.1 Polymer properties

The polymer shielding will operate as a barrier layer in selected regions during the synthesis cycle. The well-known phosphoramidite synthesis with DMT based chemistry can be performed in an automatic synthesizer with three conventional solvents: dichloromethane (CH_2Cl_2), acetonitrile (CH_3CN) and tetrahydrofuran (THF).

We have reported in the Table I the solubility parameters of different polymers. Depending on their properties, three main schemes will be considered:

3. **The polymer is insoluble in all the solvents of the synthesis cycle;** it will be eliminated after each cycle by an additional removing step with a specific solvent, outside the synthesizer chamber.

Mainly water-soluble polymers like PolyVinyl Alcohol (PVA), PolyEthyleneImine (PEI) or Poly AllylAmine (PAA) fulfill this condition.

The polymer is delivered in solution in water, and stripped off in a water wash outside the synthesizer chamber; this step will be damaging for phosphoramidite-based synthesis, which requires an anhydrous medium. In order to prevent any additional step like baking, the PVA polymer can be advantageously delivered and removed in anhydrous DMSO.

4. **The polymer is insoluble with one of the solvent used** during either detritylation or coupling steps, and soluble in the others. In this case, the polymer cap will be eliminated by a rinse step directly in the synthesizer reaction chamber.

The polymer/solvent couple will be optimised and fitted with the selected step of the cycle.

For example if we select the detritylation step (accomplished in 2% TriChloroAcetic Acid in Dichloromethane) the Poly HydroxyStyrene (PHS) or Resist XP 8843 (Shipley) are well appropriate as a protective material masking. Polymer cap is then removed into a rinse step in acetonitrile before the next coupling step, without any modification of the standard protocol.

In the same way, if we choose the coupling step as the protective barrier step, the Polystyrène (PS) insoluble in Acetonitrile is convenient; after the coupling step a standard rinse step in dichloromethane is performed in order to eliminate the polymer.

5. **The polymer is soluble whatever the three conventional solvents used**, so the process requires a particular solvent to carry out detritylation or coupling steps. We have to choose among the well-known solvents investigated for equivalent or improved coupling or detritylation efficiency. The polymer will be stripped in a wash step with an anhydrous solvent miscible with the solvents used in the protocol.

Successful methodologies have been developed for the preparation of oligonucleotides with phosphoramidite-based chemistry [15,16] with increased efficiency and purity. Comparable or improved results can be obtained at the deprotection or coupling step with toluene, xylene, glutaronitrile, dinitrile and halogenated solvents as a substitute of conventional dichloromethane and acetonitrile.

As an example PolyEhylene Oxyde (PEO) is well suited for a detritylation step in anhydrous toluene or xylene, and stripped off in acetonitrile.

POLYMER	CH_2Cl_2	CH_3CN	Protection step
PolyVinylAlcohol (PVA)	insoluble	insoluble	All
PolyStyrene (PS)	Soluble	Insoluble	Detritylation Coupling
PolyvinylCarbazole (PVK)	Soluble	Insoluble	Detritylation Coupling
PolyImide (XU 5218)	Soluble	Insoluble	Detritylation Coupling
PolyHdroxyStyrene (PHS)	Insoluble	Soluble	Detritylation
Photoresist (XP8843)	Insoluble	Soluble	Detritylation
PolyEthyleneOxide (PEO)	Soluble	Soluble	None

Table I. Solubility properties of masking polymers

An additional characteristic of the polymer is the lack of chemical reactivity towards the coupling step. If the polymer reacts, needless reagent will be consumed and the polymer solubility can be changed. However, this point was revealed of no great importance.

After screening experiments, two polymers were selected to demonstrate the potentiality of the process: PolyVinylAlcohol (PVA) and PolyHydroStyrene (PHS).
The thickness variation of the PVA polymer has been measured as a function of oligonucleotide synthesis chemicals, the results are shown in Fig.1. A solution of 10% in water of PVA with a molecular weight of 50 000-85 000 is spin coated on a silicon substrate. After baking at 80°C during one minute, the measured thickness is 750 nm. No change in thickness is observed in both CH_2Cl_2 and CH3CN solvents.

An increase of the thickness is representative of a swelling of the polymer. This phenomenon is particularly emphasized during the coupling step in presence of tetrazole with CH_3CN: a 100% thickness increase is observed.

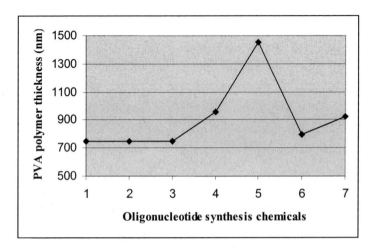

Fig.2 - PVA polymer thickness as a function of oligonucleotides synthesis chemicals With: 1) blank, 2) CH2Cl2, 3) CH3CN, 4) DCA/ CH_2Cl_2 3%, 5) tétrazole/ CH_3CN 0.45M, 6) DMAP/THF/Lutidine (6:90:6)/0.1ml ac.acétique anhydre, 7) THF/Lutidine/H2O

As depicted in Table I the PHS polymer is known to be soluble in acetonitrile. So, in this case we have checked the polymer film behaviour as a function of the detritylation time (30, 90, 180 and 360 seconds) with different polymer concentrations in DMSO, 5, 10, 20, 50% in weight and baking time ,30 seconds, 1 and 5minutes. Whatever the conditions are, a slight film consumption is observed. A thickness above 0.5µm is necessary to ensure a consistent polymer film after the detrytilation step.

3 In Situ synthesis of oligonucleotide probes

After selecting appropriate masking polymer, we performed in situ synthesis of oligonucleotides on structured silicon wafers.

3.1 Substrate cleaning, silanization and derivatizing linker

Silicon substrates were engraved using microelectronics techniques to perform microwells of 650x650x20µm³ size. The resulting surface was thermally oxidized to produce a 0.5 µm thick SiO_2.

The structured substrates were cleaned by immersion in an alkaline solution and rinsed in D.I.water. Substrate silanization is performed using a solution of glycidyloxypropyl trimethoxysilane (1ml) in toluene (3.5 ml) and triethylamine (0. 3 ml) for one night at 80°C, rinsed in acetone, dried and baked for 3 hours at 110°C.

Functionalised substrates were derivatised by hexaethylene glycol (HEG) linker in acid catalyse. The complete preparation of the support before synthesis is depicted in Fig.3.

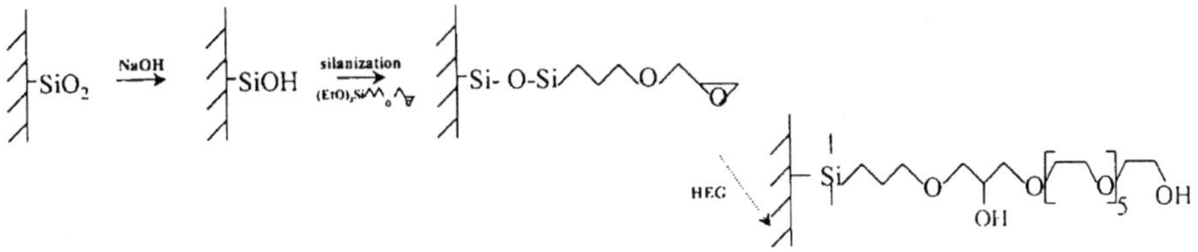

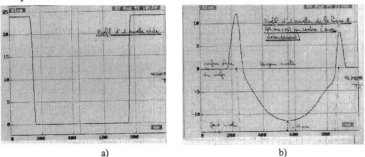

Fig.3- Chemical functionnalization of substrates for oligonucleotide in situ synthesis

A 1x1 cm^2 treated substrate can be introduced in the reaction vessel of an automatic oligonucleotide synthesizer to perform a computed sequence of DNA oligomers using the conventional phosphoramidite chemistry.

3.2 – Oligonucleotide synthesis with polymer masking

The polymer masking is realized manually by micropipeting a polymer solution in 650μm cavities. A post - bake on hot plate at 80°C for 90 seconds is made for both drying and enhancing the adhesion of the polymer. The polymer profile is shown in Fig.4. on a crossection of a microwell. Polymer removal is achieved by a solvent rinse.

Fig.4– Profile of 650μm a) empty microwells, b) after polymer masking

For example PVA will be stripped in water or DMSO but in order to avoid polymer skin deposit during development, we used a soaking in hot water (at 80°C for 1 minute). PHS is perfectly stripped by a rinse in anhydrous THF.

3.2.1 PVA masking

A process with one PVA masking step is utilized for the synthesis of two distinct 15 mers oligonucleotides with a single base change.
The same sequence 3' ATCTCAC 5' was grown up on the structured substrate

into the synthesizer. Sample is then taken out the reaction chamber and subjected to a selective polymer masking. Half of the microwells were first protected by PVA. The unmasked area of the substrate was detritylated and coupled with a C nucleotide unit. Then PVA polymer was stripped by soaking the substrate in hot water. Then, PVA polymer deposit is applied on the other half of the microwells. Coupling with an A nucleotide unit was achieved after a detritylation step. The polymer cover was removed and the end sequence CAAATAG was accomplished in the synthesizer chamber on all the microwells.

At the end of this step the resultant sample had microwell areas with two distinct sequences : 3' ATC TCA CTC AAA TAG 5' (n°1), 3' ATC TCA CCC AAA TAG 5'(n°2). The oligomers were deprotected outside the synthesizer chamber by the standard treatment with ammonia at 60°C for 45 minutes. The probes were hybridized on the substrate with the complementary biotinylated targets of probes n°2. The fluorescent detection was performed on a confocal scanner GS300, after coupling with streptavidin CY3 compound. The resulting micrographs are reported in Fig.5. A contrast value of 5 is obtained between hybridized and non hybridised probes.

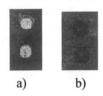

a)   b)

Fig. 5- Fluorescence signals of in situ synthesized 15mers differentiated by a single base aftet hybridisation with a) complementary target and b) non complementary target

These results show the capability of the polymer shielding to introduce a single base change during the synthesis of different probes.

3.2.2 – PHS masking

After the growing on the total a five mers with T bases only, we operate a polymer protection with PHS solution in all the microwell areas. The unmasked region was detritylated and a standard capping step was achieved. The polymer film was removed by a rinse in THF.

At first, we achieve the same sequence 3' 5T ATC TCAC ; after the seventh nucleotide base, the sites are discriminated by selective polymer masking in order to couple selectively one of the four nucleotide base A, T, G, or C (four mismatches). A 11mers sequence and a 15mers sequence were additionally obtained by respectively masking each step and no masking at all. The six different probes are grown simultaneously with the same last end of the sequence CC AAA TAG 5'. After hybridization with complementary targets of one of these probes discrimination between the four mismatches can be observed in Fig. 6. These result shows the process capability for mismatch discrimination of a 20 mers oligonucleotide in length.

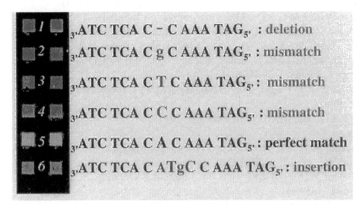

Fig.6 - Fluorescence signals of hybridised in situ synthesized oligomers with the complementary target of probe 5) for six different probes 1)14mers with a deletion in center position, 2) 15 mers with g mismatch, 3) 15 mers with T mismatch, 4) 15 mers with C mismatch, 5) 15 mers with C in center position, 6)18mers

4 Conclusion

This paper describes our attempts to develop a strategy of selective polymer masking for in situ synthesis of oligonucleotide arrays. A collection of polymers combined with a choice of suitable and compatible solvents with phosphoramidite based chemistry are described for this process. This approach provides an alternative method for the realization of high density oligonucleotides probes by combinatorial synthesis. Compared to other processes investigated for in situ synthesis oligonucleotide , involving a photo or electro addressable deprotection step, this original method offers advantage of simplicity and conditions close to those of an automatic synthesizer : efficiency of detritylation in liquid phase, a good synthesis yield and purity of oligonucleotide sequences.

References

1. " Nature genetics" supplement, vol 21 , jan.1999 (a review)
2. " DNA microarray " M. Schena P.A.S. (Oxford) 1999
3. L.Henke, U.J. Kull, Can. J. Anal. Sc. Spect. Vol 44, 2 , 61 (1999)
4. a) U. Maskos, E.M. Southern, Nucl. Acids. Res. 20 1679 (1992)
   b) M.Beeir, J.D. Hoheisel, Nucl. Acids. Res. 27,9, 1970, (1999)
5. D. Proudnikov, E. Timofeev, A. Mirzabekov, Anal. Biochem. 259, 34, (1998)

6. a) T. Livache, A.Roget, E. Dejean, C. Barthet, G. Bidan, R. Teoule, Nucl. Acids. Res. Vol. 22, 15, 2915 (1994) b) Anal. Chem. 253, 188, (1998)
7. D. Shalon, S.J . Smith, P. O. Brown Genome Resarch, 6, 639, (1996)
8. E. Southern, Patent WO 90 / 03382 (Isis Innovation Limited)
9. a) A.C. Pease, D.Solas, E. Sullivan, M. Cronin, C. Holmes, S. P.A. Fodor, Proc. Natl. Acad. Sc. USA, 91, 5022, (1994) b) S. Fodor, J.L. Read, M.C. Pirrung, L. Stryer, A. Tsilu, D. Solas, Science, Vol. 251, 770, (1991)
10. Th. Brennan, US Patent, 5,474,796 (1995)
11. A.P. Blanchard, R.J. Kaiser, L.E. Hood, Biosensors &Bioelectronics Vol. 11, 6-7, 687; (1996)
12. A. Roget, T. Livache, Mikrochim. Acta 131, 3, (1999)
13. patent cea 19 99
14. S.L. Beaucage, R. P. Iyer Tetrahedron, Vol. 49, 28, 6123, (1993)
15. Tetrahedron Vol. 48, 12, 2223 (1992
16. Krotz , Patent WO 99/ 43694 (Isis Ph.) Pattent WO 99/25724 (Protogene) ; WO 99/43694 ISSIS)
17. Th. Brennan, A. Halluin Patent WO 99/ 25724 (Protogene)

MAGNETOTRANSPORT PROPERTIES OF LA-CA-MN-O MULTILAYERS.

C. CHRISTIDES

Department of Engineering Sciences, School of Engineering, University of Patras, 26110 Patras, Greece.

Compositionally modulated structures consisting of FM $La_{1-x}Ca_xMnO_3$ layers (x=0.33, 0.4, 0.48) and $La_{1-y}Ca_yMnO_3$ antiferromagnetic (AF) layers (y=0.52, 0.67, 0.75) were grown on (001)$LaAlO_3$ by pulsed laser deposition. The effect of interfacial Ca composition on the magneto-transport properties of FM/AF multilayers is examined between 4.2 and 300 K.

Spin-engineered structures with large magnetoresistance at room temperature open up new possibilities for applications in the emerging field of magneto-electronic devices. Such devices are magnetic tunnel junctions (MTJ) that consist of a ferromagnetic (FM) top and a FM bottom electrode separated by a thin oxide (insulator) layer that defines [1] two metal-oxide interfaces (FM-I-FM). The conduction is due to quantum tunneling through the insulator. When the electrodes are FM, the tunneling of electrons across the insulating barrier is spin-polarised, and this polarization reflects that of the density of states (DOS) at the Fermi level (E_F) of the electrodes. Spin-polarized tunneling gives rise to tunnelling magnetoresistance (TMR) because the resistance of the junction depends on whether the electrodes have parallel or antiparallel magnetization: TMR=$(\Delta R)/R=(R_{AP}-R_P)/R_{AP}=(2P_1P_2)/(1+P_1P_2)$, where R_{AP} and R_P are the resistances with magnetizations of the electrodes antiparallel and parallel. The spin polarization P of tunneling electrons from a given FM electrode reflects a characteristic intrinsic spin polarisation of the DOS in the FM: $P = [N_\uparrow(E_F)-N_\downarrow(E_F)]/[N_\uparrow(E_F)+N_\downarrow(E_F)]$.

The recent interest in magnetoresistance in doped perovskite manganites was initiated by the discovery of colossal magnetoresistance (CMR) in epitaxial $La_{0.67}Ca_{0.33}MnO_3$ thin films [2]. Of great technological importance are the half-metallic $La_{2/3}A_{1/3}MnO_3$ (A=Ca, Sr) FM films where all the mobile carriers have identical spin states and the conduction band is 100% spin-polarized [3]. The full (100%) polarization of the conduction band in these FM manganites means that TMR can reach its optimum value. To test this approach several groups have prepared MTJ structures consisting of an insulating $SrTiO_3$ layer sandwiched between two layers of FM $La_{2/3}Sr_{1/3}MnO_3$. A record value of 450% TMR was observed in fields below 100 Oe at 4.2 K, corresponding to an electrode polarization of about 85% at low temperatures. However, the TMR falls with increasing temperature and becomes vanishingly small at room temperature. Since the magnetic polarization in the surface of a $La_{2/3}Sr_{1/3}MnO_3$ thin film decreases more rapidly with temperature than the bulk polarization [3], the rapid loss of TMR could

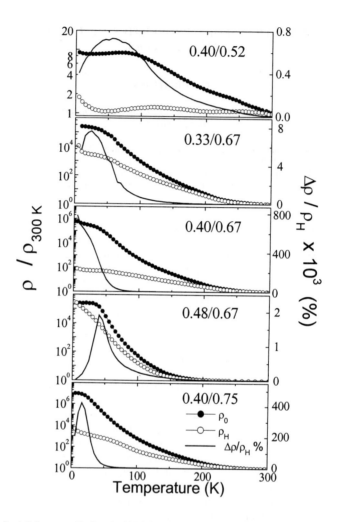

Figure 1. Resistivity, normalized to the 300 K value, as a function of temperature, measured in 50 kOe (ρ_H) and in zero applied field (ρ_0) for a series of LaAlO$_3$/AF(40 nm)/[FM(Λ/2)/AF(Λ/2)]$_{15}$ multilayers. The CMR ratio $\Delta\rho/\rho_H=[\rho_0-\rho_H]/\rho_H$ is plotted as a solid line.

be caused by charge carriers losing their spin-polarization as they cross the FM-I-FM interfaces.

The intent of this study is to investigate the effect of interfacial composition on the magneto-transport properties of FM/I manganite multilayers between 4.2 and 300 K. Two series of [(FM)La$_{1-x}$Ca$_x$MnO$_3$(Λ/2)/(AF)La$_{1-y}$Ca$_y$MnO$_3$(Λ/2)]$_{15}$ multilayers (Λ=8.2 nm is the bilayer thickness) were used to investigate the effect of Mn^{3+}:Mn^{4+} interface ratio (2-x-y)/(x+y) in the temperature dependence of the magnetoresistance (fig.1). One series is grown [4] with constant y=0.67 while x=0.33, 0.4, 0.48, and the other with constant x=0.4 while y=0.52, 0.67, 0.75. For brevity, we named the samples by the Ca^{2+} concentration ratio x/y used (fig.1).

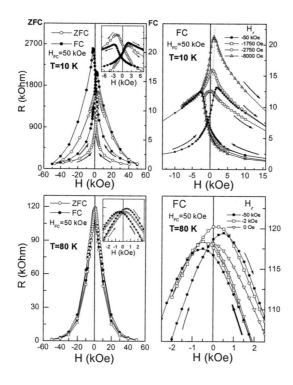

Figure 2. The magnetoresistance of the 0.40/0.67 sample at 10 K and 80 K. *Left side*: R-H loops measured at 10 K (above) and 80 K (below) after FC in 50 kOe (solid circles) and ZFC (open circles) from 300 K. The inset shows in detail the observed peaks. Arrows indicate the direction of field change. *Right side*: R-H loops measured at 10 K (above) and 80 K (below) after FC in 50 kOe from 300 K. Incomplete loops (open symbols) measured by varying the field from 50 kOe to H$_r$ and reverse to 50 kOe.

Figure 1 shows the variation of the normalized resistivity as a function of temperature, measured in 50 kOe (FC, ρ_H) and in zero applied field (ZFC, ρ_0). The resistivity increases drastically as we cool down from 300 K, spanning almost four orders of magnitude. Also, the CMR ratio becomes maximum in the temperature range below a blocking temperature [5,6] T_B ($\approx$70 K). To answer why T_B remains more or less the same it is reasonable to consider that interfacial spin ordering is confined within a few atomic planes near the AF/FM interfaces, defining an *active film volume* V_{int}. Since T_B results from a thermally activated process, its value depends on the active volume at the interfaces ($T_B \propto V_{int}$) which emerges to be similar in the examined multilayers [6,7]. The observed thermal decay of resistivity at low-T can be due to the tunneling of holes from the FM (hole-type carriers) layers [5] to AF (electron-type carriers) layers when the hopping rate between the FM and AF layer is varied [4] with temperature.

Figure 2 shows magnetoresistance loops for the 0.40/0.67 sample at 10 and 80 K. The left-plot shows saturation zero-field-cooling (ZFC) and field-cooling (FC) loops whereas the right-plot shows irreversible FC loops where the reversal (H_r) field is much lower than the saturation field of -50 kOe. Both saturation loops exhibit a large assymetry between the two branches whereas the R-maximum is shifted from the negative (ZFC) to positive field range (FC). The irreversible FC loops exhibit an R-maximum in the positive field branch. This R-maximum scales down with the magnitude of H_r, depending on whether the H_r is in the descending or ascending part of the negative field branch. Remarkably, the loop-assymetry and the irreversible behavior disappear above the T_B of about 70 K. Qualitatively similar results were observed in all the examined samples. These magnetic history depended effects indicate that the interfacial resistance becomes dominant at low-T due to [4] exchange-coupling of the AF/FM interfaces. The large magnetic irreversibilities (Fig. 2) show that [4] the origin of exchange-coupling below 80 K is interfacial magnetic disorder (like partial domain walls).

References

1. J. S. Moodera *et al*, Phys. Rev. Lett. **74**, 3273 (1995).
2. S. Jin *et al*, Science **264**, 413 (1994).
3. J. H. Park *et al*, Phys. Rev. Lett. **81**, 1953 (1998).
4. N. Moutis, C. Christides, I. Panagiotopoulos, and D. Niarchos, Phys. Rev. B **64**, (2001) *in press*.
5. I. Gordon *et al*, Phys. Rev. B **62**, 11633 (2000).
6. I. Panagiotopoulos, C. Christides, M. Pissas, and D. Niarchos, Phys. Rev. B **60**, 485 (1999).
7. I. Panagiotopoulos, C. Christides, D. Niarchos, and M. Pissas, J. Appl. Phys. **87**, 3926 (2000).

A NOVEL METHOD FOR THE CALCULATION OF THE LOCAL ELECTRIC FIELD AT THE EMITTING SURFACE OF A CARBON SINGLE-WALL NANOTUBE

G.C. KOKKORAKIS
Electrical Engineering Dept. National Technical University of Athens, Athens 15773 Greece

A. MODINOS
Physics Dept. National Technical University of Athens, Athens 15773 Greece

J.P. XANTHAKIS
Electrical Engineering Dept. National Technical University of Athens, Athens 15773 Greece
E-mail: jxanthak@central.ntua.gr

We present a novel method for the calculation of the local electric field at the emitting surface of a carbon nanotube. The nanotube is simulated by a cylindrical array of touching spheres, each sphere representing an atom of the tube. The electrostatic potential is then a linear combination of the potentials produced by each sphere. We calculate the local electric fields and the corresponding enhancement factors of both open and closed nanotubes. From the comparison we give a possible explanation as to why in some experiments the closed nanotubes emit more current than the open ones while in other experiments the opposite holds true.

1 Introduction

Among the many appealing properties of carbon nanotubes (NT) [1-2] their field emission properties are the ones promising immediate technological application i.e. flat panel displays [3]. The nanometric scale of their diameters $(20-150$ A$^\circ)$ allows the achievement of very high enhancement factors of the applied electric field and makes them ideal natural protrusions for the creation of microscopic electron guns.

The experimental evidence regarding field emission from carbon NT is far from a clear and complete picture of the mechanisms involved in the emission process [4-5]. Rinzler et al [4], for example, have observed an emission current from open NT which is 2 orders of magnitude greater than that from closed NT at elevated temperatures (1500K) and 6 orders of magnitude greater at room temperature. On the other hand Bonard et al [5] have observed that emission from closed NT is greater than that from open nanotubes: it takes a voltage of 200V to produce a current of 10^{-12} A from open NT whereas it takes 100V to produce the same current from closed NT. In both experiments the tubes were of the multi-wall type and of roughly the same diameter (20nm). Furthermore Rinzler et al claim that at room temperature a structure on the surface of the open NT arises which resembles the unraveling of a sleeve of a sweater.

It seems therefore that the emission from carbon NT is a complicated process involving such factors as the detailed geometrical structure of the top surface, the electronic structure of the terminated tube and the enhancement of the applied field. In this paper we consider the enhancement of the electrostatic field due to the geometrical structure of the emitting surface. Our choice is partly dictated by the fact that the traditional formulae for the enhancement factor at the surface of a macroscopic emitter are not applicable in the present case.

2 Method

A carbon NT could be simulated by a cylinder with a wall of finite but sharp thickness but then the edges which would appear at the top surface will make slow the convergence of any method at these points where exactly a high accuracy is required. Rounding off the edges would destroy the advantages of a simple geometry. Our method is a good compromise of the above conflicting demands. We simulate the nanotube as a cylindrical array of touching spheres, each sphere representing an atom. This structure is supposed to be perfectly conducting and is sitting on a cathode which is taken to be a sphere of radius much larger than the anode – cathode separation to simulate the planar cathode of the experiments, see figure 1. What matters here is that the applied field is screened within a distance smaller than the radius of a carbon atom [6]. The surface of the nanotube and the

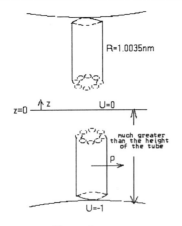

Figure 1
Simulation of the emitting carbon nanotube and origin of axes.

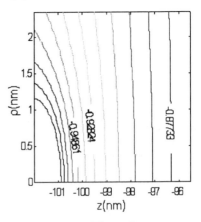

Figure 2
Equipotential lines near a closed nanotube with R=1nm and h=10nm

cathode constitute an equipotential surface of value U(x,y,z) = - V = the applied voltage=-1 unit. The anode is at z=0 where U=0. The field distribution for this structure is the same as the field calculated from a composite structure which consists of the said one (tube – cathode) and a mirror image structure (with respect to the U = 0 plane) at a constant potential +1. The total potential U (x,y,z) at an arbitrary point (x,y,z) can be written as

$$U(x,y,z) = \sum_{v=1}^{N} \sum_{L=0}^{\infty} \sum_{M=0}^{M=L} A_{L,M}^{v} \cdot P_{L}^{M}(\cos\theta_{v}) \cdot \frac{1}{r_{v}^{L+1}} \cdot \cos(M\varphi_{v}) \quad (1)$$

where the vth term represents a spherical expansion of the potential field due to the vth sphere alone [7] and N is the total number of spheres.

In the above equation P_L^M are the Legendre polynomials of order (L, M) and r_v, θ_v, φ_v, are the spherical coordinates of the point (x,y,z) with respect to a local coordinate system located on the centre of the v th sphere. The $A_{L,M}^{v}$ are coefficients to be determined. The sum over v is reduced to a sum over the real spheres by exploitation of the mirror symmetry of the system. Furthermore, because of the quasicylindrical symmetry, all coefficients $A_{L,M}^{v}$, which reside on the same horizontal plane are equal. If we truncate the infinite sum over L in equation 1 then the number of unknown coefficients will be finite, say K. These are calculated by the point matching method, i.e. by judiciously choosing K points on the equipotential surface U (x,y,z) = - V of the nanotube and the cathode and thus constructing a linear system of K equations with K unknowns. More details will be published elsewhere.

3 Results

The calculated equipotential surfaces for a closed nanotube of aspect ratio h/R =7, where h is the height and R is the radius of the tube, is shown in Figure 2. The vertical axis denotes the radial distance from the axis of the tube. The numbers on the equipotential lines denote the value of the potential in units of the applied potential V. It can be seen that at large z the field is uniform whereas near the tube the equipotential lines bend round it, thus enhancing the local field.

The field produced by the open nanotubes is radically different in that the field inside the tube and in particular near its axis remains uniform whereas near the top of the walls it is highly bent producing a significant enhancement. This is shown in Figure 3 where we plot the vertical component of the electric field as a function of the distance z from the anode, for three values of the radial distance ρ from the axis of the tube. It can be seen that for ρ=R the electric field E_z at the top of the tube is significantly higher than its far field value, while for ρ<<R it does not change significantly from its uniform value. The enhancement factor β of both open and closed nanotubes , defined as the ratio of the field at the surface to the far-field

value, is shown in Figure 4. The enhancement factor of an open NT remains roughly 3 times greater than that of a closed NT for heights up to h=100nm. Now the average h in Bonard et al was 2-3μm. It is hard at the moment to extend the calculations to this order of h due to limitations of computer time and memory. However it is not unreasonable to expect that the ratio of the β's will not increase at higher heights. Furthermore this ratio will be reduced by the multi-wall character of the NT of [5]. In fact Bonard et al find a β of 1600 and 1100 for closed and open NT respectively by arbitrarily inserting a workfunction of 5eV for both NT. Given that the closed NT contain at the top surface localized states due to the pentagons of the hemispherical cap [5] -which emit more than the extended states - it is not unreasonable to suggest that the contribution to the current of these states could compensate the lowering of the current due to a smaller (but not very smaller) local field. Finally we believe that the higher current from open NT found by Rinzler et al is likely to derive from the linear structure that appears on open NTs.

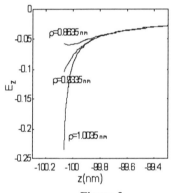
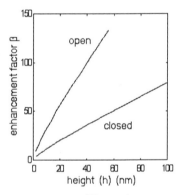

Figure 3
Vertical component of the local electric field near the surface of an open nanotube with h/R=7.

Figure 4
Variation of the enhancement factor with the height of the tube.

References

(1) T.W. Ebbeson (ed), Carbon Nanotubes: Preparation and Properties, CRC Press Princeton NJ (1997).
(2) E.W. Wong, P.E.Sheehan and C. Lieber, *Science* **277**, 1971, (1997).
(3) W.A. de Heer and R. Martel, *Physics World* **13**, 49 (2000).
(4) A.G. Rinzler et al, *Science* **269**, 1550 (1995).
(5) J.M. Bonard et al, *Appl. Phys.*, **A69**, 245 (1999).
(6) Lang and Kohn, *Phys. Rev. B* **1**, 4555 (1970).
(7) J. D. Jackson, Classical Electrodynamics 3rd edition 1999 ch. 3.

STUDY OF PHOTOLUMINESCENCE AND MICRO-PHOTOLUMINESCENCE OF V-SHAPED QUANTUM WIRES

M.TSETSERI, G.P.TRIBERIS
University of Athens, Physics Department, Solid State Section,Panepistimiopolis, 15784 Zografos, Athens, Greece

V.VOLIOTIS AND R.GROUSSON
Groupe de Physique des Solides, CNRS Universite Paris VI et Paris VII-2,place Jussieu,75251 Paris Cedex 05,France

We present a systematic theoretical study of the photoluminescence and micro-photoluminescence spectra of an $Al_{0.3}Ga_{0.7}As/GaAs$ V-shaped quantum wire.The model used reproduces the observed characteristics based on the existence of microscopic compositional fluctuations at the interfaces.

We investigate V-shaped quantum wires (QWRs) of GaAs/AlGaAs with characteristics described elsewhere[1].The intrinsic photoluminescence (PL) spectrum is assigned to the radiative recombination of excitons.At low temperatures the inhomogeneous broadening of the PL spectrum of the QWR is attributed to the interface quality since at low temperatures the principal broadening mechanism is the interface roughness. As the excitons move along the direction of free motion of the wire, y,they are trapped in interface fluctuations,become localized and they recombine radiatively.This idea has been originally used by Singh[2] for the interpretation of the PL linewidth of quantum wells.We extend this model for the case of QWRs.The energy of the excitation is given by

$$E_{exc}(\mathbf{r}, W_z, W_x) = E_e(\mathbf{r}, W_z, W_x) + E_h(\mathbf{r}, W_z, W_x) - E_b(\mathbf{r}, W_z, W_x) + E_g \quad (1)$$

where **r** denotes the exciton position , W_z the wire width on the growth direction,W_x the lateral wire width,E_e and E_h denote the energies of the electron and hole subbands respectively,E_g is the band gap of the material(GaAs) and E_b is the binding energy of the exciton.

The exciton wave function extends over a region $\approx D_0$. During the growth of GaAs on AlGaAs there are fluctuations in the wire size mainly in the direction of growth around the mean value W_{0z}.Interface fluctuations in a QWR should lead to the formation of islands elongated in the wire direction.We have assumed that from the two interfaces ($z = 0, z = W_{0z}$) only one exhibits roughness (that at $z = 0$) due to growth conditions. Fluctuations will extend in a distance (height) $z = \pm 1ML$ from $z = 0$. A parameter δ_1 describes the number of monolayers creating the fluctuations. According to the above description z,takes the values $W_{0,z} \pm \delta_1$ or $W_{0,z}$ for the "deep" (+), for the "shallow" (-),or for the "flat" regions, respectively. The extent of the island along the "free" y-direction is characterized by the smallest width δ_2. In this way, in the y-direction quantum boxes (QBs) are formed. Their dimensions are $W_{0,z} \pm \delta_1$ or $W_{0,z}, W_x$ and W_y. The smallest W_y is δ_2. The interface is described on a global scale by parameters C_a^0, C_b^0 and C_c^0 representing the mean concentrations of islands penetrating AlGaAs,GaAs the flat regions, respectively.

The energy emitted is distributed according to the extent of the fluctuations of wire widths "seen" by an exciton. An exciton in point r will "see" fluctuations in a region comparable with the exciton size D_0. The shape of the PL line is determined by the probability distribution $P(D_0, W_z)$ of finding fluctuations in the wire width extended over the exciton size. The probability of finding fluctuations C_a, C_b, C_c over a region of size D_0 when the average global concentrations are C_a^0, C_b^0, C_c^0, to first order is given by[2]

$$P(C_a, C_b, C_c; D_0) = \exp\{-(\frac{D_0^2}{\delta_{2a}^2} C_a \ln \frac{C_a}{C_a^0} + \frac{D_0^2}{\delta_{2b}^2} C_b \ln \frac{C_b}{C_b^0} + \frac{D_0^2}{\delta_{2c}^2} C_a \ln \frac{C_c}{C_c^0})\} \quad (2)$$

The wire width is given by[2]

$$W_z = W_{0z} + \delta_1[(C_a - C_a^0) - (C_b - C_b^0)] \quad (3)$$

We assume that:(i) The confinement along the growth axis is predominant and the lateral confinement is only an additional effect (ii) The confining potential along the x-axis is of a form given by[3]

$$U(x) = E_{el}(0) + \Delta E_{el} \tanh^2(x/2W_x) \quad (4)$$

where $E_{el}(0)$ is the confinement energy due to the transverse band-gap variation at the center of the crescent, $\Delta E_{el} \equiv E_{el}(\infty) - E_{el}(0)$, with $E_{el}(\infty)$ being the effective conduction band edge far away from the center and W_x is the width of the potential well.

Assuming that the wire has a width $W_{0z} = 45 \text{\AA}^1$ and $W_{0x} = 200 \text{\AA}^1$ using Eq.(1) we calculate the first electron-heavy hole excitation to occur at an energy $E_{x,0}^0 = 1.645 eV$ (using for the case of electrons a shooting method while for the case of holes the 4×4 Luttinger-Kohn Hamiltonian).

The excitation energy for $W_{0z} \pm 1ML$ occurs at $E_{x,+1}^0 = 1.635eV$ and at $E_{x,-1}^0 = 1.656eV$,respectively.Thus, the fine structure is attributed to $\pm 1ML$ fluctuations,since these energies provide an upper and a lower limit for the PL spectrum, consistent with the 20meV experimental spectrum spread.

To reproduce the experimental PL signal centered (cnt) at an energy $E_{exc}^{cnt} =$

Table 1. Full Width at Half Maximum for three different values of δ_2

width	$\delta_2 = R_0$	$\delta_2 = \frac{R_0}{2}$	$\delta_2 = \frac{R_0}{4}$
FWHM(meV)	7.10 ± 0.93	6.81 ± 0.52	4.77 ± 0.04

Table 2. Comparison between theory and experiment

PEAKS(exper)(eV)	PEAKS(theo)(eV)	W_y (Å)	z (ML)
1.6425	1.6424	150	17
1.6443	1.644	120	17
1.646	1.6454	100	17
1.6485	1.648;1.6486	70; 280	17;16
1.649	1.649;1.6494	60; 240	17;16
1.6505	1.6502;1.6505	50;200	17;16
1.652	1.6524	150	16
1.6535	1.6534	130	16
1.654	1.654	120	16

1.646eV and with full width at half maximum (FWHM) 8meV we calculate the probability $P(D_0,E)$ for three different values of $\delta_2(\delta_2 = D_0, \delta_2 = \frac{D_0}{2}, \delta_2 = \frac{D_0}{4})$. The FWHM of the PL for these values is shown in Table 1.
From the calculated FWHM we see that the experimental values fit the theoretical calculations for $\delta_2 \approx D_0$ i.e. the smallest width must have a size comparable with the size of the exciton.
A micro-PL (μ-PL) signal consists of sharp peaks assigned to fundamental states of the QBs of a width,W_y.The QWR under study has $W_{0z} = 45\text{Å}$ i.e. it consists of 16 ML ($1ML = 2.83\text{Å}$ in GaAs).A $\delta_1 = \pm 1ML$ fluctuation will create regions of height 17ML("deep"),16ML(flat) and 15ML ("shallow").Thus, in the "free" y-direction,wells of widths W_y are created and an exciton feels an additional confinement of 10 meV.
To find the sizes of the wells ,W_y,we consider them as rectangular quantum wells of finite depth (10meV) with a size distributed in a range [50Å,1000Å].In regions where W_z is 17ML the excitation energy is $E^0_{x,+1} = 1.635eV$.To this we add the confinement energies coming from the distribution of wells W_y.We do the same for 16ML where $E^0_{x,0} = 1.645eV$.We compare our theoretical results with a specific μ-PL[1].The comparison identifies the width of the boxes W_y.The experimental and the calculated peaks are shown in Table 2.With semicolumns we distinguish the peaks coming from the corresponding values of W_y and the height z. From the study of Table 2 we conclude the following:

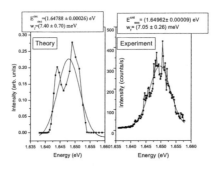

Figure 1. The theoretical and the experimental PL-spectrum. W_f is the FWHM and E_{exc}^{cnt} is the center of the peaks.(solid line:Gaussian fit, line with squares:calculated and experimentally observed peaks.)

(i) We estimate that the size of the smallest width δ_2 corresponds to a well of width $W_y = 60\text{Å}$, with a dispersion around it. (ii) With a dispersion of 20Å around $\delta_2 \approx 60\text{Å}$ in regions consisting of 17ML the peaks originate from islands with $\delta_2 \approx D_0$ and $2\delta_2$. In the regions of 16 ML the islands have widths $3\delta_2$ and $4\delta_2$. Thus, in the "deep" regions (17ML) islands have smaller width than the "flat" (16ML). We can associate the concentration of islands C_a, C_b, C_c with the intensity of the peaks, to reproduce the spatial variation of the μ-PL. The total theoretical PL signal is derived by assuming Gaussian curves centered around the energy corresponding to island of widths δ_2 and $2\delta_2$ for the case of 17 ML and Gaussian curves centered around the energy corresponding to islands of widths $3\delta_2$ and $4\delta_2$ for the case of 16 ML. The final spectrum can be fitted by a Gaussian curve that is centered round 1.648 eV and has FWHM 7.4 ± 0.7 meV (Fig. 1). This is in a very good agreement with the experiment[1] (Fig. 1). The model developed allows us to perform the cartography of the interface describing in a simple way the microscopic nature of the interface roughness. Although the real interface is likely to be considerably more complicated, the model reproduces successfully the experimental data, the main features of the interface and the physics of the problem.

References

1. J.Bellessa, V.Voliotis et al. Appl.Phys.Lett.**71**, 2481 (1997)
2. J.Singh, K.K.Bajaj Appl.Phys.Lett. **44**, 805 (1984)
3. E.Kapon, D.M.Hwang and R.Bhat *Phys.Rev.Lett.* **63**, 430 (1989)

CATALYTIC ACTION OF NI ATOMS IN THE FORMATION OF CARBON NANOTUBES: A COMBINED AB-INITIO AND MOLECULAR DYNAMICS STUDY

ANTONIS ANDRIOTIS

Institute of Electronic Structure and Laser, Foundation for Research and Technology-Hellas, P.O. Box 1527, 71110 Heraklio, Crete, Greece

MADHU MENON

Department of Physics and Astronomy, University of Kentucky, Lexington, Kentucky 40506-0055 and Center for Computational Sciences, University of Kentucky, Lexington, Kentucky 40506-0045

GEORGE FROUDAKIS

Department of Chemistry, University of Crete, P.O. Box 1470, 71409 Heraklio, Crete, GREECE

E-mail: frudakis@iesl.forth.gr

Catalytic action of Ni atoms in the growth of single-wall carbon nanotubes is investigated using tight-binding molecular dynamics and *ab-initio* methods. Our results demonstrate this to be a two step process in which the Ni atom first creates and stabilizes defects in nanotubes. The subsequent incorporation of incoming carbon atoms anneals the Ni-stabilized defects freeing the Ni atom to repeat the catalytic process.

1 Introduction

Transition metal catalysts such as Ni are known to play a key role in the production of single-wall carbon nanotubes (SWCN). Microscopic details of their interactions with SWCN are, therefore, expected to shed light on the SWCN growth process. While it is tempting to extrapolate known results of interaction of Ni with graphite to the SWCN case by drawing on the similarities between graphite and SWCN, it should be noted that the curvature could be expected to have nontrivial consequences. Indeed, our recent works have shown that such a simple extrapolation can lead to misleading results for bonding geometries, magnetic moments, and other physical properties [1,2].

In this work, we present results of a detailed theoretical study of the dynamical interaction between the Ni catalyst and SWCN in the presence of additional C atoms with a view towards an understanding of the nanotube growth mechanism.

2 Methods

Our calculations are performed using the tight-binding molecular dynamics (TBMD) method as well as accurate *ab-initio* methods [1,2]. The details of our TBMD formulation can be found elsewhere [2]. The TBMD scheme allows us to employ fully symmetry-unconstrained optimisation for all geometries considered. The TBMD calculations are further complemented by accurate *ab-initio* methods [3]. The *ab-initio* total energy calculations were performed using the GAUSSIAN-98 program package and includes density functional theory calculations with the three-parameter hybrid functional of Becke using the Lee-Yang-Parr correlation functional [3]. The atomic basis set used is of double zeta quality and includes relativistic effects for heavy atoms. Our recent works dealing with Ni chemisorption on graphite, C60, and SWCN have demonstrated the profound influence exerted by the curvature of the substrate on Ni bonding properties [1,2]. The curvature effect shows up as a re-hybridisation of the graphitic-C sp2 orbitals which in turn bond to the adsorbed Ni d orbitals. A portion of a graphene sheet consisting of 128 carbon atoms simulates the graphite. The SWCNs used in our simulations consisted of (5,5) and (10,10) types containing 150 and 320 atoms, respectively.

3 Results and Discussion

As a first step of this study, we replaced a C atom of a graphene sheet by a Ni atom and allowed the system to relax using the TBMD method [3]. The relaxation resulted in the Ni atom moving slightly outward of the graphene plane with minimal distortions to the rest of the graphene lattice. The Ni-C bonds were found to be 1.63O and the C-Ni-C angle 95–96°. These results are in very good agreement with the experimental and theoretical results. After replacing one C atom in a graphene sheet by a Ni atom we placed one extra C atom above the Ni atom (fig.1).

Figure 1. An extra C atom (blank circle) is placed above one substitutional Ni atom (black circle) in a graphene plane (left). The relaxation with TBMD resulted in the extra C atom taking the place of Ni, with the released Ni atom chemisorbing below the plane of graphite (right)

The relaxation with TBMD resulted in the extra C atom taking the place of Ni, with the released Ni atom chemisorbing below the plane of graphite. Our *ab-initio* calculations indicate this process to be energetically very favourable with an energy gain of 21eV.

After the study in graphene, we repeat the two simulations analysed above in a SWCN. The TBMD simulation of a substitutional Ni atom in a (5,5) resulted in the Ni atom moving into the interior of the nanotube leaving a C vacancy on the wall. Thus, the presence of the Ni atom appears to stabilise the C vacancy in the SWCN. This result is supported by *ab-initio* calculations. We calculate the total energy of the system starting from the TBMD relaxed Ni position and then by moving the Ni atom radically outward to an exterior position in small increments. This calculation make clear that the Ni atom is, in fact, more stable either outside or inside the tube, while its substitutional position is a transition state.

In the second step the simulations described above were repeated by including an additional C atom within bonding distance of the substitutional Ni atom in a (5,5) tube. As can be seen in fig.2 the substitutional Ni atom was replaced by the exterior C atom, while the Ni atom broke free and was left encapsulated inside the tube with no bonding to any of the C atoms. In addition, our *ab-initio* calculations showed that the substitution of a Ni atom on the wall of the carbon nanotube in the presence of an extra C atom is not energetically favourable and that a considerable energy barrier (10.5eV) separates the two stable sites. In those minima the C atom is located on the wall of the tube and the Ni atom is inside or outside of the tube.

Figure 2. Initial, intermediate and final configuration of the TBMD simulations of a (5,5) C nanotube containing a substitutional Ni atom in the presence of an incoming C atom.

These results clearly demonstrate a contrasting dynamic behaviour of the substitutional Ni atoms in graphite and in SWCNs. In particular, they show that, contrary to their behaviour in graphite, substitutional Ni atoms are not stable on the SWCNs.

Nevertheless, at the end of the tube, the substitutional Ni atom can remain stable, forming part of the hexagonal ring, although with considerable distortions. However, when new hexagons start to form by the incorporation of additional carbon atoms in the atmosphere, the Ni atom exchanges its position with an incoming carbon atom and resets itself by occupying a substitutional position on an exterior ring (fig.3). Our *ab-initio* total energy calculations support the TBMD results and show that this process is energetically very favourable with an exothermic release of 7.25 eV.

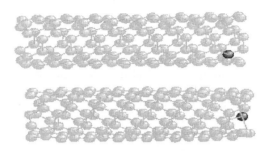

Figure 3. TBMD simulations shows that when new hexagons start to form by the incorporation of additional carbon atoms in the atmosphere, the Ni atom (black) exchanges its position with an incoming carbon atom and resets itself by occupying a substitutional position on an exterior ring

Summarising, our dynamical simulation results allow us to propose a reasonable explanation for the catalytic role of the Ni atoms in the growth process of carbon nanotubes. The Ni atom act as stabilising agents of the structural defects created in SWCN during the growth process. When other carbon atoms in the atmosphere come within bonding distance of such a Ni-stabilised defect, the defect is annealed while the freed Ni atom diffuses along the length of the tube either outside or inside. At the end of the tube, the substitutional Ni atom can remain stable, forming part of the hexagonal ring, or exchange its position with an incoming carbon atom and resets itself by occupying a substitutional position on an exterior ring.

The encapsulation tendency we find in the case of substitutional Ni in the presence of an incoming C atom is supported by the recent experimental work reported in Ref. [4] where encapsulation of Ni/Co is suggested as the most probable processes taking place during the tube growth. Our simulations, thus, give a detailed description of the possible swapping processes involved in the Ni assisted catalytic growth of nanotubes.

References

1. A. N. Andriotis, M. Menon, G. Froudakis, and J. E. Lowther, Chem. Phys. Lett. **301**, 503 (1999).
2. M. Menon, A. N. Andriotis, and G. E. Froudakis, Chem. Phys. Lett. **320**, 425 (2000).
3. M. J. Frisch *et al.*, GAUSSIAN 98, Revision A.6, Gaussian, Inc., 1998.
4. E. Dujardin, C. Meny, P. Panissod, J.-P. Kintzinger, N. Yao, and T. W. Ebbesen, Solid State Commun. **114**, 543 (2000).

Si NANOCRYSTAL MOS MEMORY OBTAINED BY LOW-ENERGY ION BEAM SYNTHESIS

E. KAPETANAKIS, P. NORMAND AND D. TSOUKALAS
Institute of Microelectronics, IMEL, NCSR 'Demokritos',
15310 Aghia Paraskevi, Greece,
E-mail: kapetan@imel.demokritos.gr

K. BELTSIOS
Institute of Microelectronics, IMEL, NCSR 'Demokritos',
15310 Aghia Paraskevi, Greece

S. ZHANG AND J. VAN DEN BERG
Department of Physics, Maxwell Building, University of Salford, England

J. STOEMENOS
Department of Physics, University of Thessaloniki, Thessaloniki, Greece

Very-low energy (1 keV) Si implantation and subsequent thermal annealing are used for the fabrication of a narrow charge storage layer within the gate oxide of metal-oxide-semiconductor capacitors and transistors. Capacitance and channel current measurements are performed to investigate the charging effects of the Si-implanted oxides as a function of Si fluence. Clear memory characteristics are observed for a dose of 1×10^{16} cm^{-2} or lower. The device electrical characteristics are found compatible with the spatial arrangement and structural state of implanted Si as well as with the presence of interface states and traps that originate from the nanocrystal formation process.

1 Introduction

Recently, memory cells consisting of a metal-oxide-semiconductor field-effect transistor (MOSFET) with a charge-storage floating-gate made of nanocrystals or excess silicon have been demonstrated [1-3]. These single-transistor memory cells are intended for low-power ultra-dense dynamic memory applications. The fabrication of semiconductor island-based floating gates through ion implantation and annealing is very promising because of its manufacturing advantages but faces the issue of generating nanocrystals close to the transistor channel without compromising the integrity of the gate oxide and the SiO_2/Si interface. For that reason, the ion implantation at a very-low energy, combined with proper annealing, is particularly attractive, as it leads to the formation of two-dimensional (2-D) arrays of semiconductor nanocrystals at relatively low doses and at a tunneling distance from the SiO_2 surface and/or the SiO_2/Si interface [3,4]. In this work we extend our recent studies [3] on the memory characteristics of Si-nanocrystal floating-gate MOSFETs obtained by low-energy ion beam synthesis. We discuss the effects, on the device operation, of the implanted dose as well as of the various traps and defects that originate from the nanocrystal fabrication process.

2 Devices

MOS capacitors with 6.4×10^{-4} cm^{-2} gate area and nMOS transistors with a gate length ranging from 2 to 10 µm and 100 µm gate width have been fabricated following a process similar to that used for conventional MOS devices. The main changes are related to the introduction of Si with a narrow distribution, peaking at a tunnelling distance from the SiO$_2$/Si interface. For this purpose, Si implantation is carried out into 8 nm thick thermally grown oxide at an energy of 1 keV and at doses ranging from 5×10^{15} to 5×10^{16} cm^{-2}. Subsequently, a 30 nm thick control oxide is deposited, followed by a 30 min nitrogen annealing at 950°C, aiming at the precipitation of Si nanocrystals. The nanocrystal characteristics (size, spatial distribution, and degree of crystallinity) as a function of the implantation dose are reported in [3,4]. In brief, a low dose (5×10^{15} cm^{-2}) leads to scattered ill-defined Si-clusters, a medium dose (1×10^{16} cm^{-2}) leads to a 3 nm-thick band of ill-crystallized 3-8 nm long Si-grains located at a tunneling distance from the SiO$_2$/Si interface and a high dose ($\geq 2 \times 10^{16}$ cm^{-2}) leads to a 4nm-thick dense band of well crystallized platelet-like 4-20nm long Si grains.

3 Results and Discussion

Figure 1 shows high-frequency (H-F) capacitance-voltage (C-V) characteristics for non-implanted and Si-implanted SiO$_2$ gate capacitors. The gate voltage is swept from −15V to +15V and back to −15V. A clear hysteresis due to the trapping of electrons/holes in the insulator during the positive/negative gate voltage sweep is found for the 5×10^{15} and 1×10^{16} cm^{-2} cases. In contrast, devices implanted with a dose of 2×10^{16} cm^{-2} or higher do not exhibit any clear memory effect. In this case the C-V curves are significantly stretched out and the capacitance in the accumulation regime increases continuously with the gate bias. Non-implanted devices do not exhibit any shift in the C-V curves, showing that the hysteresis is Si implantation related.

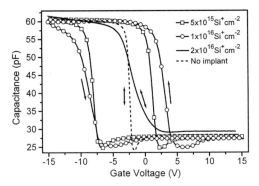

Figure 1. High-frequency C-V curves of MOS capacitors with non-implanted and Si-implanted gate oxide at three different doses.

Source-drain current (I_{DS}) versus gate voltage (V_G) characteristics of nMOS transistors with Si-implanted gate oxides appear in Fig. 2. The measurements were performed as follows [3]: (a) a load resistor (R_L=100Ω) was connected between the source electrode and ground to monitor the source-drain current through source voltage (V_S=$I_{DS}R_L$) measurements, (b) a triangular signal of +15/-15V amplitude (V_G) was applied to the gate electrode for a wide range of frequencies, (c) the V_G and V_S signals were then stored in an oscilloscope to extract the I_{DS} vs V_G curve of the transistor (Fig 2) and the corresponding threshold voltage shift (ΔV_T) value (inset of Fig. 2). Again, as in Fig.1, no memory function is observed for the 2×10^{16} implanted devices. In addition, 2×10^{16} implanted devices exhibit a strong reduction of the source-drain current. The latter reduction is due to (a) the dynamic charge exchange between nanocrystals and the inversion layer during the measurement, (b) the accumulation of charge at interface traps generated by ion implantation that reduces the mobile charge density. Effect (a) applies also as an explanation for the decrease of the threshold voltage shift observed for the 1×10^{16} cm^{-2} case in the low frequency regime (Fig.2, inset). This implies a thermal activation process as an electron injected into a nanocrystal is localized at an interface defect. Such a mechanism is supported by our measurements at 77 K (Fig.2, inset, 1×10^{16} case) where ΔV_T increases monotonically as the frequency decreases.

On the basis of the above and earlier findings [3] we can correlate the appearance of memory characteristics with the structure of the implanted oxide as follows: The flat band and threshold voltage shifts shown in figures 1 and 2 reflect the trap-like behavior of excess Si (case of 5×10^{15} cm^{-2}) and/or the presence of electronic states at the surface of uncoupled Si grains (case of 1×10^{16} cm^{-2}). For the 2×10^{16} and 5×10^{16} Si cm^{-2} implanted devices, the presence of well-crystallised Si islands that appear to be coupled laterally and the increased defect state density in the injection oxide, lead to a strong coupling between the accumulation layer and the nanocrystals at the expense of charge storage time.

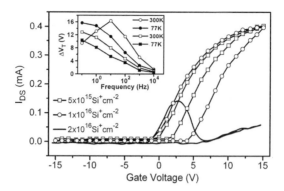

Figure 2. Source-drain current versus gate bias (1 kHz triangular signal) of transistors with 4 μm gate length for three Si implantation doses. The device threshold voltage (ΔV_T) versus frequency of +15V/-15V triangular signal for 5×10^{15} and 1×10^{16} Si$^+$cm^{-2} implanted devices at 300K and 77K is shown in the inset. The drain voltage and the load resistor are fixed at 0.1V and 100 Ω respectively.

In conclusion, we have investigated the memory characteristics of floating-gate MOS devices obtained by 1keV Si$^+$ implantation. Electrical characteristics of the devices at room and liquid nitrogen temperatures correlate strongly with the structural state of implanted Si as well as with the presence of various defects/traps in the injection oxide and at the Si nanocrystal/SiO$_2$ and SiO$_2$/Si substrate interfaces. Clear memory characteristics are observed for doses of 1.10^{16} cm^{-2} and lower.

References

1. Hanafi H. I., Tiwari S. and Khan I., Fast and long retention-time nano-crystal memory. *IEEE Trans. Electron Devices* **ED-43** (1996) pp. 1553-1558.
2. King Y. C., King T. J. and Hu C., A long-refresh dynamic/quasi-nonvolatile memory device with 2-nm tunneling oxide. *IEEE Electron Dev. Lett.* **20** (1999) pp. 409-411.
3. Kapetanakis E., Normand P., Tsoukalas D., Beltsios K., Stoemenos J., Zhang S. and Van den Berg J., Charge storage and interface states effects in Si-nanocrystal memory obtained using low-energy Si$^+$ implantation and annealing. *Appl. Phys. Lett.* **77** (2000) pp. 3450-3452.
4. Normand P., Tsoukalas D., Kapetanakis E., Van den Berg J., Armour D. G., Stoemenos J. and Vieu C., Formation of 2-D arrays of silicon nanocrystals in thin SiO$_2$ films by very-low energy Si$^+$ ion implantation, *Electrochem. Sol. State Lett.* **1** (1998) pp. 88-90.

CHARGE EFFECTS AND RELATED TRANSPORT PHENOMENA IN NANOSIZE SILICON/INSULATOR STRUCTURES

J.A. BERASHEVICH,[1] A.L. DANILYUK,[1] A.N. KHOLOD,[2] F. ARNAUD D'AVITAYA,[2] V.E. BORISENKO [1]

[1] *Belarusian State University of Informatics and Radioelectronics*
P. Browka 6, 220013 Minsk, Belarus
[2] *Centre de Recherche sur les Mécanismes de la Croissance Cristalline*
Campus de Luminy - Case 913, 13288 Marseille cedex 9, France

A kinetic model of carrier transport via traps in dielectric has been developed for quantum well Si/CaF_2 structures. Relay tunneling via traps and thermoelectron emission from the traps have been considered as possible mechanisms of the carrier transfer in the dielectric. Charge accumulation, its polarization and relaxation under external applied bias are found to result in the shift of current origin and appearance of the negative differential resistance region on the current-voltage characteristics of the structures. The effects of temperature, measurement regimes and structure geometry have been as well analyzed.

1 Introduction

Electronic devices with a negative differential resistance (NDR) behavior attract considerable attention as functional elements for circuit applications. An extensive experimental and theoretical research has been performed on NDR devices based on resonant tunneling in $A^{III}B^{V}$ quantum wells. Looking for room-temperature operation and compatibility with silicon technology there is an interest in fabrication and study of Si/CaF_2 nanostructures, in large part motivated by the impressive progress achieved in molecular beam epitaxy [1]. Recently, an NDR in perpendicular carrier transport across such structures has been observed [2]. However, the origin of the effect has not been clearly established.

In this paper we present a model of carrier transport in Si/CaF_2 quantum well structures governed by the charge capture to the traps in the dielectric and the charge release. The work has been done in order to find the parameters that condition the transport phenomena observed experimentally.

2 Model

The carrier transport is considered for the periodical structure consisting of N Si quantum wells separated by $N+1$ CaF_2 potential barriers. The wells are assumed to be undoped and trap states are supposed to exist within dielectric barriers. The following carrier transport processes are analyzed: carrier injection from the contacts; carrier tunneling via traps; carrier capture to trap states; thermoelectric carrier emission from the trap states; carrier recombination in the quantum wells.

The kinetic approach is used to describe the carrier transport. In this case the rates of carrier concentrations (n,p) in the i-th semiconductor layer are written as [3]

$$dn_i/dt = g_{i-1}^{trap}(n_{i-1}, n_i, \vec{F}_{SEM,i}) - g_i^{trap}(n_i, n_{i+1}, \vec{F}_{SEM,i+1}) - \gamma_i n_i p_i, \quad (1)$$

$$dp_i/dt = g_{i-1}^{trap}(p_{i+1}, p_i, \vec{F}_{SEM,i+1}) - g_i^{trap}(p_i, p_{i-1}, \vec{F}_{SEM,i}) - \gamma_i n_i p_i \quad (2)$$

and on the trap states (n^t, p^t) in the i-th dielectric layer as

$$dn_i^t/dt = \left(D_{i(n)}^{trap} n_i/\tau_0 - n_i^t/\tau_0\right)\exp(\alpha E_i/k_b T), \quad (3)$$

$$dp_i^t/dt = \left(D_{i(p)}^{trap} p_i/\tau_0 - p_i^t/\tau_0\right)\exp(\alpha E_i/k_b T). \quad (4)$$

Here $\vec{F}_{SEM,i}$ is the electric field in the i-th well; γ_i is the electron and hole recombination coefficient in silicon; k_b is the Boltzmann's constant; T is the absolute temperature; τ_0 is the carrier life time at the trap state given by

$$\tau_0 = v_0^{-1} \exp(E_a/k_b T), \quad (5)$$

where v_0 is the frequency factor; E_a is the activation energy of the trap. The value of τ_0 is estimated to be 5 s at the room temperature.

The electron (hole) transition rate g^{trap} is determined by the barrier transparency, which in the case of transport via traps includes the trap assisted tunneling transparency $D_{n(p),i}^{trap}$ [4] and thermoelectric emission from the traps:

$$D_{n(p),i}^{hop} = v_0 \exp\left(-(E_a - \alpha \vec{F}_{INS,i})/k_b T\right) D_{n(p),i}^{trap}, \quad (6)$$

where $\alpha \vec{F}_{INS,i}$ is the reduction of the i-th barrier by the electric field;.

Depending on the thermal ionization energy the carrier lifetime at the trap can reach seconds (as confirmed experimentally [4]). Therefore a charge accumulation occurs within dielectric barriers. This charge is polarized by the external electric field. To account for these effects discrete Poisson equations should be written as:

$$\vec{F}_{INS,i} = q \cdot d_{INS} \cdot (n_i^t - p_i^t)/\varepsilon_0 \varepsilon_{INS}, \quad (7)$$

$$\vec{F}_{SEM,i} = q \cdot d_{SEM} \cdot (n_i - p_i)/\varepsilon_0 \varepsilon_{SEM}, \quad (8)$$

for the semiconductor and insulator layers, respectively. Here q is the electron charge; ε_0 is the permittivity of vacuum; ε_{INS}, ε_{SEM} is the relative permittivity of the insulator and semiconductor, respectively; d_{SEM}, d_{INS} is the thickness of the semiconductor and insulator layers, respectively.

One additional equation is the bias condition

$$\sum_{i=1}^{N+1} \vec{F}_{SEM,i} d_{SEM} + \vec{F}_{INS,i} d_{INS} = V, \quad (9)$$

where V is the external voltage applied to the structure.

3 Results and discussion

Calculated current-voltage (*I-V*) characteristics of the 10 period Si/CaF$_2$ structure are shown in Fig. 1 for different voltage step delay time. The striking feature we want to stress is the appearance of the NDR region on the *I-V* curves. Moreover, the current origin is shifted towards negative voltage values. Both effects tend to smooth as the delay time between two consecutive voltage steps is increased. Hence, the phenomena observed are related to charging and discharging processes. Clearly, when the bias is applied the traps start to capture the carriers. As the capture time can be of the order of the step delay time the charge is accumulated in the dielectric. Polarization of this charge by the external bias causes an appearance of the internal electric field in the structure. There is a moment when both external and internal fields compensate each other and therefore the total current is vanished to zero. As the bias applied continues to decrease the reverse current due to the internal electric field arises. It reaches a peak value until the trap discharge current starts to give an important contribution. Thus, the total current decreases resulting in the NDR region in the I-V curve. There is only the discharge current when the applied bias is zero. The smoothing of the NDR and current origin shift with an increase of the step delay time is explained by the reduction of the internal electric field owing to release the accumulated charge during this period.

The influence of the parameter α, which directly relates to the nature of the trap states, has been investigated. The results are presented in the Fig. 2. According to (3) and (4), an increase of α leads to the raise of the tunneling rate through the dielectric as well as to the speeding up the discharge of the traps, when the current changes its direction. The faster the discharging processes the faster is the decrease in the internal electric field and therefore the more pronounced is the NDR region.

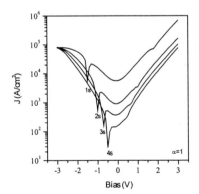
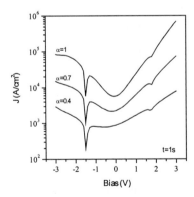

Figure 1. *I-V* curves of 10 periods Si/CaF$_2$ structure calculated for different step delay time.

Figure 2. *I-V* curves of 10 periods Si/CaF$_2$ structure calculated for different α values.

The capacitance of the structure to accumulate the charge is increased with additional number of periods. As a result, the current hysteresis effect becomes more pronounced, as it is demonstrated in Fig. 3. However, the voltage drop across one structure period is decreased in this case and therefore the rate of the trap discharge falls down. Hence the NDR region tends to disappear (Fig. 3).

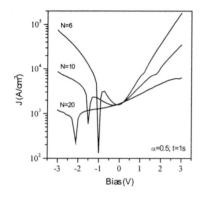

Figure 3. *I-V* curves of Si/CaF$_2$ structure calculated for different number of periods.

Another interesting result concerns the dependence of the peak-to-valley current ratio on the dielectric thickness. It is found that the NDR effect is the best pronounced at a certain thickness which is 2.0 nm of CaF$_2$ in our case. Both the NDR effect and the current origin sift have been observed to vanish rapidly as the temperature is decreased from 300 K to 200 K. This is a result of the exponential reduction of the tunneling probability via traps and charge accumulation at these traps as the temperature goes down.

4 Conclusion

The carrier transport in Si/CaF$_2$ multiquantum well structures has been studied by mean of the kinetic model assuming tunneling via traps and charge accumulation at the traps. Dynamics of charging/discharging of the traps has been found to control current hysteresis and appearance of the NDR region on the *I-V* characteristics.

References

1. F. Bassani, L. Vervoort et al, J. Appl. Phys. **79**, 4066 (1996).
2. S. Ménard, A. N. Kholod et al, phys. stat. sol.. **181** (2000), pp. 561-568.
3. A. N. Kholod, V. E. Borisenko et al, J. Appl. Phys. **85** (1999), pp. 7219-7223.
4. V. Yu. Kirpichenkov, JETF **113** (1998), pp. 1522-1530.

RADIATIVE RECOMBINATION FROM SILICON QUANTUM DOTS IN Si/SiO$_2$ SUPERLATTICES

P. PHOTOPOULOS, T. OUISSE, D.N. KOUVATSOS AND A.G. NASSIOPOULOU
Institute of Microelectronics, NCSR "Demokritos", POB 60228, 15310 Aghia Paraskevi, Greece.

In this work we have investigated radiative recombination from silicon quantum dots in Si/SiO$_2$ multiquantum wells fabricated by successive cycles of silicon deposition by Low Pressure Chemical Vapor Deposition and oxidation. The variation of PL and EL peak intensity and peak wavelength with nc-Si layer thickness, applied electric field and temperature was studied. PL appeared for silicon layer thickness smaller than 5-6 nm. EL exhibited a reversible blueshift with voltage, indicating voltage-controlled tunability of the peak wavelength. The effect of the applied electric field on the radiative recombination lifetime in silicon quantum dots embedded in SiO$_2$ was investigated theoretically.

1 Introduction

The light emitting properties of low-dimensional quantum-confined silicon nanostructures have been investigated in recent years [1-4]. Confined states and states at the Si/SiO$_2$ interface have been theoretically investigated and have both been shown to play a key role in light emission. In this work we have studied the effects of crystallite size, number of periods, electric field and temperature on the photoluminescence (PL) and electroluminescence (EL) emission from silicon quantum dots in (nc-Si/SiO$_2$)$_n$ superlattices.

2 Experimental

Si/SiO$_2$ bilayers and superlattices were fabricated by successive cycles of silicon deposition by Low Pressure Chemical Vapor Deposition, oxidation and oxide etching. In each cycle, 12 nm thick amorphous Si layers were deposited at 580°C and 300 mTorr on an initial 5 nm oxide, followed by dry oxidation at 900°C for thinning of the Si layers; then a dilute HF partial etch left about 5 nm of the formed oxide. PL and EL measurements were carried out at room and low temperatures as well as under applied electric fields.

3 Results and discussion

In single layer nc Si/SiO$_2$ structures, by crystallizing and thinning the deposited Si layer using oxidation, strong PL appears when the nominal Si thickness, as calculated by considering the oxidation rate of an equivalent polycrystalline silicon layer, goes below 5-6 nm; this is attributed to the formation of silicon nanocrystals passivated by oxygen. As observed in fig. 1, the PL intensity initially increases with oxidation time and reaches a maximum at times of 40 to 50 min. The PL peak wavelength showed an initial redshift from 650 nm to 800 - 900 nm and a subsequent blueshift to 680 - 700 nm, reaching a maximum for the same time, following an initial increase and a subsequent decrease in crystallite size with oxidation time. The PL peak blueshift stops at 680 nm, suggesting that localized states at Si/SiO_2 interfaces trap free carriers and play a key role in recombination when crystallite sizes are reduced, agreeing with a theoretical model [5]; according to this model, recombination involving localized states rather than recombination via free excitons occurs for crystallite sizes below ~ 2.8 nm. In multilayers fabricated using the optimum conditions for maximum PL obtained from a single nc-Si/SiO$_2$ bilayer, and having up to 5 periods, a superlinear PL increase is observed by increasing the number of layers, as it can be seen in fig. 2. The increase in PL intensity exceeding linearity can be attributed to better crystallization, more efficient crystallite passivation and reduction of crystallite defect density due to prolonged thermal treatment.

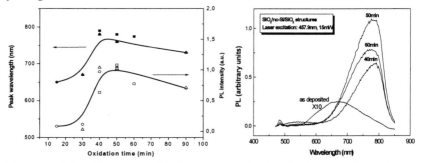

Figure 1. Variation of the peak PL and the PL intensity as a function of oxidation time for samples with one nc-Si layer (left). PL spectra from a 12 nm silicon film oxidized at 900°C for different times (right).

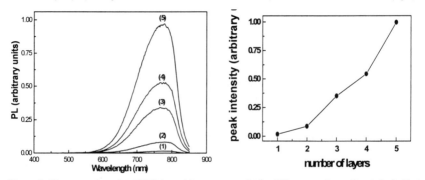

Figure 2. PL spectra from (nc-Si/SiO$_2$)$_n$ multi-quantum wells for different numbers of periods (left). A superlinear increase in PL intensity with increasing number of periods is observed (right).

Diode structures with a single nc-Si layer exhibited EL in the visible range at room and low temperatures. The EL peak wavelength, which is the same with the PL one for low excitation voltage, shows a blueshift with increasing voltage as shown in fig. 3. The blueshift is reversible, therefore it is not related to structural changes; thus the structures exhibit voltage-controlled EL peak tunability from 800 to 600 nm. The possible reasons of this blueshift are: a) field-induced quenching of the luminescence from larger crystallites (quantum-confined Stark effect), b) Auger quenching: at high fields and injection currents, more than one electron-hole pairs are likely to exist in larger crystallites, making non-radiative Auger recombination dominant, and c) the reduced injection probability into smaller nanocrystals. These factors induce a quenching of luminescence which is more effective for larger nanocrystals. Furthermore, for a given excitation voltage the EL intensity increases while the EL peak is blueshifted by decreasing the temperature; this blueshift is much larger than the corresponding PL blueshift.

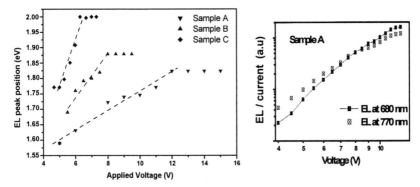

Figure 3. Variation of the EL peak energy with voltage for samples A, B and C, having nc-Si sizes of ~ 2, 1.5 and 1 nm, respectively (left). Variation of the EL efficiency with voltage, measured at 770 nm and 680 nm (sample A) (right).

We have theoretically determined that a large electric field leads to an increase in radiation lifetime and a decrease in transition energy; the energy shift is proportional to the square of the electric field for moderate fields while the field-induced increase in radiation lifetime increases rapidly as the crystallite size increases. This results in a quenching of the radiative recombination that becomes more pronounced in larger crystallites, as well as to a redshift in luminescence from each crystallite, which, for a distribution of crystallites, leads to a shift in the spectrum. However, because the contribution of larger crystallites, which emit light of longer wavelength, is effectively quenched, the luminescence of the overall crystallite distribution is found to be blueshifted despite the redshift of the emission from individual crystallites. The calculated direct and TO+LO phonon-assisted radiative recombination lifetimes as a function of crystallite size and electric field are shown in fig. 4.

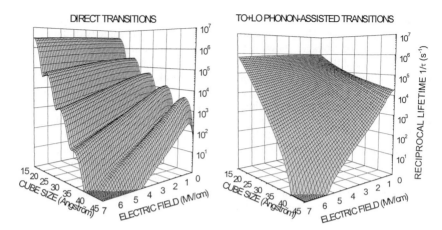

Figure 4. Direct (left) and TO+LO phonon-assisted (right) radiative recombination reciprocal lifetime as a function of crystallite size and electric field along a (100) direction (lower valleys) at T=300°K.

4 Conclusions

Fabricated nc-Si/SiO$_2$ bilayers and superlattices exhibit strong PL for silicon layer thickness below 5 nm with an initial redshift and then blueshift corresponding to increase and decrease of crystallite size. A superlinear PL increase with the number of periods was observed. EL blueshifted with increasing voltage and decreasing temperature; voltage-controlled EL peak tunability was obtained in single nc-Si layer structures. The electric field was theoretically shown to quench the radiative recombination rate more effectively in large crystallites, consistently with the experimental results.

References

1. A.G. Cullis, L.T. Canham and P.D.J. Calcott, *J. Appl. Phys.* **82**(3), 909 (1997).
2. D. Kovalev, H. Heckler, G. Polisski and F. Koch, *Phys. Stat. Sol. (b)* **215**, 871 (1999).
3. P. Fauchet, *IEEE Journal on Sel. Top. In Quant. Electr.* **4**(6), 1020 (1998).
4. A.G. Nassiopoulou, S. Grigoropoulos, D. Papadimitriou and E. Gogolides, *Appl. Phys. Lett.* **66**(9), 1114 (1995).
5. W.V. Workin, J. Jorne, P.M. Fauchet, G. Allan and C. Delerue, *Phys. Rev. Lett.* **82**(1) 197 (1999).

AVALANCHE POROUS SILICON LIGHT EMITTING DIODES FOR OPTICAL INTRA-CHIP INTERCONNECTS

S. K. LAZAROUK, P. V. JAGUIRO, A. A. LESHOK, V. E. BORISENKO

Belarusian State University of Informatics and Radioelectronics
P. Browka 6, Minsk 220013, Belarus

Recent progress in the development of integrated optoelectronic units including porous Si light emitting diode connected with a photodetector by an alumina waveguide is reported. Main attention has been devoted to the enhancement of the light emitting diode parameters. The value of quantum efficiency has been estimated to be 0.19 %. The data obtained from the transient electroluminescent wave form gave the delay time of 1.2 ns and the rise time of 1.5 ns for developed diodes. Possible methods of further device optimization are discussed.

1 Introduction

Recent developments in communication systems and computer technology make increasingly attractive the substitution of electrons with photons in transmission and processing of information. Thus, optoelectronic interconnects are required for the next generation of integrated systems. Since the discovery of efficient light emission from porous silicon, this material is considered to be promising for integrated silicon based optoelectronic systems able to emit, transmit, and detect light in the visible range.

We reported on an aluminum/porous silicon junction, which operates like a light emitting diode (LED) when biased above the avalanche breakdown [1], and on an integrated optoelectronic unit including the porous silicon LED connected with a photodetector by an alumina waveguide [2,3]. The developed avalanche LED can be integrated with conventional CMOS silicon devices on the same substrate. In this paper we present recent extended data characterizing operation of the integrated optoelectronic unit with the avalanche porous silicon LED.

2 Experimental

The main technological steps are described in [3]. The fabricated devices were characterized with time response, transient electroluminescence and delay time techniques. The time response of the avalanche LEDs was measured by a short pulse generator (I1-14) operation. The light emitted by the device passes through a monochromator and then it was detected by a fast photomultiplier. The current through the device was in the range of 10-240 mA. The transient electroluminescence was recorded between the leading and trailing edges of the current pulse measured at 10 % and 90 % of the current amplitude corresponding to 1.0 ns, which are dominated by RC time constant of the pulse generator. The delay time was measured by comparison of the recorded transient characteristics with a transient characteristic of the reference tunnel GaAs light emitting diode. The delay time of the tunnel GaAs light emitting diode was less than 0.1 ns.

3 Results and discussion

An equivalent scheme of the developed silicon integrated optoelectronic unit is shown in Fig.1. It includes two aluminum/porous silicon Schottky junctions, with an alumina layer between them. One of the junctions operates as a LED, and the other as a photodetector (PD). The distance between them is 10 μm. The anodic aluminum oxide (alumina) protects the porous silicon surface. Moreover, it plays another important role in the device. The light emitted by one of the Schottky junctions is transmitted through the alumina layer as in an optical waveguide. Since the refractive index of porous silicon (1.3-1.6) is lower than that of alumina (1.65-1.77) [2], the anodic alumina layer provides an appropriate light guiding effect.

When the bias of the left Schottky junction approaches to the avalanche breakdown and increased reverse current (I_{LED}) passes through it, light emission from the LED can be visually detected. Meantime reverse current appears in the right Schottky junction operating in a photodetector mode. The current through this junction is increased with an increase of I_{LED} as it is depicted in Fig. 2. Similar behavior is observed when light from an external source is directed to the PD. So, we conclude that the measured current is a photo response of the right Schottky junction.

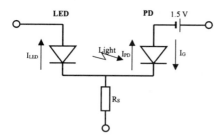 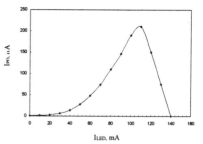

Figure 1. Equivalent electric scheme of integrated porous silicon optoelectronic unit.

Figure 2. PD current versus LED current in the porous silicon based optoelectronic unit with 10 μm alumina waveguide between LED and PD.

There is a galvanic link between the LED and the PD (see Fig. 1). However the direction of the galvanic current (I_G) is opposite to the measured PD current (I_{PD}). In order to reduce the influence of the galvanic current, we have used an additional

1.5 V battery connected with the PD, as it is shown in the Fig. 1. Moreover, the substrate resistance (R_S) is about 10 Ω, and simple calculation shows, that for I_{LED} less than 100 mA I_G could not change the PD bias. But for I_{LED} > 100 mA I_G affects the PD bias, thus resulting in the decrease of the I_{PD} values. The relationship between I_{PD} and I_{LED} is close to a quadratic dependence. The relationship between LED electroluminescence and I_{LED} is also quadratic [5] confirming that the PD response originates from LED light emission.

The ratio of I_{PD} to I_{LED} reaches 0.19 %. This value can be considered as the minimum quantum efficiency of the developed LED. The external quantum efficiency measured by an external PD was 0.01 %. This means, that the most of the light from the LED propagates through the alumina waveguide.

Special attention has been paid to the time response of the LED. The transient electroluminescence wave form with the minimized time response is shown in Fig. 3. This curve corresponds to the lowest of series resistance and capacitance in the developed LED. The basic feature of the transient electroluminescence can be characterized by the delay time (time between the application of the drive pulse and the start of the light response is visible) and the rise time. The delay time of 1.2 ns and the rise time of 1.5 ns were registered for the voltage pulse of 12 V.

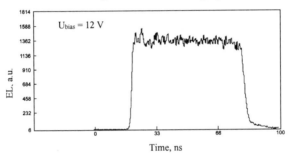

Figure 3. Electroluminescence time response of the porous silicon avalanche LED.

The shortest time response is observed for the maximum bias applied, as it was also described in [6] for forward biased porous silicon LEDs. However, our electroluminescence devices are faster in comparison with the forward biased porous silicon LEDs, because in our case there is no injection capacitance, which limits the time response of the light emission. During the light emission from reverse biased junctions the main mechanism of minor carrier generation is impact ionization at avalanche breakdown. For the avalanche breakdown it is necessary to achieve a high value of electric field. The regular columnar structure of porous silicon promotes the avalanche breakdown due to unequal electric field distribution inside the porous layer [7]. The effect of impact ionization at avalanche breakdown is very fast. For example the time of the avalanche response are estimated to be about 10 ps [8]. The faster carrier generation mechanism explains the shorter time response of our devices in comparison with forward biased porous silicon LEDs.

4 Conclusion

We have developed an effective avalanche porous silicon light emitting diode, which can be integrated with other optoelectronic components, including monocrystalline or porous silicon photodetectors and thin film alumina waveguide, in order to fabricate optoelectronic intra-chip interconnects in silicon substrates. We have demonstrated that the developed LED can operate in the nanosecond range. By further technology optimization we hope to reach the subnanosecond range. Moreover, the proposed optoelectronic unit can be also optimized in order to reduce optical losses in the waveguide and to increase the PD efficiency. It opens new possibilities for integration of electronic and optoelectronic devices. This work is in progress.

References

1. Lazarouk S., Jaguiro P., Katsouba S., Masini G., Monica S.La, Maiello G., Ferrari A., Stable electroluminescence from reverse biased n-type porous silicon-aluminum Schottky junction device, *Appl. Phys. Lett.* **68**, (1996) pp. 2108–2110.
2. Lazarouk S., Jaguiro P., Borisenko V., Integrated optoelectronic unit based on porous silicon, *Phys. Stat. Sol. (a)* **165** (1998) pp. 87-90.
3. Lazarouk S. K., Leshok A. A., Borisenko V. E., Mazzoleni C., Pavesi L., On the route towards Si-based optical interconnects, *Microelectron. Eng.* **50** (2000) pp. 81-86.
4. Bertoloti M., Carassiti F., Fazio E., Ferrari A., Monica S. La, Lazarouk S., Liakhou G., Maello G., Proverbio E., Schirone L., Porous silicon obtained by anodization in the transition regime, *Thin Solid Films* **255** (1995) pp. 152-154.
5. Lazarouk S., Katsouba S., Tomlinson A., Benedetti S., Mazzoleni C., Mulloni V., Mariotto G., Pavesi L., Optical characterization of reverse biased porous silicon light emitting diode, *Mater. Sci. Eng. B* **69-70** (2000) pp. 114-117.
6. Cox T. I., Simons A. J., Loni A., Calcott P. D. J., Canham L. T., Uren M. J. and Nash K. J., Modulation speed of an efficient porous silicon light emitting device, *J. Appl. Phys.* **86** (1999) pp. 2764-2773.
7. Lazarouk S. K. and Tomlinson A. A. G., Formation of pillared arrays by anodization of silicon in the boundary transition region: an AFM and XRD study, *J. Mat. Chem.* **7** (1997) pp. 667-673.
8. Sze S. M. Semiconductor Devices: Physics and Technology (A Wiley-Interscience publication, New York, 1985).

INFRARED ABSORPTION IN STRAINED $Si/Si_{1-x}Ge_x/Si$ QUANTUM WELLS

G. HIONIS AND G. P. TRIBERIS

University of Athens, Physics Department, Solid State Section,
Panepistimiopolis 157 84 Zografos, Athens, Greece
E-mail: ghionis@cc.uoa.gr

The energy subbands in pseudomorphic p-type $Si/Si_{1-x}Ge_x/Si$ quantum wells are calculated within the multiband effective-mass approximation that describes the heavy, light and split-off hole valence bands. We examine the infrared intersubband absorption in this system. The selection rules are obtained for light polarization vector parallel or perpendicular to the growth direction. Comparison is made with other theories and experiment.

1 Introduction

Infrared absorption in heterostructures can be exploited for technological applications, such as infrared photodetectors (IRP)[1]. These devices have tremendous applications in satellite resource mapping, night vision, medical thermography and optical communications. In particular the Si/SiGe system has attracted the researchers for potential applications in IRPs. This stems from the potential subsequent integration of Si/SiGe quantum wells on Si-based circuits. Integration of IRPs is desired (e.g. in silica-based optical fibers and in imaging applications that call for high resolution and thus for complex detector arrays) since it is promising for low cost and miniaturization of the relevant devices. On the other hand the band offset of these systems is significant in the valence band. Valence band quantum wells are especially attractive since intersubband absorption is allowed for light polarized both parallel and perpendicular to the growth direction[2]. Thus, normal-incidence quantum well IRPs can be realized without the need of a grating coupler that is used in III-V systems.

In the precent work we study the infrared intersubband absorption in p-type Si/SiGe quantum wells using a multiband effective-mass equation, concentrating on the selection rules governing the absorption.

2 Theory

The system we study is a single pseudomorphic $p - Si/Si_{1-x}Ge_x/p - Si$ symmetrically modulated doped quantum well with growth direction $< 100 >$, which we take as the z axis. The $Si_{1-x}Ge_x$ layer extends from $-L$ to L, while in the Si layer there is a spacer of thickness D_s at both sides of the well.

We work in the envelope function approximation (EFA). For the description of the quasi two-dimensional hole gas (Q2DHG) present in the pseudomorphic Si/SiGe system the full 6×6 Luttinger-Kohn (L–K) Hamiltonian has to be employed including the heavy hole (HH), the light hole (LH) and the split-off hole (SO) valence bands [3]. Strain effects are easily included in the picture of deformation potential theory. The full theory has been published elsewhere [3].

For the solution of the multiband effective mass equation we use the Fourier Grid Hamiltonian Method (FGHM) [4], properly modified.

Absorption spectra can be deduced from free carrier and intersubband absorption. Here, we are interested on the selection rules that govern the lateral. The absorption coefficient for inter-sub-valence-band transition between the subband n and the subband n' is given by[5]

$$\alpha_{n,n'}(\omega) = \frac{4\pi^2 e^2}{\sqrt{\epsilon} m_0^2 \omega c} \sum_{\mathbf{k}_\parallel} [f_n^0(\mathbf{k}_\parallel) - f_{n'}^0(\mathbf{k}_\parallel)] \left| \vec{\varepsilon} \vec{P}_{nn'} \right|^2 \frac{\hbar \Gamma_{nn'}/\pi}{[\Delta_{nn'} - \hbar\omega]^2 + [\hbar\Gamma_{nn'}]^2} \quad (1)$$

where ϵ is the dielectric constant, m_0 the free electron mass, $\Delta_{nn'}(\mathbf{k}_\parallel) \equiv |E_n(\mathbf{k}_\parallel) - E_{n'}(\mathbf{k}_\parallel)|$, $\vec{P}_{nn'}(\mathbf{k}_\parallel)$ the momentum matrix element, $E_n(\mathbf{k}_\parallel)$ and $E_{n'}(\mathbf{k}_\parallel)$ the in-plane dispersion relations, $\vec{\varepsilon}$ the light polarization vector $\Gamma_{nn'}(\mathbf{k}_\parallel)$ the scattering rates between n and n' subbands and $\mathbf{k}_\parallel$ is the wavelength parallel to the plane of the well.

We focus on $\vec{\varepsilon} \vec{P}_{nn'}(\mathbf{k}_\parallel)$ term. This can be expressed in terms of the coefficients of the inverse effective-mass tensor of the valence band $D^{\alpha,\beta}_{n,n'}$ ($\alpha, \beta = x, y, z$), and the holes envelope function $f_\nu(n, \mathbf{k}_\parallel; z)$. It expresses the selection rules imposed by the active material of the heterostructure as well as by the envelope function of the localized holes.

In the L-K set of basis states ($\nu = 1, .., 6$), the $\vec{\varepsilon} \vec{P}_{nn'}$ term is given by[5]

$$\vec{\varepsilon} \vec{P}_{nn'} = \left(\frac{2m_0}{\hbar^2} \right) \left[\hbar \vec{\varepsilon} \sum_{\nu,\nu'} (\vec{\mathcal{R}}_{\nu\nu'} \mathcal{O}_{\nu\nu'}^{nn'} + \vec{\mathcal{Q}}_{\nu\nu'} \mathcal{D}_{\nu\nu'}^{nn'}) \right] \quad (2)$$

where $\vec{\varepsilon} \vec{\mathcal{R}}_{\nu\nu'} \equiv \sum_i \varepsilon_i \left(D_{\nu\nu'}^{xi} k_x + D_{\nu\nu'}^{yi} k_y \right)$ and $\vec{\varepsilon} \vec{\mathcal{Q}}_{\nu\nu'} \equiv \sum_i \varepsilon_i D_{\nu\nu'}^{zi}$. $\mathcal{O}_{\nu\nu'}^{nn'}$ is the element of the overlap matrix $\mathcal{O}^{nn'}$: $\mathcal{O}_{\nu\nu'}^{nn'} = \int F_\nu^*(n, \mathbf{k}_\parallel; z) F_{\nu'}(n', \mathbf{k}_\parallel; z) dz$ and $\mathcal{D}_{\nu\nu'}^{nn'}$ is the element of the dipole matrix $\mathcal{D}^{nn'}$: $\mathcal{D}_{\nu\nu'}^{nn'} = i \int F_\nu^*(n, \mathbf{k}_\parallel; z) \frac{\partial}{\partial z} F_{\nu'}(n', \mathbf{k}_\parallel; z) dz$

3 Results and Discussion

We examine the case of a $Si/Si_{0.75}Ge_{0.25}/Si$ quantum well with $2L = 47\text{Å}$, $D_s = 20\text{Å}$ and acceptor concentration $2.2 \times 10^{18} cm^{-3}$, in order to compare our results with experiment[6] and other theoretical investigations[6].

The subbands are labeled from the corresponding dominant spinor component at $\mathbf{k}_\parallel \sim 0$. The successive subbands of the same kind (for example HH) have alternately different parity. Only the first subband ($n = 1$ HH1) is occupied, with $k_F \sim 0.02\text{Å}^{-1}$, thus we study the optical transitions for $\mathbf{k}_\parallel \sim 0$.

By $\mathcal{M}$ we denote a concrete form of the transition matrix in order to have a unified picture to apply the various polarizations. This is given by

$$\mathcal{M} = \begin{bmatrix} dh & A & B & 0 & (i/\sqrt{2})A & -i\sqrt{2}B \\ A^\dagger & dl & 0 & B & iC & i\sqrt{3/2}A \\ B^\dagger & 0 & dl & -A & -i\sqrt{3/2}A^\dagger & iC \\ 0 & B^\dagger & -A^\dagger & dh & -i\sqrt{2}B^\dagger & -(i/\sqrt{2})A^\dagger \\ -(i/\sqrt{2})A^\dagger & -iC & i\sqrt{3/2}A & i\sqrt{2}B & ds & 0 \\ i\sqrt{2}B^\dagger & -i\sqrt{3/2}A^\dagger & -iC & (i/\sqrt{2})A & 0 & ds \end{bmatrix} \quad (3)$$

thus $(\mathcal{R} or \mathcal{Q})^i = -\frac{\hbar^2}{2m_0}\mathcal{M}$, $i = x, y, z$

We distinguish two cases for the polarization of the incident field. For $\vec{\epsilon} \parallel \vec{z}$ the matrix $\mathcal{R}^z$ for $\mathbf{k}_\parallel = 0$ is zero and for $\mathbf{k}_\parallel \sim 0$ is non-zero only with very small but non-vanishing elements A ($A \equiv -\sqrt{3}\gamma_3(k_y + ik_x)$). In the $\mathcal{Q}^z$ matrix only the diagonal terms survive ($dh = \gamma_1 - 2\gamma_2$, $dl = \gamma_1 + 2\gamma_2$, $ds = \gamma_1$, and $C = 2\sqrt{2}\gamma_2$).

Therefore transitions are allowed between states of the same kind i.e. HH-HH and LH-LH and also of different kind i.e. HH-LH (although weaker), for $\mathbf{k}_\parallel \sim 0$. According to Eq. (2), for transitions between states of the same kind we must also have $\mathcal{D}_{nn'} \neq 0$. Thus only transitions between states with the different parity are allowed. Considering these selection rules, at $\mathbf{k}_\parallel \sim 0$, the most important allowed transition is between HH_1 and HH_2.

For $\vec{\epsilon} \perp \vec{z}$ e.g. $\vec{\epsilon} \parallel \vec{x}$, the matrix $\mathcal{R}^x$ is zero while in $\mathcal{Q}^x$ only $A \neq 0$ ($A \equiv -i\sqrt{3}\gamma_3$). Therefore only the transitions between states of different kind are allowed (i.e. HH-LH). According to Eq. (2) we must have $\mathcal{D}_{nn'} \neq 0$. Thus only transitions between states with different parity are allowed. Considering these selection rules, at $\mathbf{k}_\parallel \sim 0$, the most important allowed transition is between HH_1 and LH_2.

Detailed numerical calculations (Fig.1) verify the conclusions from the above analysis of $\mathcal{R}$ and $\mathcal{Q}$.

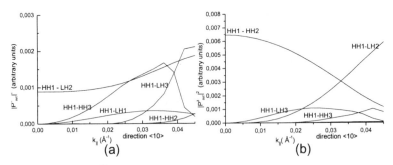

Figure 1. $\left|\vec{\varepsilon}\vec{P}_{nn'}(\mathbf{k}_{\|})\right|^2$ from the first to lower subband as a function of $k_{//}$ in the $<10>$ direction, (a) for polarization $\vec{\epsilon} \parallel \vec{z}$ and (b) for polarization $\vec{\epsilon} \perp \vec{z}$.

For $\mathbf{k}_{\|}$ away from zero the selection rules discussed above become less strict. Nevertheless the transitions we concluded remain predominant, for every kind of polarization, up to k_F. The possibility of absorption for $\vec{\epsilon} \perp \vec{z}$ is a consequence of the band mixing and characteristic of the quantum wells of holes. Compared with the absorption for polarization $\vec{\epsilon} \parallel \vec{z}$ it is less intense.

In the work of Fromherz et al.[6] the observation of absorption for polarization $\vec{\epsilon} \perp \vec{z}$ was not possible. For polarization $\vec{\epsilon} \parallel \vec{z}$ they observed maximum absorption at wavenumber $750 cm^{-1}$. They proposed a theory based on a self-consistent one-subband model and they attribute this line to the transition between states of the same kind and different parity. Our multiband approach is in agreement with their results and we have a very good coincidence with experiment as our line for the absorption according to our calculations is found at a wavenumber $755 cm^{-1}$.

References

1. B.F. Levine, J.Appl.Phys.**74**, R1 (1993).
2. R. People, J.C. Bean, C.G. Bethea, S.K. Sputz and L.J. Peticolas, Appl.Phys.Lett.**61**, 1122 (1992).
3. G. Hionis and G.P. Triberis, Superlatt. Microstruct. **24** 399 (1998).
4. C.C. Marston and G.G. Balint-Kurti, J.Chem.Phys.**91** 3571 (1989).
5. C.Chang and R.B. James, Phys.Rev.B **39** 12672 (1989).
6. T. Fromherz, E. Koppensteiner, M. Helm and G. Bauer, Phys.Rev.B **50** 15073 (1994).

THERMOPOWER CALCULATIONS AT FILLING FACTOR 3/2 AND 1/2 FOR TWO-DIMENSIONAL SYSTEMS

V. C. KARAVOLAS AND G.P. TRIBERIS

*University of Athens, Physics Department, Solid State Section,
Panepistimiopolis, 15784 Zografos, Athens, Greece
E-mail: vkaravol@cc.uoa.gr*

The diffusion and the phonon-drag thermopower are calculated for a two-dimensional electron and hole gas at low temperatures in the Fractional Quantum Hall Effect (FQHE) regime at filling factors $\nu = 3/2$ and $\nu=1/2$. The composite fermions (CFs) picture enables us to use the Integer Quantum Hall Effect (IQHE) and Shubnikov de Haas (SdH) conductivity models for a quantitative comparison with experiment. We use the edge states model to calculate the thermopower of the system. We use the idea of parallel conduction of two gases. One gas, composed of electrons (holes) fully occupies one of the two spin levels of the lowest Landau level, and a second composed of composite fermions partially occupies the other spin level. Comparison is given with experiment.

The relation of the transport coefficients such as the thermopower in the fractional quantum Hall effect regime at different filling factors is still an open question. Experiments for electron gases [1] show that for the phonon-drag (S^g) diagonal components at filling factors $\nu = 3/2$ and $\nu=1/2$ are equal i.e. $S^g_{xx,\nu=1/2} = S^g_{xx,\nu=3/2}$. For the case of diffusive transport experimental data [2] show that the diagonal components of the diffusion thermopower (S^d) at filling factors $\nu = 3/2$ and $\nu=1/2$ are related as $S^d_{xx,\nu=1/2} = 0.7\, S^d_{xx,\nu=3/2}$. We present a theoretical study for both cases. The system under study consists of N carriers moving on a plane (x,y) in the presence of an external magnetic field $B = (0,0,B_z)$ perpendicular to the plane. We will consider only the case when the magnetic field is so high that all the carriers populate the two lowest spin-levels of the lowest Landau level.
The transport coefficients are given by

$$\sigma_{ij} = \int_{-\infty}^{\infty}(-\frac{\partial f(E)}{\partial E})\sigma_{ij}(E)dE \quad , \quad L_{ij} = \frac{1}{eT}\int_{-\infty}^{\infty}(-\frac{\partial f(E)}{\partial E})(E-E_F)\sigma_{ij}(E)dE \quad (1)$$

where $\sigma_{ij}(E)$ is the conductivity at T=0 K, for E=E_F, and f(E) is the Fermi-Dirac distribution function.
<u>Filling factor $\nu=1/2$</u>: We calculate the resistivity and thermopower using the transport theory of low dimensional structures based on the edge states of composite fermions [1]. The electrochemical potential of the electrons μ and the composite fermions' μ^* are connected through the following equation [1]

$$\mu_\alpha - \mu_\beta = \mu_\alpha^* - \mu_\beta^* + \int_\beta^\alpha (2h/e)\mathbf{J}\cdot\hat{\mathbf{B}}\times \mathbf{dr} \Rightarrow \Delta\mu = \Delta\mu^* + \int_\beta^\alpha (2h/e)\mathbf{J}\cdot\hat{\mathbf{B}}\times \mathbf{dr} \quad (2)$$

where the indices α and β indicate the edges of the sample. $\int_{\beta}^{\alpha}(2h/e)\mathbf{J}\cdot\hat{\mathbf{B}}\times d\mathbf{r}$ is the net electric current flowing across the path connecting the two edges. It is simply (2h/e)I, where I=-eJ$_N$ the total current and J$_N$ the particle current. Eq. (2) gives

$$\mu_\alpha^* - \mu_\beta^* = \mu_\alpha - \mu_\beta + \frac{2h}{e}I \quad \text{or} \quad \Delta\mu = \Delta\mu^* + 2hJ_N \qquad (3)$$

The resistivity and the thermopower are given by

$$\rho = \frac{\nabla\mu}{e^2 J_N} = \frac{\Delta\mu^* + \int_{\beta}^{\alpha}(2h/e)\mathbf{J}\cdot\hat{\mathbf{B}}\times d\mathbf{r}}{e^2 J_N} = \frac{\Delta\mu^*}{e^2 J_N} + \frac{2hJ_N}{e^2 J_N} = \rho^{QP} + \rho_{CS} \,, \quad (\nabla T = 0) \qquad (4)$$

$$S = -\frac{1}{e}\frac{\nabla\mu}{\nabla T} = -\frac{1}{e}\frac{\Delta\mu^* + \int_{\beta}^{\alpha}(2h/e)\mathbf{J}\cdot\hat{\mathbf{B}}\times d\mathbf{r}}{\nabla T} = -\frac{1}{e}\frac{\Delta\mu^*}{\nabla T} + \frac{1}{e}\frac{2hJ_N}{\nabla T} = S^{QP} \,, \quad (J_N = 0)\,, (5)$$

where $\rho^{QP} = \frac{\Delta\mu^*}{e^2 J_N}$, and $S^{QP} = \frac{1}{e}\frac{\Delta\mu^*}{\nabla T}$, are the quasi-particle components of the resistivity and the thermopower. The conductivities for the carriers in the FQHE and for the CFs in the IQHE are added in parallel. A very important result is that in the thermopower tensor there is not the Chern-Simmons term which appears in the resistivity[4]. Thus the total thermopower of the system is given only by the quasi-particle term. The analytical expressions for ρ and S appear elsewhere [4,5]. The conductivity of the carriers is calculated from the inversion of the resistivity matrix. At high magnetic fields applied to 2D systems, ρ_{xx} becomes vanishingly small and ρ_{xy} shows plateaus, in finite ranges of the magnetic field, when E$_F$ lies between two separated Landau levels. Englert [6] proposed a model that explains satisfactory the above behavior. We use this model taking gaussian distributions for the total density of states, D_N, and for the extended states, $D_{\lambda N}$ [2].

Filling factor ν=3/2: The ν=3/2 case is qualitatively different from the ν=1/2 case due to the fact that the magnetic field is inadequate to transform all the carriers to composite fermions, as in the ν=1/2 case. Our analysis is based on the discrimination of the carriers in two different gases, showing parallel conduction [5]. The first consists of carriers which fully occupies one of the two spin levels of the lowest Landau level (ν=1). Its transport coefficients are σ^1_{xx}=0, L^1_{xx}=0 και σ^1_{xy}=e^2/h, L^1_{xy}=0. The second consists of carriers, which have been transformed to CFs, partially occupying the other spin level (ν=1/2). As the actual magnetic field changes, the concentration of each gas changes. Due to the fact that the carrier concentration of the fully occupied spin level is field dependent, the field resulting in a filling factor ν= 1/2, for the composite fermion gas, will also be field dependent. Thus the effective magnetic field will be $B_{eff} = B_{tot} - B_{\nu=1/2}$ where B$_{tot}$ is

the total magnetic field. The evaluation of the total conductivity and the resistivity of the system is presented elsewhere[5].
The thermopower of the system is given by

$$S = -L\rho \qquad (6)$$

where L has matrix elements given by Eq.(1). For $B_{eff}=0$ $\rho_{xx} \ll h/e^2$ and thus it can be neglected. For $B_{eff}=0$, also $\rho_{xy}=0$, $S_{xy}^{QP}=0$. Eq. (6) after detailed manipulations results in

$$S_{xx} = S_{xx}^{QP}/3 \qquad (7)$$

Fig.1 shows S_{xx}^{QP}/S_{xx} for a wide range of magnetic field around v=3/2. With decreasing B_{eff} the ratio S_{xx}^{QP}/S_{xx} decreases. The peaks observed are due to the contribution of the S_{xy}^{QP} to the S_{xx}.

Fig.1. Theoretical results for S_{xx}^{QP}/S_{xx} around filling factor v=3/2.

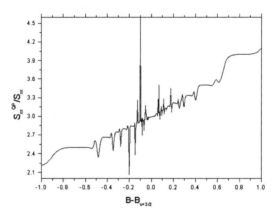

A) Diffusion
It can be shown using Sommerfeld analysis [7] that the L coefficients for the fully occupied landau level are zero. Using Mott's formula for the diffusion thermopower,

$$S_{xx}^d = -\frac{\pi^2 kT}{3eE_F}(p+1) \qquad (8)$$

and taking into account the fact that the number of carriers that are transformed to CFs at v=3/2 is the one third of those transformed to CFs at v=1/2 and that the effective mass of the CFs is proportional to the square root of the CF concentration [8] we find that

$$S_{xx,v=1/2}^d = 1.73 S_{xx,v=3/2}^d \qquad (9)$$

The experimental data of Ying et al. [9] show a ratio $S^d_{xx,\nu=1/2}=0.71\ S^d_{xx,\nu=3/2}$ However, their accuracy around $\nu=3/2$ is limited due to the difficulty to clearly discriminate the oscillations of the thermopower. The deviation of our theoretical results is due to the value for the effective mass used[4] in our calculations.

B) Phonon-Drag
Experimentally it is well known that for a fully occupied Landau level $S^g_{xx}=0$, $S^g_{xy}=0$ and $\rho_{xx}=0$. Using equations

$$S^g_{xy} = \rho_{xy} L_{xx} + \rho_{xx} L_{xy} \quad , \quad S^g_{xx} = \rho_{xx} L_{xx} + \rho_{xy} L_{xy} \qquad (10)$$

we conclude $L_{xx}=0$ και $L_{xy}=0$. Also experimentally it has been found that [1] the phonon drag component of the thermopower is proportional to the inverse of the CFs concentration at T=440 mK. Using the above, from Eq. (7) we find that for the phonon-drag component

$$S^g_{xx,\nu=3/2} = S^g_{xx,\nu=1/2} \qquad (11)$$

at least for the temperature values of Tieke's et al. experiment [1]. This is in absolute agreement with Tieke et al. experimental data.

V.C.K. acknowledges the support of the Public Benefit Foundation "Alexander S. Onassis".

References

1. B. Tieke, U. Zeitler, R. Fletcher, S.A. Wiegers, A.K. Geim, J.C. Maan and M. Hennini, Phys. Rev. Lett. **76** (3630) 1996.
2. X. Ying, V. Bayot, M.B. Sandos and M. Shaeygan, Phys.Rev.B **50**, 4969 (1994)
3. G. Kirczenow and B.L. Johnson, Phys.Rev.B **51**, 17579 (1995)
4. V.C. Karavolas, G.P. Triberis and F.M. Peeters, Phys.Rev.B **56**, 15289 (1997)
5. V.C. Karavolas and G.P. Triberis, Phys.Rev.B **63**, 035313 (2001)
6. Th. Englert, in *Application of High Magnetic Fields in Semiconductor Physics II*, edited by G. Landwehr (Springer, Berlin, 1983) p.165.
7. V.C. Karavolas and G.P. Triberis, Phys.Rev.B **59**, 7590 (1999)
8. N. d' Ambrumenil and R. Morf, Surf.Sci. **361-362**, 92 (1996).

THERMOELECTRIC PROPERTIES OF COMPOSITE FERMIONS

M. TSAOUSIDOU[1,2] AND G. P. TRIBERIS[1]

[1] *University of Athens, Physics Department, Section of Solid State Physics, Zografos 157 84, Greece.*
[2] *University of Warwick, Physics Department, Coventry CV4 7AL, United Kingdom.*

We explain thermopower measurements at a half filled Landau level in GaAs/AlGaAs structures. The phonon-drag thermopower, S^g, dominates in the whole temperature range we examine here. We describe S^g by using the standard Cantrell-Butcher theory for S^g at zero magnetic field within the composite fermion framework.

Significant progress for the understanding of the fractional quantum Hall effect (FQHE) at Landau levels with even denominator filling factors, $\nu = 1/2m$, has been made with the appearance of a novel model known as the composite fermion (CF) model [1,2]. A composite fermion consists of an electron (or hole) with an even number ($2m$) of attached magnetic flux quanta. In mean field theory the external magnetic field at $\nu = 1/2m$ is canceled by the fictitious field associated with the attached flux quanta [1]. The CFs form a well defined Fermi surface of radius $k_f = (4\pi n_i)^{1/2}$, where n_i is the carrier sheet density [1].

In the present work we adopt the CF approach in order to explain the measured thermopower at $\nu = 1/2$ in two-dimensional electron [3] and hole [4] gases confined in AlGaAs/GaAs heterostructures. The measurements span a temperature range between 0.2 – 1.2 K. At those temperatures the phonon-drag contribution, S^g, to the thermopower dominates over the diffusion contribution [3].

The origin of the phonon-drag thermopower of a 2D carrier gas imbedded in a bath of 3D acoustic phonons is very simple. Applying a small thermal gradient ∇T along the plane of the 2D gas phonons flow in the direction $-\nabla T$. This leads to momentum transfer from the perturbed phonons to the carriers through the carrier-phonon interaction. In other words, the carriers are *dragged* by phonons producing a phonon-drag contribution $\mathbf{J}^g$ to the carrier current density. The electric field required to stop $\mathbf{J}^g$ is: $\mathbf{E} = S^g \nabla T$, where S^g is the phonon-drag thermopower. The theoretical model for the description of S^g at zero magnetic field has been proposed by Cantrell and Butcher [5].

The only unknown parameter in our calculations of S^g is the value of the composite fermion mass, m_{cf}. Here we take $m_{cf} = 1.5\,m_e$, where m_e is the free electron mass. This value is consistent with the theoretical estimation of Park and Jain [6] and with the experimental values suggested by Du et al [7] and Manoharan et al [8] when $\nu \to 1/2$.

1 Theory

We apply the Cantrell-Butcher theory to the case of CFs by replacing the carrier band mass by the CF mass, m_{cf}, and by removing the spin degeneracy. Then we take [9]

$$S^g = \frac{\pm \Lambda m_{cf}}{8\pi^2 k_B T^2 n_i e} \sum_{\mathbf{Q}s} \frac{\omega_{\mathbf{Q}} q^2}{Q} \int_0^{2\pi} d\theta \int d\epsilon_{\mathbf{k}} f^0(\epsilon_{\mathbf{k}})[1-f^0(\epsilon_{\mathbf{k}}+\hbar\omega_{\mathbf{Q}})] P_{\mathbf{Q}}^a(\mathbf{k},\mathbf{k}+\mathbf{q}) \quad (1)$$

where the minus sign corresponds to electrons and the plus sign to holes. Here, Λ is the phonon mean free path, n_i is the carrier sheet density, and e is the magnitude of the electronic charge. Moreover, $\omega_{\mathbf{Q}}$ is the frequency of the acoustic phonon with wave vector $\mathbf{Q} = (\mathbf{q}, q_z)$ and mode s, $\mathbf{k} = (k_x, k_y)$ is the CF wave vector, $\epsilon_{\mathbf{k}} = \hbar^2 k^2 / 2m_{cf}$ is the CF energy, f^0 is the Fermi-Dirac distribution function and θ is the angle between $\mathbf{k}$ and $\mathbf{q}$. Finally, $P_{\mathbf{Q}}^a(\mathbf{k},\mathbf{k}+\mathbf{q})$ is the transition rate at which the CF transfers from $\mathbf{k}$ to $\mathbf{k}+\mathbf{q}$ by absorbing one phonon with wave vector $\mathbf{Q}$. The full expression for $P_{\mathbf{Q}}^a(\mathbf{k},\mathbf{k}+\mathbf{q})$ is given in Refs [5,9]. The transition rate is proportional to the square of the matrix element of the screened carrier-phonon interaction which is given by $|M^s|^2 = |\Xi^s|^2 / |\epsilon(\omega_{\mathbf{Q}}, q)|^2$, where $\epsilon(\omega_{\mathbf{Q}}, q)$ is the 2D dielectric function for the ground state [10] and $|\Xi^s|^2$ is the square of the matrix element of the 'effective' carrier-phonon interaction. Since GaAs is a polar material $|\Xi^s|^2$ accounts for both deformation potential coupling and piezoelectric coupling. For the longitudinal acoustic branch (l) and for each of the transverse branches (t) we obtain [11] $|\Xi^l|^2 = \Xi_d^2 + |\Xi^l_{piez}|^2$ and $|\Xi^t|^2 = |\Xi^t_{piez}|^2$ where Ξ_d is the deformation potential constant. The expressions for $|\Xi^s_{piez}|^2$ are given by Lyo [11].

At temperatures lower that 1 K we need to consider the effects of the Chern-Simons gauge field on the CF-phonon piezoelectric coupling as it is proposed by Khveshchenko and Reizer [12]. Then the square of the matrix element for the screened piezoelectric interaction is [9]

$$|M^{cf,s}_{piez}|^2 = \frac{|\Xi^s_{piez}|^2}{|\epsilon_{cf}(\omega_{\mathbf{Q}}, q)|^2}[1 + \Phi^2 F^2(q)\frac{k^2}{q^2}\sin^2\theta], \quad (2)$$

where $\epsilon_{cf}(\omega_{\mathbf{Q}}, q)$ is the dielectric function at $\nu = 1/2$ [9,12]. The second term in the right hand side of Eq. (2) is the contribution that accounts for the gauge field. Here, $\Phi = 2$ is the number of the magnetic flux quanta attached to each carrier at $\nu = 1/2$ and $F(q)$ is the dielectric form factor [10].

2 Results and Discussion

We use the above theoretical formalism and we calculate the phonon-drag thermopower at $\nu = 1/2$ for a two-dimensional electron gas (2DEG) and a two-dimensional hole gas (2DHG) confined in AlGaAs/GaAs. Our results are compared with the experimental data obtained by Tieke et al [3] in a 2DEG with sheet density $n_e = 1.75 \times 10^{15}$ m^{-2} (full details are given in Ref. [9]) and the data of Crump et al [4] in a 2DHG with sheet density $n_h = 1.25 \times 10^{15}$ m^{-2}. The values of the parameters used are the standard ones for GaAs [11]. The deformation potential constant is taken to be $\Xi_d = 9.3$ eV for the conduction band [11] and $\Xi_d = 9.2$ eV for the valence band [13]. The phonon mean free path Λ is extracted from the measured thermal conductivity.

Fig. 1 shows the theoretical and the experimental values of thermopower for the 2DEG (Fig.1a) and the 2DHG (Fig.2b) we consider here. The circles are the experimental data and the solid lines are our calculations. The dot lines correspond to the theoretical estimation of S^g when the effect of the Chern-Simons gauge field on the CF-phonon piezoelectric coupling is neglected. We see that at low temperatures the gauge field plays important role in the CF-phonon interaction.

We should stress here that the only uncertain parameter in our calculations is the value of the CF mass. Here we take $m_{cf} = 1.5 m_e$ [6,7,8,9]. However, at low T where screening is severe S^g becomes independent of the value of the mass [9]. Thus, practically speaking, our calculation of S^g does not involve any adjustable parameter.

Acknowledgments

The authors wish to dedicate this paper in the memory of Professor Paul N. Butcher who will be greatly missed. His contribution to this work was invaluable and decisive. MT acknowledges the UK Engineering and Physical Sciences Research Council for funding this research. Also, the authors wish to thank the University of Athens for financial support.

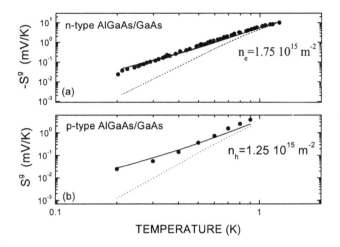

Figure 1. The thermopower as a function of T at $\nu = 1/2$ for (a) a 2DEG and (b) a 2DHG in AlGaAs/GaAs. The circles are the experimental data and the solid lines are the theoretical values of S^g. The dot lines correspond to the theoretical prediction of S^g without taking into account the effect of the gauge field on the CF-phonon piezoelectric interaction.

References

1. B.I. Halperin, P.A. Lee, and N. Read, *Phys. Rev.* B **47**, 7312 (1993).
2. J.K. Jain, *Phys. Rev. Lett.* **63**, 199 (1989).
3. B. Tieke et al, *Phys. Rev.* B **58**, 2017 (1998).
4. P.A. Crump et al, *Sur. Sci.* **361-362**, 50 (1996).
5. D.G. Cantrell and P.N. Butcher, *J. Phys.* C **20**, 1985 (1987); 1993 (1987).
6. K. Park and J.K. Jain, *Phys. Rev. Lett.* **81**, 4200 (1998).
7. R.R. Du et al, *Phys. Rev. Lett.* **73**, 3274 (1994).
8. H.C. Manoharan et al, *Phys. Rev. Lett.* **73**, 3270 (1994)
9. M. Tsaousidou et al, *Phys. Rev. Lett.* **83**, 4820 (1999).
10. A. Gold and V.T. Dolgopolov, *Phys. Rev.* B **33**, 1076 (1986).
11. S.K. Lyo, *Phys. Rev.* B **38**, 6345 (1988).
12. D.V. Khveshchenko and M. Yu. Reizer, *Phys. Rev. Lett.* **78**, 3531 (1997).
13. B.K. Ridley in *Quantum Processes in Semiconductors* (Clarendon Press, Oxford, 1988).

DESIGN AND FABRICATION OF SUPPORTED-METAL CATALYSTS THROUGH NANOTECHNOLOGY

I. ZUBURTIKUDIS

Dept. of Industrial Design Engineering, TEI of West Macedonia, 50100 Kozani, GREECE
E-mail: izub@kozani.teikoz.gr

A design and a solid state fabrication scheme are proposed for the production of supported-metal catalysts micrometer-long and nanometer-wide. Ni catalysts supported on SiO_2 have been fabricated according to the proposed scheme. The catalytic testing of the model Ni/ SiO_2 system revealed that it can reproduce the behavior of traditional Ni on silica clusters. Thus, the model system is an adequate catalyst and the proposed scheme is appropriate for the manufacturing of supported-metal catalysts.

1 Introduction

The need for maximizing the effective surface area of catalytically active metals in heterogeneous systems led to the extensive use of highly dispersed metals in the form of clusters supported on a variety of carrier materials. Clusters with diameters up to 20 nm are of catalytic interest.

Traditionally, supported metal catalysts are produced through chemical routes. A variety of techniques such as coprecipitation and impregnation are used [1]. However, these preparation techniques do not yield uniformly-shaped and -sized clusters. Full reduction of the metal is not assured nor is complete absence of impurities guaranteed. The characterization of the clusters is difficult, since they are within the porous matrix of the support and certain assumptions must be made.

These drawbacks of the chemically-produced supported-metal clusters necessitate the production of model catalytic systems that are carefully prepared, have well-defined geometrical features in the nanoscale, can be characterized relatively easy and have enough surface area for catalytic reaction studies at realistic conditions [2]. The model metal clusters should be uniform in shape, monodisperse, fully reduced, clean or with controllable impurities and with controllable supports.

2 The Design Idea

The advent of modern solid state microfabrication techniques makes possible the construction of new model supported metal catalysts with such features [2]. By alternatively depositing layers of catalyst and support on a suitable substrate, one can form a catalyst/support Layered Synthetic Microstructure (LSM), such as the one shown in Figure 1 below.

The edge of the LSM exhibits a compositionally modulated surface that consists of catalyst and support. The metal catalyst is in the form of strips between those of the support. The width of the catalyst strips is in the nm scale and comparable to the size of supported clusters, while their length is macroscopic.

If the top most layer of the LSM is made of support, the only exposed catalyst surfaces are on the edges of the LSM. In order to provide sufficient metal surface area lithographic and etching techniques can be used to carve the initial LSM in a repetitive pattern in the direction perpendicular to the top surface of the layers. A forest of micrometer-sized towers of given shapes is produced, as it shown below:

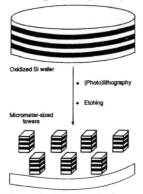

Figure 1. The initial LSM and the forest of micrometer-sized towers. Shaded areas are metal catalyst layers. Ideal picture; dimensions not to scale.

Each tower is essentially identical to every other and has a lateral surface that is also identical to the edge of the initial LSM. The metal catalyst is present as micrometer-long, nanometer-wide metal strips on the edges of each tower. Because all edges are identical, all metal strips are identical and non-uniformities in shape and size of the catalysts are reduced to a minimum. For more, see references [2,3].

3 Fabrication of the Model Catalysts

We used e-beam evaporation and a conventional vacuum system for wafer coating to prepare four different Ni/SiO$_2$ LSMs on top of oxidized 3-inch Si wafers [2]. The Ni layer thickness ranged between 2 and 10 nm, whereas the SiO$_2$ thickness was fixed at 4.1 nm. Photolithography and chemical etching resulted in the creation of ~10^7 towers per Si wafer with a total Ni surface of 1 to 2 cm^2 depending on the specimen. A laser interferometric micrograph of the forest with the 2-nm-thick Ni and the 4.1-nm SiO$_2$ layers (specimen 1) is shown in Figure 2 below:

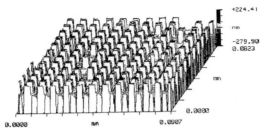

Figure 2. Laser inteferometric micrograph of the manufactured forest for specimen 1.

The Ni catalyst strips are on the edges of these towers, but they are too thin to be resolved optically. Details about the actual fabrication of the Ni/ SiO$_2$ LSMs, the forest of the micrometer-sized towers, the micrometer-long, nanometer-wide Ni strips and their characterization can be found in references [2,3].

4 Catalytic Testing of the Fabricated Model System

We used hydrogenolysis and hydrogenation reactions to test the catalytic activity of the fabricated Ni LSMs. The rate of ethane hydrogenolysis ($C_2H_6 + H_2 =$ 2CH$_4$) and the rate of propane hygrodenolysis ($C_3H_8 + 2H_2 = 3CH_4$ and $C_3H_8 + H_2 =$ CH$_4$+ C$_2$H$_6$) over supported-Ni on SiO$_2$, when normalized to unit surface area of Ni, show a dependency on the Ni-cluster size [1], whereas the normalized rate of ethylene hydrogenation ($C_2H_4 + H_2 = C_2H_6$) over the same system does not show any size dependence [1]. Therefore, these reactions are ideal for revealing whether a model system such as the one we fabricated is functioning properly.

Here, the results for ethane and propane hydrogenolysis over the Ni LSMs are summarized in Fig. 3(A&B) (curve 1 in both A&B) and compared with results over cluster data (curves 2 and 3) and Ni foil (curve 4 only for ethane hydrogenolysis).

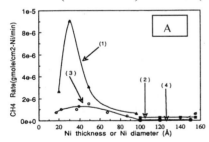
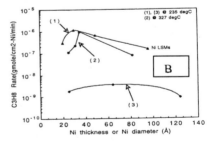

Figure 3(A&B). Ethane (A) and Propane (B) hydrogenolysis over Ni LSMs and traditional Ni catalysts

The observed rates over the Ni LSMs, corrected for background and normalized to the calculated geometrical exposed area of the LSM edge, exhibited a maximum as a function of the Ni layer thickness. The maximum occurred at a thickness very close to the diameter reported for a rate maximum on traditional Ni on silica clusters.

As it is shown in Fig. 4, the normalized rate of ethylene hydrogenation on the Ni LSMs (curve 1) showed no thickness dependence, in agreement with observations made with traditional Ni on silica clusters (curve 2).

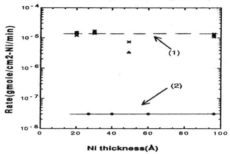

Figure 4. Ethylene hydrogenation over Ni LSMs and traditional Ni catalysts

5 Conclusions

All the above experimental results indicate that the Ni LSMs reproduce the behavior of the traditional Ni on silica clusters. As a result, the model Ni LSM system is an adequate catalyst. Thus, using solid state microfabrication techniques, we developed a scheme for producing supported metal catalysts micrometer-long and nanometer-wide. This fabrication scheme is very promising in producing enough nanometer-sized entities for various chemical studies and very interesting as a potential catalyst-manufacturing route. (This research was possible only after the support of the Univ. of Rochester (Dept. of Chemical Eng.) and RIT (Dept. of Microelectronics Eng.). Their help is gratefully acknowledged.)

References

1. Che, M., Bennett, C. O., The Influence of Particle Size on the Catalytic Properties of Supported Metals, *Advances in Catalysis* **36** (1989) pp. 55-172.
2. Zuburtikudis, I., Layered Synthetic Microstructures and Size Effects of Supported Metal Catalysts,, PhD thesis, Univ. of Rochester, 1992.
3. Zuburtikudis, I., Saltsburg, H., Linear Metal Nanostructures and Size Effects of Supported Metal Catalysts, *Science* **258** (1992) pp. 1337-1339.

CALCULATED SPONTANEOUS EMISSION RATES IN SILICON QUANTUM WIRES GROWN IN {100} PLANE

X. ZIANNI AND A. G. NASSIOPOULOU
IMEL/ NCSR "Demokritos", 153 10 Aghia Paraskevi, Attiki, Greece
E-mail: xzianni@imel.demokritos.gr

We discuss the dependence of the calculated spontaneous emission rates of Si rectangular quantum wires grown in {100} plane on the growth direction. The eigenstates of electrons and holes are calculated within the effective mass approximation. The directional dependence of the electron states is reflected on the spontaneous emission of the quantum wires. It becomes obvious that light emission from quantum wires of diameters of a few nanometers has strong direct transition character, if the wires direction is along one of the main crystallographic directions or along [110]. Phonon assisted transitions are in these cases also present, but their intensity is three orders of magnitude smaller than the intensity of direct transitions.

1 Introduction

The optical properties of Si nanostructures have attracted the interest of research in the last decade [1]. This research has been stimulated by the effort to develop new Si-based devices. Zero-, one- and two- dimensional structures were fabricated and their properties were widely studied. Silicon quantum wires were fabricated by using lithography and highly anisotropic silicon etching [2] and light emission was observed from those wires at room temperature in the visible range both under optical and under electrical excitation due to quantum confinement [3]. Theoretical calculations predict discrete quantum confined states in the conduction and valence band and a size dependent band gap opening responsible for light emission in the visible range [4,5]. Here, we discuss the dependence of the spontaneous emission of Si quantum wires grown in {100} plane on the wire growth direction.

2 Model

We consider uniform quantum wires of infinite length. The direction of the wires is in the {110} plane, at an angle θ with respect to the [010] crystallographic direction. The confinement dimensions of the quantum wires are L_x and L_z. The eigenstates of electrons and holes are calculated within the effective mass approximation [4,5]. In references [4,5] it has been verified that the effective mass theory gives a good description of the electronic states in quantum wires down to at least 1nm wide by comparing their results with those of first principle calculations. For holes, the minimum of the one-dimensional sub-bands is at the Γ point for the three holes bands (heavy hole, light hole and split-off band). Here, the holes bands are approximated as independent. For electrons, the six anisotropic valleys at the minimum of the bulk Si conduction band are taken into account. The six valleys are not equivalent in the case of the quantum wire and so the minima of the electron energy sub-bands depend on the growth direction of the wire. For some angles in the {100} plane, the minimum of the sub-bands is at the Γ-point. The transition rates are given by Fermi's Golden Rule for direct and indirect transitions [4]. The transition rates depend on the position of the energy sub-band minima and the overlap integrals. The overlap integrals exhibit directional dependence, because the phases of the electron wave functions depend on the wire direction.

3 Results and discussion

For illustration purposes we consider two representative cases of quantum wires that have the same cross section but different confinement dimension ratios L_z/L_x. Wire A: L_x=2.5 nm and L_z=2.0 nm with L_z/L_x=0.8 . Wire B: L_x=2.0 nm and L_z=2.5 nm with L_z/L_x=1.25. The results for the energy eigenstates of the two wires are discussed in detail in [4]. For θ=0°, the lowest electron sub-band corresponds either to [001] or to [100] valleys and the wires have direct band gap. Moreover, the spectra are independent of the ratio L_z/L_x of the two wires. For orientations away from the crystallographic axes (θ?0°), the states exhibit different directional dependence

The calculated spontaneous emission spectra for wire A are shown in figure 1.

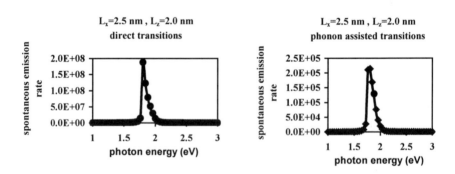

Figure 1. Wire A. Spontaneous emission rate for 0° (solid line) and 30° (dots).

Wire A has a direct gap irrespectively of the wire direction since the ground electron sub-band corresponds to valleys [001] and it is direction independent. A strong direct transition peak is seen at photon energies 1.8 eV that corresponds to the effective band gap of the wire. In the phonon assisted transitions, the dominant contribution is from transverse optical phonons. We find that phonon assisted transitions exhibit a peak at the same energy as direct transitions. This is because the relevant photon energy is in both cases of transitions the effective band gap at the Γ point [4]. Phonon assisted transitions are three orders of magnitude weaker than direct transitions and they are consequently screened by the direct transitions in the emission spectrum.

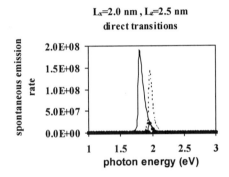

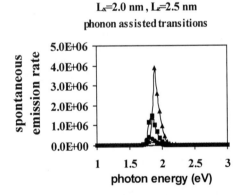

Figure 2. Wire B. Spontaneous emission rate for 0° (solid line), 10° (dots), 20° (squares), 30° (triangles) and 45° (dashed line).

For wire B, the electron ground sub-band corresponds to valleys [100] for most wire directions, namely $0°<\theta<35°$. For these wire directions, the quantum wire has indirect gap. A cross-over occurs at higher angles, $35°<\theta<45°$, where the electron ground sub-band corresponds to valleys [001] and the wires have direct gap. These features are reflected in the emission spectra of the wires. In figure 2, it can be seen that a small deviation by 10° of the wire direction from the main crystallographic directions, shifts the peak to higher photon energies because of the shift of the minimum of the gap away from Γ-point. Now, direct transitions are allowed due to thermal broadening of the electrons energy distribution. Further increase of the angle θ, causes bigger shift of the band gap minimum away from the Γ-point and consequent elimination of direct emission rates (figure 2). As θ increases, the phonon assisted emission rates become stronger, as it can be seen in figure 2. Phonons involved in these transitions have bigger wave numbers that correspond to the position of the energy gap minimum. For θ=45°, i.e. growth along [110], the electron ground sub-band corresponds to the [001] valley and the wire has direct gap. Now, a strong direct emission peak is found, that is shifted to bigger photon energies that correspond to the actual effective energy gap. Phonon assisted transitions are of the same order of magnitude as for θ=0°.

4 Conclusion

From the above discussion, it becomes obvious that light emission from quantum wires of diameters of a few nanometers has strong direct transition character, if the wires direction is along one of the main crystallographic directions (θ=0°) or along [110] (θ=45°). Phonon assisted transitions are in these cases also present, but their intensity is three orders of magnitude smaller than the intensity of direct transitions. A comparison with experimental results is not attempted here because of the lack of data for the luminescence of Si nanowires in which the effect of confinement dominates.

References

1. Canham L.T., Appl.Phys.Lett.**57** (1990) pp1046-1048.
2. Nassiopoulou A. G., Grigoropoulos S., Papadimitriou P. and Gogolides E., *Applied Physics Letters* **66** (1995) pp 1114.
3. Nassiopoulou A. G., Grigoropoulos S. and Papadimitriou P., *Applied Physics Letters* **69** (1996) pp 2267.
4. Zianni X. and Nassiopoulou A.G., unpublished.
5. Horigushi S., Conditions for a direct band gap in Si quantum wires, *Superlattices and Microstructures* **23** (1998) pp. 355–364.

ELECTRICAL MODELING AND CHARACTERIZATION OF Si/SiO$_2$ SUPERLATTICES

T. OUISSE, A.G. NASSIOPOULOU AND D.N. KOUVATSOS
Institute of Microelectronics, NCSR "Demokritos", POB 60228, 15310 Aghia Paraskevi, Greece.

In this work we have characterized diode structures containing a single nc-Si layer between SiO$_2$ layers and modeled their electrical behavior as a double RC circuit. The free carrier concentrations and the current-voltage characteristics of Si/SiO$_2$ superlattices were numerically computed as a function of parameters such as the number of Si wells, the Si and SiO$_2$ film thicknesses and the temperature.

1 Introduction

Confined 2-D systems are essential in silicon technology as evidenced by the field effect transistor itself [1]. Enhancement of the confinement by fabricating low-dimensional structures can result in novel applications from single electron devices to luminescent devices [2-4]; we have recently demonstrated high PL efficiency in nc-Si/SiO$_2$ superlattices [5] and voltage-tunable EL in single nc-Si layer diodes [6]. Thus the characterization and modeling of such 2-D structures is of great interest.

2 Experimental

Single nanocrystalline Si layers in-between SiO$_2$ layers were fabricated by utilizing silicon deposition by LPCVD, oxidation and oxide etching. First, 12 nm thick amorphous Si layers were deposited at 580°C and 300 mTorr on an initial 5 nm pad oxide, followed by dry oxidation at 900°C, for 50 min to 120 min, for crystallization and thinning of the Si layer; then a diluted HF (1%) partial oxide etch left about 5 nm of the formed oxide. I-V and C-V electrical characterization as well as capacitance and conductance frequency dependence measurements were performed on diode structures containing such SiO$_2$/nc-Si/SiO$_2$ stacks as dielectric.

3 Results and discussion

The fabricated diode structures were characterized by I-V and C-V measurements and were found to exhibit high gate currents. However, control structures having only the 5 nm pad oxide had much lower leakage currents, indicating that the leakage originated in the subsequent CVD and etching process. Fig. 1 shows a typical static I-V characteristic for a structure with one nc-Si layer oxidized for 50 min.; the dispersion of values for structures across a wafer was approximately one order of magnitude. As shown in fig. 2, this leakage current induced a decrease of the capacitance with increasing frequency with one or two cutoff frequencies, which was modeled as a double RC circuit and the data were fitted assuming different leakage properties of the top and bottom oxides and easier conduction inside the polysilicon (nc-Si) layer. The high frequency cutoff is due to the series resistance R_s while the low frequency cutoff, when visible, is mainly dependent on the lower of the two oxide resistance values and on the corresponding capacitance. No C-V hysteresis was observed since the injection current was high.

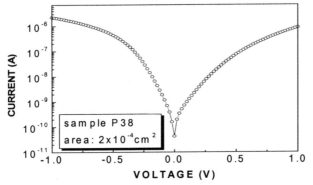

Figure 1. Typical I-V characteristic of a SiO_2/nc-Si/SiO_2 diode structure.

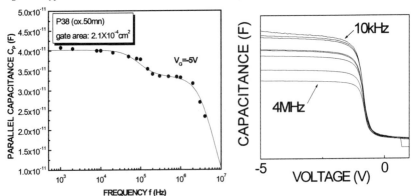

Figure 2. Parallel capacitance versus frequency for a structure with 50 min oxidation time (left) and C-V curves of the same device with the frequency as a parameter (right).

The parallel conductance for the same device against the frequency, as well as the RC equivalent circuit used for fitting the data, are shown in fig. 3. For structures oxidized for longer times (90-120 min) the nc-Si layer is not continuous, as verified by TEM observations, and the dependence of capacitance on voltage and frequency is more intricate, exhibiting peaks before decreasing at high accumulation voltage.

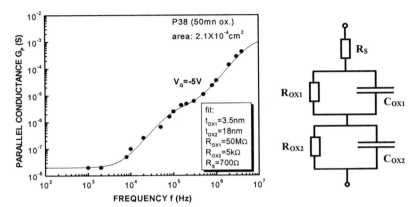

Figure 3. Parallel conductance of the same device as in the previous figure (left) and the equivalent circuit used for fitting the data (right).

A theoretical fully self-consistent electrical model of the general diode structure containing the Si/SiO$_2$ superlattice between the metal and silicon electrodes was developed using a double iteration numerical solving procedure, assuming that the nc-Si wells are thin enough to form locally two-dimensional systems. The square potential well approximation was used. The free carrier concentrations and the current-voltage characteristics of the structure were numerically computed as a function of parameters such as the number of Si wells, the Si and SiO$_2$ film thicknesses and the temperature. In contrast to superlattices based on III-V semiconductors, due to the large barrier heights between the wells, current by Schottky emission over the barrier is limited and only present at very high electric fields and thus carrier confinement is achieved, with charge accumulated inside most of the wells. Thus, the Si/SiO$_2$ superlattices behave as an ensemble of MOS capacitors in series submitted to high electric fields and connected to one another by tunnel barriers, with no quantum coherence between adjacent wells.

As shown in fig. 4, it is found that for a single well configuration with negligible recombination through defect centers thin top and bottom oxides with thickness of about 2.5 nm maximize the carrier concentrations, and thus the currents and the radiative recombination, while in the presence of significant recombination through defect centers it is not possible to have high concentrations of both holes and electrons in the well. In contrast to the case of PL, no gain in EL properties is expected from the use of many layers, because all injected holes recombine in the well closest to the substrate no matter what the number of layers is; no holes reach the metal gate, while a non-negligible proportion of electrons reaches the Si substrate. Thus the recombination current density for the same field remains the same regardless of the number of layers, and so does the expected EL. The total density of electrons trapped inside the structure increases markedly with increasing number of layers. Moreover, an increase in temperature leads to a drastic increase of the current drive capability and thus of the expected EL properties of the structure.

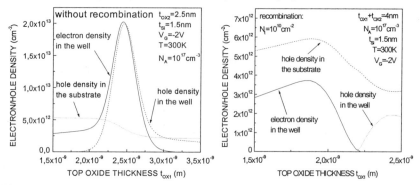

Figure 4. Variation of the electron and hole concentrations in an Al/SiO$_2$/Si/SiO$_2$ structure versus the top oxide thickness, with constant bottom oxide thickness and voltage, assuming that no recombination occurs in the single Si well (left) or that trap recombination occurs (right).

4 Conclusions

MOS structures containing one nc-Si layer sandwiched between two thin oxides were characterized and found to exhibit high gate current. This leakage current induced a decrease of capacitance with increasing frequency with one or two cutoff frequencies, explained by modeling the structure as a double RC circuit. A general structure containing a Si/SiO$_2$ superlattice between metal and silicon electrodes was modeled using a double iteration numerical solving procedure. For a single well configuration, thin top and bottom oxides (2.5 nm thick) maximize the currents and the recombination. Contrary to the PL case, no gain in EL is expected from the use of a multiple layer configuration, as all injected holes recombine in the well closest to the substrate; the current density and the EL for the same field remain the same.

5 References

1. T. Ando, A.B. Fowler and F. Stern, *Rev. Mod. Phys.* **54**, 437 (1982).
2. U. Meirav and E.B. Foxman, *Semicond. Sci. Technol.* **10**, 255 (1995).
3. S. Tiwari, F. Rana, K. Chan, L. Shi and H. Hanafi, *Appl. Phys. Lett.* **69**, 1232 (1996).
4. D.J. Lockwood, Z.H. Lu and J.-M. Baribeau, *Phys. Rev. Lett.* **76**(3), 539 (1996).
5. P. Photopoulos, A.G. Nassiopoulou, D.N. Kouvatsos and A. Travlos, *Appl. Phys. Lett.* **76**(24), 3588 (2000).
6. P. Photopoulos and A.G. Nassiopoulou, *Appl. Phys. Lett.* **77**, 1816 (2000).

Ge/SiO$_2$ THIN LAYERS THROUGH LOW-ENERGY Ge$^+$ IMPLANTATION AND ANNEALING: NANOSTRUCTURE EVOLUTION AND ELECTRICAL CHARACTERISTICS

K. BELTSIOS, P. NORMAND, E. KAPETANAKIS AND D. TSOUKALAS
Institute of Microelectronics, IMEL, NCSR 'Demokritos',
15310 Aghia Paraskevi, Greece,
E-mail: kgbelt@mail.demokritos.gr

A. TRAVLOS
Institute of Materials Science, NCSR 'Demokritos',
15310 Aghia Paraskevi, Greece

J. GAUTIER, F. JOURDAN AND P. HOLLIGER
LETI/CEA, 17 Avenue des Martyrs, 38054 Cedex 9, Grenoble, France

The structure and electrical characteristics of low-energy Ge-implanted SiO$_2$ thin layers are investigated. Glass transition and Ge-species relocation lead to an arrest of phase separation in the annealed oxide, which, nevertheless, can phase separate following e-irradiation during transmission electron microscopy observation. The electrical characteristics of the oxide layers annealed at different temperatures are studied through high-frequency capacitance-voltage measurements of metal-oxide-semiconductor capacitors before and after the application of constant voltage stress. Measurements indicate that Ge-implanted-SiO$_2$ undergoes a major structural change after annealing at temperatures > 910°C. The latter change is identified as a relocation of Ge towards the upper portion of the oxide with a significant fraction of them leaving the oxide.

1 Introduction

During the last few years, formation of charge-storage materials such as nanocrystals or excess silicon in the gate oxide of metal-oxide-semiconductor field-effect-transistors (MOSFET) has received considerable attention for low-power ultra-dense memory applications [1,2]. The structure of these materials strongly affects the charge storage mechanism [3, 4]; electron confinement is expected in the case of phase separated Si or Ge islands, while trap-like behavior is expected in the case of Ge or Si/SiO$_2$ non-separated glass. Here we are concerned with the structural and electrical properties of annealed low-energy Ge-implanted thin SiO$_2$ layers. In particular, the charge storage characteristics as a function of the thermal budget are studied through capacitance-voltage measurements of metal gate capacitors and related to the structural evolution of the Ge-implanted oxides. In addition, we site and discuss observations pertinent to structure formation under annealing and e-irradiation conditions.

2 Experimental

Thin SiO_2 films with a 9 nm thickness were grown on (100), 8″ n-type 3-6 Ω-cm silicon wafers and subsequently implanted with 3 keV Ge^+ at a dose of 1×10^{16} cm^{-2}. Following implantation, the wafers were cut in samples and annealed at temperatures ranging from 500°C to 950°C for 30 min in nitrogen environment. Finally, the oxide on the backside of the samples was etched away and 10^{-4} cm^{-2} aluminum (Al) gate and Al substrate contacts were formed by thermal evaporation. For comparison, control samples without Ge implantation were processed in the same manner. Structural investigation of the implanted films was carried out by transmission electron microscopy (TEM) using a 200 kV acceleration voltage.

3 Results and Discussion

3.1 TEM observations and Structure

Phase diagram /glass transition considerations [5] suggest that the single phase SiO_2/Ge matrix in consideration should be heated at temperatures in excess of 900°C for extensive phase separation to occur. At the same time secondary ion mass spectroscopy (SIMS) compositional profiles suggest that annealing in the range of 900-950°C leads to partial escape of Ge, which in turn raises T_g and, again, phase separation fails to take place. Nevertheless, phase separation does take place during prolonged TEM observation. This can be due to rapid heating of the sample and/or the enhancement of the matrix mobility as a result of e-irradiation. Another factor favoring phase separation is the fact that the thinning of the oxide (for cross-sectional view) or of the Si support (for top view) leads to a substantial T_g depression. The progress of phase separation under e-irradiation is shown in figure 1.

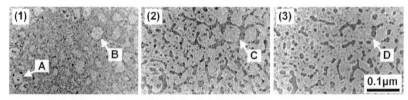

Figure 1. Progress of phase separation under e-irradiation. (1):60 sec, (2): 120 sec, (3): 300 sec

Dark gray corresponds to compositions close to those of the original matrix. Black corresponds to Ge pockets and light gray to the Ge-depleted (rich in SiO_2) matrix. In figure 1.1 depletion areas appear with a central Ge pocket (A) or, more often, with Ge pockets only in the periphery (B). Configuration B may reflect nucleation and growth situations typical of the nanoscale: a central Ge particle is not favored as it contributes (at an early stage) to a high density of compositional gradients. As separation progresses (Fig.1: B, C, D), the connectivity of dark gray areas is reduced as the original matrix gradually disappears and the structure tends towards a SiO_2 matrix with a dispersion of quasi-spherical Ge pockets.

3.2 Electrical characterization

As the implantation affects the integrity of the oxide and the quality of the Si/SiO_2 interface, a thermal treatment is required for the healing of the matrix and the interface. The effect of the annealing temperature on the electrical characteristics of the Ge-implanted oxides was examined through high-frequency capacitance-voltage (C-V) measurements of MOS capacitors before and after the application of constant voltage stress (V_S) for 1 s in accumulation regime. High positive V_s [4] induces a flat-band voltage shift (ΔV_{FB}) in the positive direction of the bias axis of the C-V curve (Fig. 2 and inset) indicating that electrons are trapped into the oxide. Non-implanted capacitors do not exhibit any shift in the C-V curves, showing that the hysteresis effect due to electron trapping is Ge implantation related. The density of traps at the Si/SiO_2 interface as well as the density and distribution over the matrix of bulk traps generated by implantation depend on the annealing temperature and strongly affect the injection and the storage of electrons in the oxide, as depicted by the ΔV_{FB} versus stress-voltage (ΔV_{FB}-V_S) plots of figure 2.

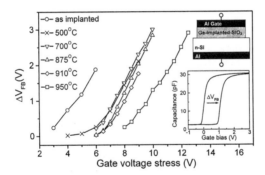

Figure 2. Flat-band voltage shift of 1×10^{16} Ge cm^{-2} implanted MOS capacitors as a function of 1 s. voltage stress. A positive stress voltage shifts the C-V curve in the positive direction of the gate bias (inset).

Substantial local-scale annealing of implantation-related-defects such as broken bonds, or displaced Si and O atoms, occurs at temperatures as low as 500°C leading to a reduction of the bulk and interface state density and to an enhancement of the voltage-stress required for effective electron injection to occur. The small shift of the ΔV_{FB}-V_S plots along of the stress-voltage axis as the temperature increases from

500 to 910°C can be related to the decrease of the interface trap level density (D_{it}) with the annealing temperature (3×10^{13}, 4×10^{11} and 1×10^{11} cm^{-2} eV^{-1} for 500, 875 and 910°C). Still, overall, the structure changes little for annealing temperatures between 500 and 910°C (comparable ΔV_{FB}-V_S plots), while subsequently (950°C) Ge gains enough mobility for relocalization towards the oxide-air interface (with part of it fully escaping from the oxide, according to SIMS measurements). As a result, ΔV_{FB}-V_S plots for 950°C are distinct from those for annealing temperatures up to 910°C (though $D_{it} = 5 \times 10^{10}$ cm^{-2} eV^{-1}, i.e. comparable to that for 910°C).

In conclusion, we have shown that the implantation of thin SiO$_2$ films with 1×10^{16} cm^{-2} Ge ions at an energy of 3 keV and subsequent 950°C annealing generates a single-phase Ge-SiO$_2$ glassy layer, which, nevertheless, can phase separate following e-irradiation. MOS capacitor characteristics suggest that the oxide undergoes a major structural change after annealing at temperatures > 910°C, a finding in harmony with compositional profiles showing that at such temperatures Ge species relocate to the upper portion of the oxide and partially escape.

References

1. Hanafi H. I., Tiwari S. and Khan I., Fast and long retention-time nano-crystal memory. *IEEE Trans. Electron Devices* **ED-43** (1996) pp. 1553-1558.
2. King Y. C., King T. J. and Hu C., A long-refresh dynamic/quasi-nonvolatile memory device with 2-nm tunneling oxide. *IEEE Electron Dev. Lett.* **20** (1999) pp. 409-411.
3. Kapetanakis E., Normand P., Tsoukalas D., Beltsios K., Stoemenos J., Zhang S. and Van den Berg J., Charge storage and interface states effects in Si-nanocrystal memory obtained using low-energy Si$^+$ implantation and annealing. *Appl. Phys. Lett.* **77** (2000) pp. 3450-3452.
4. Kapetanakis E., Normand P., Tsoukalas D., Beltsios K., Travlos T., Gautier J., Palun L. and F. Jourdan, Structure and memory effects of low energy Ge-implanted thin SiO$_2$ films. In *Proc. 29th European Solid-State Device Research Conference ESSDERC'99*, ed. by Maes H. E., Mertens R. P., Declerck G. and Grunbacher H. (Editions Frontieres, Leuven, Belgium, 1999) pp. 432-435.
5. Normand P., Beltsios K., Kapetanakis E., Tsoukalas D., Travlos A., Stoemenos J., Van Den Berg J., Zhang S., Vieu C., Launois H., Gautier J., Jourdan F. and Palun L., Formation of 2-D arrays of semiconductor nanocrystals or semiconductor-rich monolayers by very-low energy Si or Ge ion implantation in silicon oxide films. To appear in *Nucl. Instrum. Meth. Phys. Res. B*.

VERTICAL TRANSPORT MECHANISMS IN nc-Si/CaF$_2$ MULTI-LAYERS

V.IOANNOU-SOUGLERIDIS, AND A.G. NASSIOPOULOU
Institute of Microelectronics NCSR "Demokritos"
PO Box 60228 153-10 Aghia Paraskevi ,Greece
E-mail: v.ioannou@imel.demokritos.gr

F. BASSANI AND F. ARNAUD D'AVITAYA
CRMC2 CNRS Campus de Luminy
Case 913, F-13288 Marseille Cedex 9 France
E-mail: bassani@crmc2.univ-mrs.fr

An investigation of the electrical transport properties of Si/CaF$_2$ multi-layers is presented. These multi-layers were synthesized by MBE at room temperature. The current-voltage and current temperature characteristics showed that the conduction mechanism was thermally activated. In particular structures having CaF$_2$ thickness in each bilayer larger than 1nm showed the presence of a continuous distribution of defects with activation energies within the range of 0.3-0.8eV. On the other hand MLs with smaller CaF$_2$ thickness are characterised by a constant activation energy of 0.3eV. In those latter structures the dominant transport mechanism follows the Poole-Frenkel model.

1 Introduction

The unique physical properties of silicon nanocrystallites triggered an extensive investigation in order to exploit potential application fields. Among the various structures containing synthesized low dimensional silicon are the nc-Si/CaF$_2$ multi-layers (MLs). Structures consisting of several bilayers of low dimensional Si and CaF$_2$ (typical values 50 or 100 bilayers) have been successfully fabricated by Molecular Beam Epitaxy at room temperature [1]. The end result of this low temperature deposition process is that both CaF$_2$ and Si layers are nanocrystalline. Previous studies on the electrical properties of these structures showed that an important parameter controlling their vertical transport characteristics is the CaF$_2$ layer thickness. In particular, MLs with CaF$_2$ thickness larger than 1nm exhibit extended slow trapping effects, manifested as hysteresis in the C-V and I-V characteristics [2,3]. This work investigates the vertical transport properties of these ML structures and their dependence on the CaF$_2$ layer thickness.

2 Experimental

The study of the transport characteristics of nc-Si/CaF$_2$ multi-layers was carried out using I-V characteristics in the temperature range 80-300K and the temperature dependence of the dark current. Fig.1 shows the Arrhenius plot of the temperature dependence of dark current from a sample, consisting of 100 bilayers with t_{CaF2}= 0.56nm and t_{Si}=1.6nm, for three gate voltages. This behaviour is typical for structures with thin CaF$_2$ layers. The Arrhenius plot is a straight line indicating that the vertical transport of carriers through the MLs is determined by traps with an activation energy E_α in the range of 0.28-0.3eV, in the temperature range 100-300K. E_α decreases as the gate voltage increases and it follows a law $E_\alpha \propto \sqrt{E}$ where E is the applied electric field [4]. Identical results were obtained for positive gate voltages. By combing I-V measurements in the temperature range 100-280K it was verified that conduction in these samples follows the Poole-Frenkel model, corresponding to electric field enhanced thermal excitation of carriers from Coulombic traps [4]. This process was found to be characterized by an activation energy of 0.35eV at zero electric field, in close agreement with the results presented in fig.1.

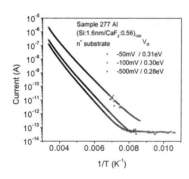

Figure 1. Temperature dependence of the dark conductivity of a sample composed of 100 bilayers, with t_{CaF2}=0.56nm and t_{Si}=1.6nm.deposited on an n$^+$ substrate for three values of the gate bias.

The case of MLs with CaF$_2$ thickness in each bilayer larger than ≈1nm is illustrated in fig.2, in which the obtained Arrhenius plot of the temperature dependence of the dark current from a structure consisting of 50 bilayers with t_{Si}=1.5nm and t_{CaF2}=3nm for several gate voltages is shown. In this case, the behaviour of the structure is distinctly different from that of MLs with smaller CaF$_2$ thickness in each bilayer. The current is very low, below 10^{-12} A and it starts to increase above 180K. The Arrhenius plots are non-linear, especially in the temperature range 180-250K. It is also evident that a narrow hump exists around 250K, separating the Arrhenius plot in two regions. From the slopes of the Arrhenius plot the activation energy was found to increase with temperature, having values in the range of 0.3-0.7eV. It has been suggested that the non-linearity of the Arrhenius plot indicates a distribution of trapping levels within the structure and/or a tunneling process between traps [5]. However the quite large estimated activation energies rule out tunneling process, and the existence of a distribution of trapping levels within the material seems more probable. Almost the same results were found in all samples examined, with CaF_2 layer thickness in each bilayer larger than 1nm.

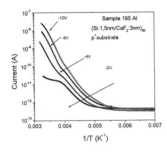

Figure 2. Temperature dependence of the dark conductivity of a sample composed of 50 bilayers, with $t_{CaF2}=3$nm and $t_{Si}=1.5$nm, deposited on a p^+ substrate for several values of the gate voltage.

The thermally stimulated detrapping current (TSDC) was also employed for samples with CaF_2 thickness in each bilayer above 1nm. The samples were cooled down to 100K under bias in order to fill the traps with carriers. At 100K the bias was switched-off and a heat-up temperature ramp of 3K/min was applied, while the gate was connected to the substrate via an electrometer. Therefore, when trapped carriers acquire enough thermal energy, they are thermally emitted, producing in this way a discharging current measured by the electrometer. This is shown in fig.3a for a series of negative voltages. From the increasing part of the peak the activation energy can be extracted. From fig.3a an activation energy of 0.3eV was found. However the peaks were extremely broad indicating the existence of a continuous distribution of trapping levels. This point was further investigated by the fractional heating method. This is achieved following the same experimental procedure as in the TSDC method, except that the heating is stopped near the peak, the temperature is lowered again to 80K without biasing and the heating-up cycle is repeated again. In this way the charge from the trap states is partly collected, since the heat-up cycle is stopped at certain temperatures. The remaining trapped charge is thermally emitted in the next heat-up cycle. The results of the fractional heating method are shown in fig.3. Activation energies are extracted in the same way as in TSDC. The most important result is the continuous increase of activation energies from 0.3 to 0.5 and finally to 0.8eV, consistent with the above hypothesis that a distribution of defects exists within the material.

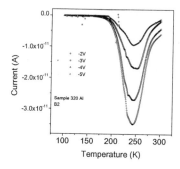 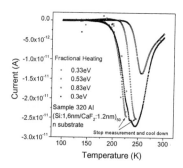

Figure 3. a)TSDC measurements for a series of negative gate voltages from a sample consisting of 50 bilayers with $t_{CaF2}=1.2$nm and $t_{Si}=1.6$nm. b) TSDC fractional heating method. The method reveals that the TSDC peak is due to carriers trapped in a continuous distribution of states.

3 Conclusion

The presented experimental results indicate that the vertical transport properties of nc-Si/CaF_2 MLs, are temperature dependent and they are different in samples with CaF_2 thickness in each bilayer below≈1nm, compared with samples with thicker CaF_2 layer. Samples with CaF_2 below≈1nm are more conductive and their electrical characteristics are dominated by a single activation energy of the order of 0.3eV. On the other hand MLs with CaF_2 thickness in each bilayer above≈1nm behave almost as insulators and the vertical current through the structure is limited. MLs with thicker CaF_2 are characterized by a continuous distribution of deep trapping states in the range of 0.3-0.7eV. The dependence of transport properties on CaF_2 thickness indicates further that the majority of trapping levels are situated within CaF_2. These trapping states are responsible for the formation of extensive reversible space charge regions near the injecting electrode.

References

1. F. Bassani, L. Vervoot, I. Mihalcescu, J.C. Vial and F. Arnaud d'Avitaya J. Appl. Phys. **79**, 4066, (1996).
2. V. Ioannou-Sougleridis, V. Tsakiri, A. G. Nassiopoulou, F. Bassani, S. Menard, F. Arnaud d'Avitaya, Mat. Sci. Eng. B69-70 (2000) pp 309-313.
3. A.G. Nassiopoulou, V. Tsakiri, V. Ioannou-Sougleridis, P. Photopoulos, S. Menard, F. Bassani and F. Arnaud d'Avitaya, J. Lumin. 80 (1999) pp.81-89
4. V. Ioannou-Sougleridis, T.Ouisse, A. G. Nassiopoulou, F. Bassani, and F. Arnaud d'Avitaya, Journ. Appl. Phys.89 (2001) pp610- 614.
5. A. Jauhiainen, S. Bengsson, O. Engstrom, Journ. Appl. Phys. 82 (1997) pp.4966-4973

PHOTO- AND ELECTROLUMINESCENCE FROM nc-Si/CaF$_2$ SUPERLATTICES

V.IOANNOU-SOUGLERIDIS, T. OUISSE AND A.G. NASSIOPOULOU
Institute of Microelectronics NCSR "Demokritos"
PO Box 60228 153-10 Aghia Paraskevi, Greece
E-mail: v.ioannou@imel.demokritos.gr

F. BASSANI AND F. ARNAUD D'AVITAYA
CRMC2 CNRS Campus de Luminy
Case 913, F-13288 Marseille Cedex 9 France
E-mail: bassani@crmc2.univ-mrs.fr

The photoluminescence and electroluminescence characteristics of nanocrystalline silicon in Si/CaF$_2$ superlattices synthesized by Molecular Beam Epitaxy on Si (111) substrates at room temperature were investigated. Light emission from these structures was in the visible range and the spectra were broad. Simple light emitting devices were fabricated with semitransparent gold or ITO as gate metal and an aluminium back ohmic contact, so as to get current in the vertical direction. Photo- and electroluminescence peaks were at the same wavelength. Voltage–tunable electroluminescence in the red, similar to that obtained in porous silicon and Si/SiO$_2$ superlattices was observed. The observed blue shift by increasing the applied voltage was limited to ≈650-700nm, suggesting that some states within the gap were involved in radiative recombination from the silicon crystallites. The spatial distribution of the emitted light from the device area was also investigated.

1 Introduction

Nanocrystalline Si/CaF$_2$ superlattices synthesized by Molecular Beam Epitaxy were proposed as alternative structures for the study of the properties of silicon nanocrystallites [1]. Multi-layer (ML) structures consisting of many alternating layers (typically 50 or 100 bilayers) of nanocrystalline Si and CaF$_2$ have been successfully synthesized by MBE at room temperature [1]. The low temperature growth process favorizes the formation of nanocrystalline layers, which exhibit visible photoluminescence at ambient temperature, if the Si layer thickness is below 2.5-3nm. In this work we examine the spectral characteristics of the photo- and electroluminescence from these structures.

2 Experimental

Fig.1 shows a typical photoluminescence (PL) spectrum from an as-grown sample, consisting of 100 bilayers with t_{CaF_2}= 0.56nm and t_{Si}=1.6nm. As shown in fig.1 the PL spectrum is composed of three Gaussians peaked at 700nm (1.77eV), 780nm (1.59eV) and 530nm (2.34eV), named A, B and C respectively.

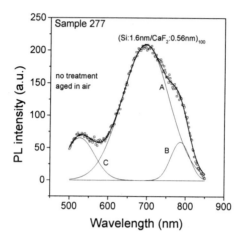

Figure 1. Room temperature of a sample consisting of 100 bilayers, with $t_{CaF2}=0.56$nm and $t_{Si}=1.6$nm. The experimental points (open squares) can be fitted assuming the existence of three gaussian peaks.

Peak C was present in all samples and its position, as well as that of peak B did not depend on the silicon nanocrystal size. These peaks were attributed to defects within the material. Peak A was tunable with silicon grain size and was attributed to quantum confinement. Electroluminescence from nc-Si/CaF$_2$ superlattices was observed above a threshold of few volts. Below this threshold the structures behaved as insulators and the current through the MLs was small. When the gate exceeded a threshold voltage, an instability mechanism was initiated and the current started to increase although the voltage was constant. This peculiar effect was quite extended and it resulted in five orders of magnitude increase in the current. Around 0.1-1Acm^{-2} the current stabilized and a low level red-orange electroluminescence was clearly observed by naked eye. The EL spectra of the sample of fig.1 are shown in fig.2, for several current levels. EL and PL signals show identical characteristics in the low energy region. Peak C was not observed in the EL spectrum. The EL signal increased with current until device burnout, with no saturation tendency. As the current increased, the red peak of the EL spectra showed a clear blue shift. This shift concerned only peak A, while peak B was practically constant at the same wavelength, and independed of the injected current as shown in fig.2b. A current or voltage tunability of electroluminescence was also observed in the case of Si crystallites in the Si/SiO$_2$ superlattices [2] and was ascribed to three different effects: a) size-dependent carrier injection, [3] b) Auger recombination which quenches radiative recombination when more than one electron-hole pair is present within a nanocrystal [3] and c) the effect of a high electric field [4]. This last effect is less likely to occur in the present case, of Si/CaF$_2$ superlattices since the applied electric fields are not so important, due to the small thickness and nature of CaF$_2$ in each bilayer.

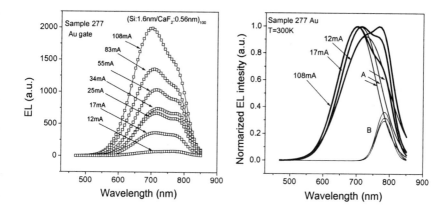

Figure 2. a) Electroluminescence spectra for several current levels from the same structure as in fig.1 (left). b) The spectra have been resolved into two peaks. While peak A is clearly blue shifting with increasing current level, peak B is not. Spectra have been normalized (right)

The homogeneity of the light emission from the device area has been examined by using CCD imaging with a resolution of 0.1 μm. It was found that light was emitted from a number of discrete spots on the device surface. This suggests that the current passes only through a determined number of paths. As the voltage increased the total number of spots increased, the older ones continuing to emit light exactly at the same position on the device area, while new ones were added. By probing the spots individually we observed that the light intensity of one spot did not exhibit any kind of saturation with voltage and that the threshold voltage at which a spot begun to emit was different for the different spots.

Conclusion

The EL properties of nc-Si/CaF$_2$ superlattices were examined. It was found that EL and PL spectra have identical characteristics, which suggests that the same recombination mechanism is involved in both cases. Three peaks were observed in the spectra, two of them being attributed to defects and the red one, which depended on silicon crystallite size and was tunable with voltage, to quantum confinement. Finally light emission was found to originate from a number of discrete light spots on the device surface.

References

1. F. Bassani, L. Vervoot, I. Mihalcescu, J.C. Vial and F. Arnaud d'Avitaya J. Appl. Phys. **79**, 4066, (1996)
2. T. Ouisse and A. G. Nassiopoulou Europhys. Lett., 2000, **51**, pp.168-173
3. I. Mihalsescu, J.C. Vial, A. Bsiesy, F. Muller, R. Romenstain, E. Martin, C. Delerue M. Lannoo and G. Allan Phys. Rev. B, 1995, **51**, pp.17605-17613
4. P. Photopoulos, A.G. Nassiopoulou, D.N. Kouvatsos and A. Travlos App. Phys. Lett. 2000, 76, pp. 3588-3590

AB INITIO CALCULATION OF THE OPTICAL GAP IN SMALL SILICON NANOPARTICLES.

C.S. GAROUFALIS AND ARISTIDES D. ZDETSIS
Department of Physics, University of Patras, 26500 Patras, Greece.

We have calculated the optical gap of small silicon nanocrystals, passivated by hydrogen, with diameters up to 12 Å, using the high level Configuration Interaction Singlets (CIS) method based on both Hartree-Fock (HF) and density functional theory (DFT). The hybrid nonlocal exchange-correlation functional of Becke and Lee, Yang and Parr, including partially exact exchange (B3LYP) was used throughout the calculations.

Our results, which can be rather safely extrapolated for larger nanoparticles with diameters up to about 16 Å, clearly show that for nanocrystals of this size, quantum confinement by itself cannot explain the observed visible photoluminescence since the calculated optical gap is definitely larger than 3.0 eV.

Thus, additional factors, such as the presence of oxygen in the surface of the samples, must play a significant role. This is in agreement with other theoretical calculations, and our own preliminary work for oxygen covered nanocrystals.

The visible photoluminescence of porous Silicon and silicon nanoparticles has attracted a lot of attention in the last five years both experimentally and theoretically [1-13]. A large portion of this work has been devoted to understanding the visible photoluminescence of this material and correlating its spectrum with the diameter of the nanoparticles from which it is composed. It is widely accepted that the luminescence in the visible is mainly due to quantum confinement of the corresponding nanoparticles.

The observed inverse correlation of the optical gap with the nanoparticle size is readily explained by quantum confinement. However, models involving siloxene derivates, polysilane and hydrides on the surface of porous silicon have challenged this hypothesis [10]. Alternatively, the existence of silicon oxide layers on the surface of the crystalline Si core has been also considered [3,12] as an essential factor, for the observed optical properties of these systems.

On the theoretical side, a large part of the work deals with the calculation of the "HOMO-LUMO" gap using in most cases methods of empirical or semi-empirical character since they rely on suitably selected parameters, designed to reproduce the bulk band structure [4,5,8]. Thus, these methods are expected to be reliable only for large nanocrystals (ie for nanoparticles with a number of atoms N>>1000). For such large nanocrystals the difference between the highest occupied and lowest unoccupied molecular orbitals (the HOMO-LUMO gap) and the real excitation energy, which is measured by experiment, is expected to be relatively small [14]. For smaller size nanoparticles which contain less or about 100 Si atoms (with diameters in the range of 10 to 20 Å) for which visible photoluminescence has been suggested both theoretically and experimentally [2,7], new calculations are needed. These calculations must deal with the excitation spectrum and/or the optical (not the HOMO-LUMO) gap, with *ab initio* techniques, without "customized" adjustable parameters.

In this work we employ the *ab initio* Configuration Interaction method, including all Single substitutions (CIS) : 1st) out of Hartee – Fock (HF) determinant and 2nd) the newly developed [15] density functional theory (DFT) CIS method, based on DFT/B3LYP wave-functions.

The nanoparticles which we have considered using these techniques range in diameter from 2 to about 12 Å, which involve up to 29 Si atoms with 4 to 36 the hydrogen atoms at the surface (of total of about 65 atoms). However the conclusions about the optical gap reached here can be rather safely extrapolated to nanocrystals with diameters up to about 15-16 Å. All clusters have Td symmetry and their geometries have been fully optimized within this symmetry constrain. The DFT/CIS calculations, which are based on the hybrid B3LYP functional with the exact HF exchange, have been carried out using the method and program developed by S. Grimme [15].

In the traditional CIS method (based on a HF reference state), the wavefunction of the excited state is written as a linear combination of all possible singly excited determinants: $\Psi_{CIS} = \sum_{ar} a_a^r \psi_a^r$.

The configuration interaction coefficients are deduced as the normalized eigenvectors of the Hamiltonian matrix:

$$\langle \psi_a^r | \psi_b^s \rangle = [E_{HF} + \varepsilon_r - \varepsilon_a] \delta_{rs} \delta_{ab} - (ar \| bs) \quad (1)$$

In the DFT/CIS method shifted molecular-orbital eigenvalues from Kohn-Sham density functional theory are used in the diagonal matrix elements of the CI treatment and all Coulomb type two-electron integrals are scaled by an empirically determined factor. The Hamiltonian matrix elements are expressed as:

$$\langle \psi_a^r | H - E_0 | \psi_a^r \rangle = \varepsilon_r^{KS} - \varepsilon_a^{KS} - c_1 J_{ar} + 2K_{ar} + \Delta(K_{ar}, \varepsilon_a) \quad (2)$$

with $\langle \psi_0 | H | \psi_a^r \rangle = 0$ and $\langle \psi_a^r | H | \psi_b^s \rangle = -c_1(\varphi_a \varphi_b | \varphi_r \varphi_s) + 2(\varphi_a \varphi_r | \varphi_b \varphi_s)$

In this last relation c_1 is an empirical parameter and $\Delta(K_{ar}, \varepsilon_a)$ is an empirical shift applied in the diagonal elements.

The DFT/CIS molecular orbitals derived from the solution of the Kohn-Sham equations closely resemble the HF MOs in the occupied space but differ substantially in the virtual part. The virtual KS-MOs are more compact and they are better suited for CI calculations than their HF counterparts. Additionally the term $\varepsilon_r^{KS} - \varepsilon_a^{KS}$, although not formally proven, can be considered as a more realistic single particle excitation energy. As a result, the electron hole interaction enters in relation (2) scaled by an empirical factor c_1. This parameter can be considered to incorporate, in our treatment, the effect of electron screening in the electron-hole interaction. The value of this parameter, which has been determined by a list square fit of calculated and experimental vertical excitation energies of a selected set of molecules [15], is 0.317.

The HF, DFT, the HF/CIS calculations have been performed with the 6-31G* basis set, whereas for the DFT/CIS calculations the SVP, split valence basis set was employed. For the SiH_4 cluster, in addition to those basis sets the larger TZVP basis set was also used for comparison.

The results of our calculations are summarized in table 1. In the first two lines of table 1 we list the results for the Si_1H_4, Si_2H_6 molecules for which unambiguous experimental data exist.

It is obvious that the DFT/CIS values practically coincide with the experimental ones. As we can see the DFT/CIS approximation gives more reliable results, while at the same time its computational performance is by far superior compared to the corresponding HF/CIS. This was proven to be true even in the case of the Si_2H_6 molecule where the reference transition is to a Rydberg state.

Table 1: The optical gap calculated by the two CIS methods. The larger nanoparticle has a diameter of approximately 12Å. The numbers in brackets correspond to calculations with the TZVP basis set.

No of Si atoms	Total No Of atoms	HF/CIS (eV)	DFT/CIS (eV)	Exprt (eV)
1	5	9.2 (9.1)	9.1 (8.8)	8.8
2	8	7.3	7.5	7.6
5	17	6.5	6.0	--
17	53	5.5	5.3	--
29	65	5.4	5.0	--

The larger nanoparticle studied by the CIS method, $Si_{29}H_{36}$, for which the optical gap is in the order of 5eV has a typical diameter of approximately 12 Å. Thus nanoparticles of this size cannot produce visible photoluminescence. For larger nanocrystals, up to about 16 Å in diameter, we can extrapolate this conclusion using the relation $\varepsilon_g^{opt} = \varepsilon_g^{HL} - E_{coul}$ (3), where ε_g^{HL} is the single particle HOMO-LUMO gap and E_{coul} is the screened electron – hole interaction:

$$E_{Coul} = \int \frac{|\Psi_H(1)|^2 \cdot |\Psi_L(2)|^2}{\varepsilon_\infty^{dot}|r_1 - r_2|} dr_1 dr_2 \qquad (4)$$

Based on calculated HOMO –LUMO gaps [16] and the magnitude of the Coulomb integral in (4), which is between 0.2 and 0.5 eV, according to ref 8 or between 0.15 and 0.75 according to our calculations, we see from (3) that the optical gap for these nanocrystals cannot be smaller than 3eV. Therefore, no visible photoluminescence can be observed (if the luminescence is only due to quantum confinement effect) from these nanocrystals, as has been considered by several researchers.

In conclusion, we have found that the optical gap for nanocrystals with diameters in the range of 10 Å to 16 Å, is larger than 3eV and the reported experimental values. Thus the reported visible photoluminescence cannot be solely due to quantum confinement.

References

1. J. von Behren, T. van Buuren, M. Zacharias, E.H. Chimowitz and P.M. Fauchet *Sol. Stat. Comm.* 105, 317 (1998).
2. S. Schuppler, S.L. Friedman, M.A. Marcus, D.L. Adler, Y.H. Xie, F.M. Ross, Y.L. Chabal, T.D. Harris, L.E. Brus, W.L. Brown, E.E. Chaban, P.F. Szajowski S.B. Christman and P.H. Citrin *Phys. Rev. B52*, 4910 (1995).
3. Y. Kanemitsu and S. Okamoto, M. Otobe and S. Oda, *Phys. Rev. B* 55, 7375 (1997)
4. L.W. Wang and A. Zunger *J. Phys. Chem.* 98, 2158 (1994).
5. J.P. Proot, C. Delerue and G. Allan *Appl. Phys. Lett.* 61, 1948 (1992).
6. M. Rohlfing and S.G.Louie *Phys. Rev. Lett.* 80, 3320 (1998).
7. R.J. Baierle, M.J. Caldas, E. Molinari and S. Ossicini *Sol. Stat. Comm.* 102, 545 (1997)
8. C. Delerue, G. Allan and M. Lanoo *Phys Rev B* 48, 11024 (1993)
9. S. Ogut and J. Chelikowsky, S. G. Louie. *Phys. Rev. Lett.* 79, 1770 (1997); Phys. Rev. Lett. 80, 3162 (1998)
10. J. L. Gole and D. A. Dixon, *Phys. Rev. B* 57 12002 (1998)
11. F. A. Reboredo A. Franceschetti and A. Zunger, *Phys. Rev. B* 61 13073 (2000)
12. M. V. Wolkin, J. Jorne and P. M. Fauchet G, Allan and C. Delerue *Phys. Rev. Lett.* 82 197 (1999)
13. L.E. Brus, P.F. Szajowski, W.L. Wilson, T. D. Harris, S. Schuppler, and P.H. Citrin, *J. Am. Chem. Soc.* 117, 2915 (1995)
14. C. S. Garoufalis, A. D. Zdetsis and S.Grimme, in preparation.
15. S. Grimme *Chem. Phys. Lett.* 259 (1996) 128.
16. C. S. Garoufalis, A. D. Zdetsis and J. P. Xanthakis, submitted.

GROUND STATE ELECTRONIC STRUCTURE OF SMALL SI QUANTUM DOTS

C. S. GAROUFALIS *, ARISTIDES D. ZDETSIS*[+] AND J.P. XANTHAKIS**

*Department of Physics, University of Patras, 26500 Patras, Greece
**Electrical Engineering Department, National Technical University of Athens, Athens 15773, Greece. [+]Corresponding author.

We have calculated the electronic structure of small Si dots terminated by hydrogen atoms with diameters up to 15Å. Having in mind the extention, in the future, of the present calculations to larger dots we have used a variety of theoretical techniques ranging from semi-empirical to *ab initio* Hartree – Fock (HF) and density functional theory (DFT). Within the framework of the DFT theory, we have utilized the hybrid (HF+DFT) non-local exchange and correlation functional of Becke, Lee, Young and Parr (B3LYP), which is known to be very successful for silicon and other semiconductor clusters. The calculated band gaps are significantly larger than those suggested by recent experiments design to measure ground state band edges. We suggest that the DFT / B3LYP description is the most accurate of the methods, despite the fact that the calculated band gaps by simple LDA method, without non local corrections, and by empirical methods fitted to bulk properties are artificially closer to the values suggested by experiment. The discrepancy between experimental and theoretical values is open to various interpretations.

The electronic structure of Si quantum dots has attracted a lot of attention in the last five years both experimentally and theoretically [1]-[9]. Although the exciton binding energy is the quantity measured during the absorption process, the one-electron band gap E_g, which is a ground state property, is easier to calculate theoretically. In order to correlate with the experimentally measured exciton binding energy most of the existing theoretical work, which is semi-empirical in nature, use the dielectric screening theory to estimate the electron-hole attraction, calculated [9] to vary between 0.2 and 0.5 eV. Subtracting this quantity from the calculated band gap gives a good estimate of the exciton binding energy. Naturally, E_g is difficult to measure experimentally. Recently however, experimental data of the band gap for hydrogen-passivated dots have become available [3] by measuring independently the conduction band edge (CBE) shift and the valence band edge (VBE) shift. This has been achieved by exciting an electron from the valence band to vacuum (to measure the VBE shift) and a core electron to the conduction band by an $L_{2,3}$ absorption process (to measure the CBE shift). Therefore, we have undertaken the task to calculate as accurately as possible the electronic structure of the smaller size dots (before we extend the calculations to larger sizes) and compare with the "E_g measurements" of Van Buuren et al [3]. We will stick here only to ground state electronic properties and we shall not extrapolate our results through the dielectric screening approximation to excitonic optical properties, in order to avoid additional assumptions and approximations. Having in mind future extensions to larger dots, in addition to HF and DFT, we have also performed empirical and semi-empirical AM1 (Austin Model 1) [11] and MNDO (Modified Neglect of Differential Overlap) [11] calculations with the standard "unbiased" parameters, without any further "customization". In this respect it is worth pointing out that in almost all of the theoretical work so far, the methods used (pseudopotential or tight-binding) [4],[5],[9] are of an empirical nature, relying on values of parameters which are designed to reproduce the bulk band structure thus making them reliable only for large dots with a number of atoms N, N>>1000 .The experimental work of Van Buuren et al [3], however, contain data for quantum dots with diameters ,d, as small as d =10 to 15Å. These dots contain less or about 100 atoms, in which case any attempted comparison with existing theoretical data (with parameters fitted to bulk values) will be misleading.

Recently, large scale *ab initio* calculations have become available mostly within the local density approximation (LDA), utilizing pseudopotential methods. Such kind of calculations suffers from the well-known trend of LDA to underestimate the band gap. The addition of an arbitrary, size independent, self-energy correction of 0.68 eV, which corresponds to bulk limit, could make the agreement with experiment look better. However, this type of correction, which is frequently used in simple LDA calculations, ignores the effect of quantum confinement on the shelf-energy [10]. Our DFT/B3LYP [10] results, in which we include the partially exact Hartree-Fock exchange [12], do not suffer from this shortcoming. Also the gradient corrections, which are included in this functional, are important for the correct description of the electronic properties of these systems. At the same time the inclusion of HF exchange remedies the deficiency of pure LDA to underestimate the band gap. It should be stressed, as is well known, that the DFT / B3LYP method is comparable in accuracy to high correlation methods, such as the Coupled Cluster method including all Single and Double substitutions together with a quasiperturbative estimate of the effect of Triple excitations (CCSD(T)).

In our calculations we have used the split valence 6-31g* basis set. The sizes of silicon crystallites, which have T_d symmetry, range from 5 to 71 silicon atoms (or from 6Å to about 15Å). All surface dangling bonds are passivated with hydrogen atoms as required by experiment. The geometry of each crystallite has been optimized within the T_d symmetry constrains. These results are shown in figure 1 together with the DFT / B3LYP and the "experimental" results of reference [3]. As we can see in figure 1, these methods, without additional adjustment of the parameters, give diverse results for the HOMO-LUMO gap, with differences as large as 50-60%. The HF approximation is, as expected, inappropriate to describe the Si dots due to the lack of correlation, and the well-established trend to overestimate the HOMO-LUMO gap. On the other hand it is important to mention, as is well known, that the simple LDA without gradient corrections or hybrid functionals clearly underestimates the gap.

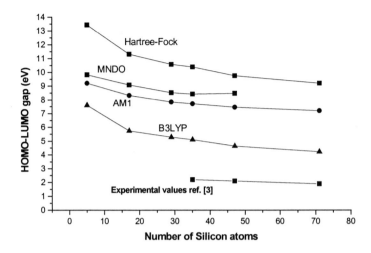

Figure 1: Comparison of the calculated band gaps by different methods as a function of the number of atoms in the dot.

Therefore, we expect that the theoretical results based on the DFT / B3LYP method, which is not suffering from these shortcomings and is comparable in accuracy to CCSD(T), represent an unbiased and realistic ab initio description of these dots. However it is clear from figure 1 that a large discrepancy between theoretical and experimental values exists. Such a large difference of about 2.5 eV, on the average, from the DFT/B3LYP results is much larger than any of the experimental error bars. We should mention that four of the existing calculations of other workers get results, which are closer than 2.5 eV to the experimental values but still beyond any possible experimental error. One of the calculations is the empirical calculation of Bairle et al [8] which employs the MNDO method with a special customized parameterization. The other calculations by Wang and Zunger [4], by Proot et al [5] and by Delerue et al [9] have in common that their parameters are deduced from bulk Si properties but their results are still away from the experimental values of van Buuren et al [3]. For reasons that we have explained before, we believe that the real discrepancy between the theoretical and the experimental values is of the order of 2.5 eV, as the B3LYP results suggest. The smaller discrepancy of the semi-empirical calculations of ref. 4, 5 and 8 is only fortuitous.

So what are the reasons for this large discrepancy? If we assume that the experimental problems with the size determination of the dots have been taken care of properly (which is very difficult for small dots) and furthermore that the measurements of ref. [3] really correspond to the exact band edges (assumption challenged recently) we are led to the conclusion that this discrepancy must be related to the presence of oxygen in the samples. As van Buuren et al have suggested, despite the care taken to prepare the samples in vacuum, at least half a monolayer of oxygen may be present on the surface. This could reduce significantly the measured band gap. We believe, in conclusion, that besides the role of oxygen, questions about the real size of the dots and the true meaning of the measured "single particle" energies must be examined. Some of these questions will be considered in forthcoming work.

References

1. J. von Behren , T. van Buuren , M. Zacharias , E.H. Chimowitz and P. M. Fauchet Sol. Stat. Comm. 105, 317 (1998)
2. S. Schuppler , S.L. Friedman , M.A. Marcus , D.L. Adler , Y.H. Xie , F.M. Ross , Y.L. Chabal , T.D. Harris , L.E. Brus , W.L. Brown , E.E. Chaban , P.F. Szajowski S.B. Christman and P.H. Citrin Phys. Rev. B52 , 4910 (1995).
3. T. van Buuren , L.N. Dinh , L.L. Chase , W.J. Siekhaus and L.J.Terminello Phys. Rev. Lett 80 , 3803 (1998).
4. L.W. Wang and A. Zunger J. Phys. Chem. 98 , 2158 (1994).
5. J.P. Proot , C. Delerue and G. Allan Appl. Phys. Lett. 61 , 1948 (1992).
6. A. Franceschetti and A. Zunger Phys. Rev. Lett. 78 , 915 (1997).
7. R.J. Baierle , M.J. Caldas , E. Molinari and S. Ossicini Sol. Stat. Comm. 102 , 545 (1997)
8. C. Delerue, G. Allan and M. Lanoo Phys Rev B 48 , 11024 (1993)
9. S. Ogut and J. Chelikowsky, S. G. Louie. Phys. Rev. Lett. 79, 1770 (1997); Phys. Rev. Lett. 80, 3162 (1998)
10. A. D. Becke, J. Chem. Phys. 98, 5648, (1993)
11. Gaussian 94, M.J. Frisch, *et al,* Gaussian inc., Pittsburgh PA, (1995)

Processing

TECHNOLOGY ROADMAP CHALLENGES FOR DEEP SUBMICRON CMOS

C. L. CLAEYS AND H. E. MAES (invited)
IMEC, Kapeldreef 75, B-3001 Leuven, Belgium
also at E.E. Depart., KU Leuven, Belgium
E-mail: Cor.claeys@imec.be

The ultimate downsizing of the minimum feature size is hampered by physical, technological and economical limitations. To ensure Moore's law below 100 nm technology nodes both front- and back-end processing has to face technological challenges as clearly stipulated by the International Technology Roadmap for Semiconductors (ITRS). Lithography, gate stack, shallow junctions, high- and low-k dielectrics and interconnect schemes are nowadays amongst the hot research issues leading to a global collaboration. This paper reviews some of the on-going research efforts to come to cost-effective solutions forming the backbone for future technology generations.

1 Introduction

The trend in scaling future technologies is governed by the International Technology Roadmap for Semiconductors (ITRS)[1], put together by industrial and academic experts from Europe, Japan, Korea and Taiwan. This ITRS roadmap is updated on a regular basis and has during the last years been accelerated several times, as illustrated in Fig. 1. This figure shows the evolution of the DRAM half pitch from 1994 to 1999, whereby it can clearly be noticed that each update of the roadmap enhances the technology node by one year. It is generally believed that 100 nm technologies will be in production in 2003-2004, while a decade later the 35 nm technology node should become available. This continuous acceleration of the ITRS roadmap also imposes several technological challenges in order to avoid the so-called "Red Brick Wall", i.e., technological areas for which no solution is available today and therefore requiring very innovative and high risk research. This will go along with the introduction of new materials, as clearly illustrated by the study of the low- and high-k materials, and with the use of less standard and new equipment approaches in the semiconductor manufacturing. Typical examples are dry cleaning, atomic layer chemical vapor deposition, electroplating techniques, plasma and spike annealing etc.

A key technological challenge is the choice of the lithography for achieving the small feature sizes. Although the overall optical lithography trend is going from 248 over 193 to 157 nm, the questions around the affordability of the lithography tools and the Cost of Ownership (CoO) remain. For sub 70 nm linewidths, several options for the Next Generation of Lithography (NGL), such as extreme UV, electron projection lithography, and X-ray lithography remain open.

Another key issue is the selection of the gate dielectric. Whereas the lifetime of SiO_2 can somewhat be extended by switching over to nitrided (NO) or reoxidized nitrided oxides (RNO), a further down scaling requires the implementation of high-k gate dielectrics such as e.g. ZrO_2, HfO_2, Al_2O_3, $SrTiO_3$, Ta_2O_5, TiO_2 and various silicate alloys. Not only the fabrication techniques, but also the reliability issues have to be investigated in detail.

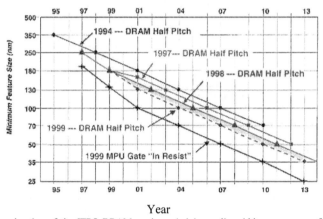

Figure 1. Acceleration of the ITRS DRAM roadmap (minimum linewidth versus year of production) according to the different updates published.

Device engineering is imposing major challenges for achieving the expected device performance. Novel concepts are introduced for controlling the channel profile, such as e.g. L-shape devices. Advanced ion implantation issues (e.g. use of BF_2 and In as dopants, new implantation techniques, reduced thermal budgets....) are required for obtaining ultra shallow junctions. Also in the area of interconnects improved or novel silicidation techniques may have to be implemented. Several of the proposed alternatives will be reviewed. An area of extensive investigations is related to the back-end processing. Here process steps and modules such as low-k dielectrics, advanced metallization schemes (e.g. dual damascene), Cu metallization, and Chemical Mechanical Polishing (CMP) are deserving full attention.

As it is not sure to what extent the anticipated technological roadblocks may be solved, investigations on alternative device approaches, such as e.g. vertical transistors and the use of silicon-on-insulator concepts, are gaining more importance.

The paper first outlines some basic concepts related to the impact of scaling on the device performance, before addressing more in detail some of the technological challenges for 100 nm and below technologies. Several of the above mentioned topics are discussed in order to give an indication of how deep submicron technologies can be realized. Finally the future outlook is addressed, including the transition area between micro- and nano-electronics, i.e., devices such as single electron transistors (SETs), quantum dots and quantum well devices, and spin components.

2 Downscaling and Device Performance

In 1970 a 10 μm PMOS technology was used for the fabrication of 1kbit DRAMS and 4 bit MPUs operating at a clock fequency of 750 kHz. Today, a 120 nm CMOS technology allows to fabricate 512 Mbit DRAMS and MPUs with a 1.13 GHz clock. Moore's law predicts a doubling of the number of transistors every 18 months and a doubling of the CPU speed every two years. The competitiveness of the IC market necessitates an increased functionality and device performance for a reduced cost per function. The scaling MPU requirements are more stringent for MPUs than for DRAMs. It is, however, expected that the aggressive downscaling trend will not continue the next decade and that several of the performance parameters will rather saturate, such as e.g. the MPU clock frequency at 3 GHz which will be limited by the on chip propagation of the electromagnetic waves and the dielectric constant of the used materials [2].

The roadmap of the device performance parameters depends on the envisaged application and is for high performance microprocessors given in Table 1. The scaling of the supply voltage has a beneficial impact on the gate dielectric leakage, the junction breakdown and the latch-up performance [3]. There is also an impact on the transistor drive current I_{DSAT}, which is given by the formula:

$$I_{DSAT} = \frac{\varepsilon_{ox}\mu W}{T_{EOT} L_{eff}} (V_{GS} - V_T)^{\gamma}$$

Table 1. Device characteristics for high performance MPUs [1].

Technology node (nm)	130	100	70	50	35
Supply voltage V_{DD} (V)	1.5-1.2	1.2-0.9	0.9-0.6	0.6-0.5	0.6-0.3
Gate delay CV/I (ps)	7.3	5.7	3.7	2.6	2.4
Maxium off current I_{OFF} (nA/μm)	10	20	40	80	160
Nominal I_{ON} (μA/μm) (nMOS/pMOS)	750/350	750/350	750/350	750/350	750/350

where ε_{ox} is the dielectric constant, μ is the carrier mobility, W is the device width, T_{EOT} is the equivalent gate oxide thickness, L_{eff} is the electrical device length, V_{GS}

is the gate voltage, V_T is the threshold voltage and γ is a parameter between 1 and 2. The scaling of the supply voltage and the V_{DD}/V_T ratio lowers the gate overdrive voltage. Also the electron mobility decreases by reducing the effective gate length and is dominated by surface scattering. On the other hand, an increase of the ε_{ox}/T_{EOT} ratio has a positive influence on the drive current. The drive current on itself is limiting the gate delay, as both parameters are linked by the expression

$$t_{pd} = C_{gate} \frac{V_{DD}}{I_{DSAT}}$$

with C_{gate} the gate capacitance. In case the original scaling method (constant electric field) proposed by Dennard et al [4] would be used, a reduction of the device length with a factor K would reduce both the drive current and the gate delay with a factor 1/K. The downscaling of the devices also requires a suppression of short channel effects and an optimization of the channel and source/drain resistances, so that in practice different scaling rules are applied [2,5].

3 Sub 100 nm Technology Challenges

3.1 Optical Lithography

For linewidths below 0.35 μm the optical lithography tools are changing the exposure wavelength from I-line (365 nm) to deep UV based on either KrF (248 nm) or ArF (193 nm) excimer lasers. By using resolution enhancement techniques such as off-axis illumination (OAI), optical proximity correction (OPC), multilayer resist schemes and/or phase shifting (PS) masks this should allow linewidths for the 100-70 nm technology nodes. The high resolution capability of projection lithography is, however, accompanied by a high cost. For sub 70 nm the use of F_2 excimer lasers (157 nm) may be the first choice [6], although alternative approaches such as extreme ultra violet lithography (EUVL), electron beam projection (EPL), ion projection lithography and X-ray are under investigation [7]. A lithography roadmap is given in Fig. 2. Beside the used exposure tool it is also essential to keep into account mask making aspects such as availability, fabrication difficulties and overall cost. In general it is not believed that sub 100 nm lithography will become the foremost showstopper, although there may be economical limitations preventing its industrial implementation. Another aspect that is often underestimated is the availability of appropriate metrology tools.

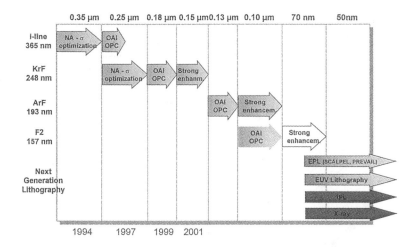

Figure 2. Lithography roadmap.

3.2 Gate Dielectrics

Scaling has direct consequences for the gate dielectric of the transistor. Standard SiO_2 as thin gate dielectric faces several difficulties such as; i) for a too low thickness the reproducibility and uniformity of the fabrication process decrease, ii) no longer sufficient resistant to the in-diffusion of boron from the p^+-doped polysilicon [8], iii) reliability restrictions, especially for slightly higher operating temperatures [9], iv) reduced hot carrier immunity, v) direct tunneling, with an exponential increase of the tunnel current [10], vi) quantum mechanical effects [11]. For the 100 nm technology node the T_{EOT} is about 1.0-1.5 nm, implying that the direct tunneling gate leakage current becomes of the same order as the device off current, thereby impacting the off-state power level of the devices. Therefore, much attention is given the use of high-k dielectrics as a replacement [12]. A first step in that direction is the implementation of nitrided oxides (NO) or reoxidized nitrided oxides (RNO). Nitrogen incorporation in the oxide film provides a good barrier against dopant diffusion, leads to an increase of the dielectric constant, and has a beneficial impact on the reliability [13-14]. However, as there is less than a factor of two increase in the dielectric constant of nitride compared to oxide, the used film thickness must remain rather thin to obtain the required capacitance. In most cases these films are operating in the direct tunneling region, whereby the tunnel current increases monotonically with the oxygen content for a given equivalent oxide thickness [15]. The feasibility of using alternative high-k gate dielectrics such as e.g. Sc_2O_3 (k>10), Ta_2O_5 (k=25), TiO_2 (k=60), and BST (k=300) is getting much attention. Although it is not clear yet who will be the winner, very promising results have already been published as illustrated for the gate current density at 1 V in HfO_2 [16] and TiO_2/Si_3N_4 dielectrics [17], respectively. It is quite important to obtain good interfacial properties between the high-k dielectric and both the silicon substrate and the polysilicon gate electrode. For the first interface a thin SiO_2 layer is often grown first. The top interface remains a challenge even in case that preference would be given to a metal gate electrode.

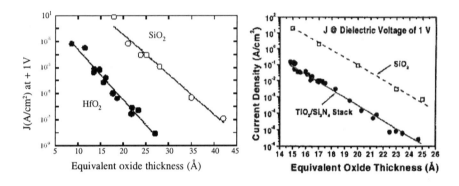

Figure 3. The gate current density at 1V in HfO_2 [16] and TiO_2/Si_3N_4 dielectrics [17], respectively

3.3 Device Engineering

The scaling of the lateral and vertical transistor dimensions imposes severe challenges to the overall device engineering approach, i.e., shallow junctions, channel profile control, spacer concepts, source/drain extensions, etc. A schematic illustration of advanced transistor concept is shown in Fig. 4.

The control of the dopant profiles requires a tight control of the overall thermal budget during processing, without penalizing the dopant activation and the elimination of the ion-implantation damage. Several optimized Rapid Thermal Anneal (RTA) schemes have been proposed [18]. However, dependent on the ramp rate and the doping type, the profiles may be degraded due to transient enhanced diffusion (TED) in addition to the possible negative impact of the defect growth during the ramp-up cycle. As TED is caused by the excess interstitials generated by the ion implantation damage, defect engineering can be used for TED suppression [19]. In this case a well controlled amount of impurities (e.g. C or O) or point defects are introduced in the material before the junction implantation. The use of lower implant energies aggravates TED and reduces the dopant activations [20]. Therefore atomic layer doping seems a very promising technique for obtaining sub 20 nm junction depths [21].

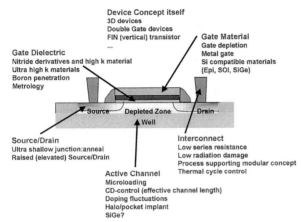

Figure 4. Schematic cross-section of advanced device concepts

Highly doped shallow junctions can be fabricated by e.g. low-energy implants, plasma doping and/or the use of selective epitaxy. Much effort is devoted to optimize the ion implantation process [22]. Device engineering based on lowly doped drain (LDD) regions, source/drain extensions, HALO or highly doped drain (HDD) regions are well-known approaches and all require a very tight control of the thermal budget [23]. Recently, a new technique based on a L-shaped spacer architecture has been proposed. As nitride L-shaped spacers are formed before extension and junction implants, the process complexity is reduced by elimination of lithographic and implantation steps and reducing the total thermal budget [24].

3.4 Interconnects and Multilvel Metallization

Improved gate delay can only beneficially be used if also the interconnect delay is minimized. Therefore the so-called back-end of line (BEOL) processing, i.e. intermediate layer dielectrics (ILDs), metallization schemes, intermetal dielectric layers (IMDs) and multi-level metallization systems, has received tremendous attention during the last years and is nowadays forming the most time consuming and expensive part of the processing. Directly associated with this issue is the use of oxide and metal CMP in order to achieve the required depth of focus (DOF) for the optical lithography. The number of publications on CMP related issues, such as modeling, system optimization, slurries, post CMP clean etc, has increased exponentially.

A large variety of low-k dielectrics are investigated as potential candidate. An extensive and complete review on this topic can be found in the handbook published by Nalwa [25]. Early winners include hydrogen silsesquionane (HSQ) and fluorinated oxides resulting in a k-value of 3 and 3.5 respectively. For lower k-values, organic polymers such as poly(arylene)ethers (PAE), benzocyclubutene (BC) and an aromatic hydorcarbon has been studied. Also silicon based CVD films are very promising. In case that ultra-low k values are needed, spin-on candidates include nanoporous silica films, porous polymers and polytetrafluoroethylene (PTFE). A schematic overview of the different candidates and their properties is summarized in Table 2 [26]. Beside the deposition technique (CVD, spin-coating...) attention has to be given to the dry etch performance and to the planarization (locally and global) aspects.

Table 2. Overview of low-k materials used for ILDs [26]

Dielectric	Dielectric constant (k)	Glass transition temperature (Tg) (°C)	Refractive Index	Water adsorption (%)	Stress (MPa)	Gap Fill (μm)	Cure temperature (°C)	Weight loss (%wt) at 450°C
FSG (silicon oxyfluoride, SixOFy)	3.4 - 4.1	> 800	1.42	< 1.5	-130	< 0.35	no issue	none
HSQ (hydrogen silsesquioxane)	2.9	> 500	1.37	< 0.5	70 - 80	< 0.10	350 - 450	< 3
Nanoporous silica	1.3 - 1.5	> 500	1.15	TBD	0	< 0.25	400	none
Fluorinated polyimide	2.6 - 2.9	> 400	ΔRI>0.15(ai)	1.5	2	< 0.5	350	< 0.1
Poly(arylene) ether	2.6 - 2.8	260 - 450	1.67	< 0.4	60	< 0.15	375 - 425	< 1.0
Paralyne AF4 (aliphatic tretrafluorinated poly-p-xylylene)	2.5	Tmelt > 510	1.548 ΔRI > 0.09(ai)		100	0.18	420 - 450	0.5
PTFE (polytrtrafluoro-ethylene)	1.9	-100	1.34	< 0.01	25 - 27	< 0.3	360 - 390	0.8
DVS-BCB (divinyl siloxane bis-benzocyclobutene)	2.65	> 350	1.561	< 0.2	30 - 35	< 0.22	300	< 1.0
Aromatic hydrocarbon	2.65	> 490	1.628	< 0.25	55 - 60	< 0.05	400 - 500	< 1.0
Hybrid-silsesquioxanes	<3.0	Tmelt > 250	1.58	0	30 - 40	< 0.1	450	6

Another important issue is the metallization approach, e.g. single or dual damascene, and the choice of the metallization system, such as e.g. hot Al or Cu. The selected solution must take into account issues such as seed layers, barrier layers (TiN, TaN, CVD-WN...), the type of fill process (Cu deposition, electroless deposition, electroplating), dry etch aspects, CMP performance, overall manufacturability and CoO.

4 Outlook

Although Moore's law with its exponential increase in density and performance has been valid for more than three decades, it is expected that the aggressive scaling will not be maintained forever. Today there is a strong interest in Silicon-on-Insulator (SOI) technologies as they offer improved device performance compared to the same technology mode for standard CMOS. As illustrated in Fig. 5, compared to bulk Si, SOI CMOS can run at 20 to 50% higher switching speeds or with a 2 to 3 times lower power consumption for the same speed [27]. Depending on the application, both fully and partially depleted technologies are used. Also double gate structures are gaining attention. In general, there is a strong tendency to increase the functionality with process options such as e.g. non volatile memories (FeRAMS and MRAMS), rf-technologies based on SiGe or SiGeC strained layers in CMOS or BiCMOS. Improved performance and functionality at a reduced cost remains the main driver.

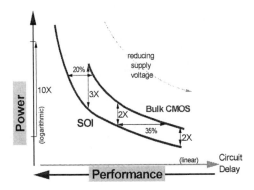

Figure 5. Power drain and access time for bulk and SOI CMOS for V_{DD} of 1.2-2.5 V [27].

The future slow down of the CMOS scaling may trigger the breakthrough of nanoelectronic devices, the feasibility of which has been demonstrated in recent years. Nanoelectronics consists of a new family of devices and components such as e.g. single electron transistors (SETs), resonant tunneling diodes (RTDs), spin controlled devices, rapid single quantum flux logic (RSQF), and intramolecular nanoelectronics. SETS can be considered as devices making a bridge between micro- and nanoelectronics. However, they are not yet a mature alternative to CMOS because of remaining challenges [28] such as i) controlling the background charges, ii) reproducibility and manufacturing tolerances, iii) electrostatic interactions between the devices, iv) control of the operating conditions such as voltage range and temperature, and v) some technological bottlenecks.

Another interesting issue is the future impact of the breakthrough of systems on a chip (SOCs), as schematically illustrated in Fig. 6, on the ITRS roadmap. Although this may relax some of the present challenges, new challenges are associated with the intra- and interchip interconnects and the package aspects

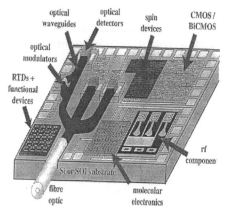

Figure 6. Schematic illustration of a system on a chip (SOC) [28].

5 Acknowledgements

The authors want to thank the members of the Silicon Process Technology and Silicon Technology and Device Integration divisions for stimulating discussions.

References

1. International Technology Roadmap for Semiconductors (ITRS), 1999 Edition, [http.www.itrs.net/1999_SIA_Roadmap/Home.htm], Semiconductor Industry Association.
2. Iwai H. and Ohmi S.-I., ULSI Process Integration for 2005 and Beyond. In *ULSI Process Integration II*, ed. by C. Claeys, F. Gonzalez, H. Murota and K. Saraswat (The Electrochemical Society, Pennington, 2001) pp. 3-33.
3. Packan P. A., Scaling Transistors into the Deep-Submicron Regime. *MRS Bulletin* (June 2000) pp. 19-21.
4. Dennard R. H., Gaensslen F.H., Yu H.-N., Rideout V. L., Bassous E. and LeBlanc A. R., Design of Ion-Implanted MOSFETs With Very Small Physical Dimensions. *IEEE Solid-State Circuits* **SC-9** (1974) pp. 256-268.
5. Dennard R. H., Scaling Challenges for DRAM and Microprocessors in the 21st Century. In *ULSI Science and Technology/1997*, ed. by H.Z. Massoud, H. Iwai, C. Claeys and R.B. Fair (The Electrochemical Society, Pennington, 1997) pp. 519-532.
6. Wakabayashi O., Enami T., Ohta T., Tanaka H., Kubo H., Suzuki T., Terashima K., Sumitani A. and Mizoguchi H., Billion Level Durable Excimer Laser With Highly Stable Energy, *Proc. SPIE* **3679** (1999).

7. Ishitani A., Future Prospects for Sub 100 nm Lithography. In *ULSI Process Integration II*, ed. by C. Claeys, F. Gonzalez, H. Murota and K. Saraswat (The Electrochemical Society, Pennington, 2001) pp. 71-79.
8. Krish K. S., Manchanda L., Bouman F. H., Green M. L., Brasen D., Feldman L. C. and Ourmazd A., Impact of Boron Diffusion Through O_2 and N_2O gate Dielectrics on the Process Margin of Dual-Poly Low-Power CMOS. In *IEDM Techn. Digest* (1994) pp. 325-328.
9. Degraeve R., Pangon N., Kaczer B., Nigam T., Groeseneken G. and Naem A., Temperate Acceleration of Oxide Breakdown and Its Impact on Ultra-Thin Gate Oxide Reliability. In *1999 Symp. on VLSI Technology* (1999) pp. 59-60.
10. Maserjian J., Tunneling in Thin MOS Structures, *J. Vac. Sci. Technol.* **11** (1974) pp. 996-1003.
11. Lo S.-H., Buchanan D. A., Taur Y. and Wang W., Quantum Mechanical Modeling of Electron Tunneling Current from the Inversion Layer of Ultra-Thin Oxide nMOSFETs. *IEEE Trans. Electron Dev.* **18** (1997) pp. 209-211.
12. Huff H. R., Brown A. and Larson L. A., The Gate Stack/Shallow Junction Challenges for Sub-100 nm Technology Generations. In *ULSI Process Integration II*, ed. by C. Claeys, F. Gonzalez, H. Murota and K. Saraswat (The Electrochemical Society, Pennington, 2001) pp. 223-249.
13. Tseng H., O'Meara D. L., Tobin P. J. and Wang V. S., Reduced Gate Leakage Current and Boron Penetration of 0.18 μm 1.5V MOSFETs Using Integrated RTCVD Oxynitride Gate Dielectric. In *IEDM Techn. Digest* (1998) 98-101.
14. Ibok E., Ahmed K. and Ogle B., Electrical Characteristics of Ultrathin [2.5 nm] Gate Quality LPCVD Nitride/Oxide Films. In *ULSI Science and Technology/1999*, ed. by C. Claeys, H. Iwai, G. Bronner and R. B. Fair (The Electrochemical Society, Pennington, 1999) pp. 181-192.
15. Guo X. and Ma T. P., Tunneling Leakage Current in Oxynitride: Dependence on Oxygen/Nitride Content. *IEEE Electron Dev. Lett.* **19** (1998) pp. 207-209.
16. Lee B. H., Qi W.-J., Nieh R., Jeon Y., Onishi K. and Lee J. C., Ultrathin Hafnium Oxide With Low Leakage and Excellent Reliability for Alternative Gate Dielectric Application. In *IEDM Techn. Digest.* (1999) pp. 133-136.
17. Guo X., Wang X., Luo Z., Ma t. P. and Tamagawa T., High Quality Ultra-thin (1.5 nm) TiO_2/Si_3N_4 Gate Dielectric for Deep Sub-micron CMOS Technology. In *IEDM Techn. Digest.* (1999) pp. 137-140.
18. Larsen L. A. and Covington B. C., Shallow Junction Challenges to Rapid Thermal processing. In *Rapid Thermal Processing and Other Short-Time Processing Technologies*, ed. by F. Roozeboom, M. C. Ozturk, J. C. Gelpey, K. G. Reid and D.-L. Kwong (The Electrochem. Soc., Pennington, 2000, pp. 129-136.
19. Privitera V., Impact of the Purity of Silicon on the Evolution of Ion Beam Generated Defects: From Research to Technology. In *High Purity Silicon VI*, ed.

by C. L. Claeys, P. Rai-Choudhury, M. Watanabe, P. Stallhofer and H. J. Dawson (The Electrochem. Soc., Pennington, 2000) pp. 606-620.
20. Nishida A., Murakami E. and Kimura S., Characteristics of Low-Energy BF_2- or As-Implanted Layers and Their Effect on the Electrical performance of 0.15 μm MOSFETs. *IEEE Trans. Electron Dev.* **45** (1998) pp. 701-709.
21. Koyanagi M., Ultra-Shallow Junction Formation by Atomic Layer Doping. In *ULSI Process Integration II*, ed. by C. Claeys, F. Gonzalez, H. Murota and K. Saraswat (The Electrochemical Society, Pennington, 2001) pp. 133-139.
22. Borland J. O., Improved Device Scaling & Process Simplification Through Advanced Ion Implantation Techniques. In *ULSI Process Integration II*, ed. by C. Claeys, F. Gonzalez, H. Murota and K. Saraswat (The Electrochemical Society, Pennington, 2001) pp. 273-288.
23. Wakabayashi H., Ueki M., Narihiro M., Uejima K., Fukai T., Togo M., Yamamoto T., Takeuchi K., Ochiai Y. and Mogami T., Process Technology for Sub 100-nm CMOS Devices. In *ULSI Process Integration II*, ed. by C. Claeys, F. Gonzalez, H. Murota and K. Saraswat (The Electrochemical Society, Pennington, 2001) pp. 34-49.
24. Augendre E., Perello C., Vandamme E., Pochet S., Rooyackers R., Beckx S., de Potter M., Lauwers A. and Badenes G., L-Shape Spacer Architecture for Low Cost, High Performance CMOS. In *ULSI Process Integration II*, ed. by C. Claeys, F. Gonzalez, H. Murota and K. Saraswat (The Electrochemical Society, Pennington, 2001) pp. 297-304.
25. Nalwa H. S., Handbook on Low and High Dielectric Constant Materials and Their Applications – Materials and Processing, Academic Press, vol 1 & 2 (1999).
26. Peters L., "Pursuing the Perfect Low-k Dielectric. *Semicond. Int.* **21** (1998) pp. 64-74.
27. Sadana D.K., SOI for CMOS Logic and Memory Applications. In *ULSI Process Integration II*, ed. by C. Claeys, F. Gonzalez, H. Murota and K. Saraswat (The Electrochemical Society, Pennington, 2001) pp. 474-488.
28. Likharev K. K., Single-Electron Devices and Their Applications. *Proc. IEEE* **87** (1999) pp. 606-632
29. Microelectronics Advanced Research Initiative (MEL-ARI) - Nanoelectronics Roadmap (ESPRIT Report - 1999).

PHOTOLITHOGRAPHIC MATERIALS FOR NOVEL BIOCOMPATIBLE LIFT OFF PROCESSES

A. DOUVAS[1,2], C.D. DIAKOUMAKOS, P. ARGITIS AND K. MISIAKOS

[1]*Institute of Microelectronics, IMEL, NCSR 'Demokritos',*
15310 Aghia Paraskevi, Greece,
E-mail: argitis@imel.demokritos.gr

D. DIMOTIKALI

[2]*Department of Chemical Engineering, Nat. Tech. Univ. of Athens,*
15780 Zografou, Athens, Greece

C. MASTIHIADIS AND S. KAKABAKOS

Institute of Radio-Isotopes and Radio-Pharmaceutical Products, NCSR 'Demokritos',
15310 Aghia Paraskevi, Greece

New photolithographic resists are being developed for use in biomolecule patterning. These resists fulfill biocompatible lithographic requirements necessary for the implementation of lift-off process for this purpose. t-Butyl acrylate homopolymer and a new synthesized copolymer having t-butyl ester groups were used in the resist formulations. The resists were chemically amplified, positive tone and had onium salts as photoacid generators. The method is general, independent of the specific type of biomolecules and it was applied on both planar and cylindrical plastic surfaces.

1 Introduction

The biosensors are used today in analytical chemistry to provide information about the presence and relative concentration of an analyte, but not about its type as do classical techniques such as GC/MS [1]. The advantages of bioanalytical methods are selectivity, sensitivity and response speed. Recently, a significant part of the research effort in this field is focused on creation of multi-analyte immunoassays or devices, which analyse more than one substance at the same time [2,3]. The biomolecule patterning is mostly based, until now, on photochemical modification of substrate for covalent binding or selective physisorption of specific types of biomolecules. This method depends strictly on the type of biomolecules and the specific substrate in a way that only one kind of biomolecules can be deposited each time on the specifically modified substrate [4]. A new approach has been introduced by our group [5], which is further explored in this paper, based on the lift-off process, a technique broadly used in semiconductor technology. The usual lithographic resists can not be used with this technique, since their processing causes complete denaturation of the biomolecules for subsequent molecular recognition.

New photoresists and processes developed in the context of our work, which do not denaturate proteins and are applicable on several types of binding biomolecules and substrate types, are discussed.

2 Experimental

Components of the biocompatible resists were: a) 35% w/w *t*-butyl-acrylate solution in toluene purchased by Poly Sciences, b) (meth)acrylate copolymer containing *t*-butyl acid labile group synthesised in our laboratory, c) triphenyl sulfonium hexafluoroantimonate salt (for exposure at 254nm) purchased by General Electric and d) 50 % w/w triaryl sulfonium hexafluoroantimonate salt, solution in propylene carbonate, purchased by Union Carbide for exposure at $\lambda > 300$nm. Ethyl lactate was used as solvent. The resist was coated either on planar substrates of Si wafers and polystyrene disks or on the cylindrical inner surfaces of polystyrene and polymethylpentene capillary tubes. The films were exposed either by Orial Hg-Xe 500 W lamp at DUV or by Carl Sus aligner at near UV. Narrow band filters (10nm bandwidth at half maximum) and broad band ones (50nm bandwidth at half maximum) were used for radiation at 254nm and pyrex cut off filters for radiation at $\lambda > 300$nm. UV and FTIR spectroscopy were used for chemical reaction studies, optical and electronic microscopy and profilometry for lithographic characterization, and fluorescence scanning for detection of fluorophore substance FITC, which is conjugated with the antibody.

3 Results and Discussion

Our approach for biopatterning is as follows: after lithographic patterning of resist, in which the exposed areas were removed and the unexposed remained, biomolecules are adsorbed on both of them. Second lithography removes the initially unexposed resist areas with the biomolecules on them, leaving a biomolecular pattern on the substrate. The requirements imposed by the biopatterning process to the lithographic resist were: positive tone lithography, thermal treatment at T<60°C, development in dilute aqueous bases and exposure preferentially at $\lambda \geq 300$nm for no loss of antibody immunoreactivity. The resist formulations that fulfill all the above requirements, are positive, chemically amplified systems based on acrylate resin and triaryl (or triphenyl) sulfonium salt as photoacid generator.

The chemical amplification mechanism of the biocompatible lithographic resist has two main steps: photochemical acid production by the sensitizer during exposure and acid catalyzed deprotection reaction of the *t*-butyl pendant group of the acrylate homopolymer or (meth)acrylate copolymer during thermal treatment after exposure.

The thermal treatment after exposure was not necessary for the (meth)acrylate copolymer, but it was for the homopolymer. The material formulation optimisation followed the chemical system selection. The combination of polymer and sensitizer concentration was optimised in order to achieve minimum chemical change of unexposed resist during development in dilute aqueous base. Then, the conditions of critical process steps were optimized. The influence of post exposure thermal treatment temperature on the homopolymer resist sensitivity was studied (Fig. 1). Also the developer concentration effect on sensitivity of this resist was examined (Fig. 2). The biocompatible lithographic process with the homopolymer resist gave on Si wafer microstructures of 6.25μm lines/spaces with exposure through pyrex filter (Fig. 3 a,b).

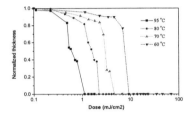 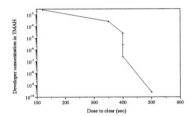

Figure 1. Influence of post exposure thermal treatment temperature on the sensitivity of the homopolymer acrylate resist. Developer: dilute aq. base 2.6×10^{-3} N. Exposure: 254nm broad band filter.

Figure 2. Influence of developer concentration on sensitivity of homopolymer acrylate resist. Thermal treatment after exposure: 50°C, 30min. Developer: aqueous TMAH at different cocentrations. Exposure through pyrex filter.

The lithographic process in the capillary tube included the following steps: two bands were created in resist film of the inner surface using the acrylate homo- or copolymer with triaryl sulfonium salt as sensitiser. Then, rabbit-IgG molecules were physisorbed on this surface by incubation with an appropriate solution. The initially unexposed resist was removed by exposure and development in dilute aq. TMAH solution (2.7×10^{-3} N). Afterwards, bovine serum albumin solution was introduced in order to cover free substrate sites where the rabbit-IgG molecules were not adsorbed (blocking). Then, a goat anti-rabbit antibody labelled with an enzyme (horse radish peroxidase) solution was injected in order to react exclusively with physisorbed rabbit-IgG. Finally, chromogenic substrate ABTS/H_2O_2 solution was introduced and two green bands were obtained indicating the rabbit-IgG patterning (Fig. 4a). In order to examine the homogeneity of the protein bands formation on a more quantitative basis, antibody labelled with fluorescent dye FITC was used instead of the enzyme labelled one and the immunoreaction of antibody-antigen was followed by fluorescence scanning (Fig. 4b). Although the detection using the enzyme label offers ten times higher sensitivity compared with the one withe the fluorescent label, the second method of detection was selected, because of the availability of appropriate instrumentation to scan the capillary containing fluorescent labelled biomolecules. The surface density of rabbit-IgG on the plain TPX inner capillary surface at saturation is 550 ng/cm² (determination using radio-iodinated rabbit-IgG). At the above surface coverage conditions 44-50 ng/cm² of fluorescently labelled antibody molecules bind onto the immobilized rabbit-IgG.

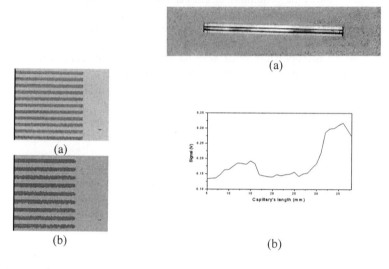

Figure 3. Structures, 6.25 μm lines/spaces, with homopolymer acrylate resist. Thermal treatment after exposure: 45 °C, developer: dilute aq. base (2.6x10⁻³ N), exposure (pyrex filter): (a) 270s and (b) 280s.

Figure 4. Two bands of rabbit-IgG patterned with copolymer acrylate resist in capillary inner surface. Visualization through: (a) enzyme label conjugated to antibody in chromogenic substrate and (b) fluorophore label conjugated to antibody and fluorescence scanning. Color formation could be observed close to capillary ends in (a), due to manual injection of the bioreagents by syringes, which destruct these areas of the resist film. The different signals presented in the specific case in (b), reveals the difficulties imposed by the capillary cylindrical geometry to the optimization of the lithographic process.

References

1. Thomson M. and Krull U.J., *Amer.. Chem. Soc.*, **63**:7 (1991).
2. Morgan C.L., Newman D.J.and Price C.P., *Clinidal Chemistry*, 193 (1996).
3. Misiakos K. and Kakabakos S.E., *Biosens. and Bioelectron.*, **13**:825 (1998).
4. Roznyai L.F., Benson D.R., Fodor S.P.A. and Schults P.G., *Angew. Chem. Int. Ed. Engl.*, **31**:6 (1992).
5. Douvas A., Argitis P., Misiakos K., Dimotikali D., Petrou P.S. and Kakabakos S.E., *Biosens. and Bioelectron.*, (2001), in press.

POLYCRYSTALLINE SILICON THIN FILM TRANSISTORS HAVING GATE OXIDES DEPOSITED USING TEOS

V. EM. VAMVAKAS, D. N. KOUVATSOS AND D. DAVAZOGLOU

Institute of Microelectronics, NCSR "Demokritos", POB 60228, 15310 Aghia Paraskevi, Greece.

In this work we investigated the use of SiO_2 dielectrics deposited by LPCVD using TEOS under various deposition temperatures and pressures as gate oxides in thin film transistors (TFTs). It was found, using FTIR spectroscopy, that at temperatures lower than 635°C carbon contamination is introduced in the oxide. The degradation of the TFT transfer characteristics under various stressing gate bias values was studied. The evolution of the electrical parameters with stressing time was determined. The degradation of both the threshold voltage and the subthreshold swing exhibits logarithmic dependence with stressing time, indicating a charge trapping process mainly occurring near and at the polysilicon / SiO_2 interface.

1 Introduction

Thin film transistors (TFTs) in polycrystalline silicon films are important because of their use in AMLCDs, in circuits on special substrates and in three-dimensional integration as in SRAMs; however, the use of thermal oxides in these devices is impractical because of their high growth temperatures. In this work we investigated the use of SiO_2 gate dielectrics deposited by LPCVD using tetraethylorthosilicate (TEOS, $Si(OC_2H_5)_4$) in TFTs. These SiO_2 films have been shown to have very good step coverage and low mechanical stress [1], but they also have a granular form, contain undesirable impurities like carbon, and are substoichiometric in oxygen [2]. The behavior of TEOS-deposited oxides under electrical stressing and the resulting effects on the performance of devices utilizing them should thus be characterized in order to gain useful understanding and facilitate their use for various applications.

2 Experimental

Amorphous silicon films 100 nm thick were deposited by LPCVD using silane, at 550°C and 230 mTorr, on oxidized silicon wafers and were crystallized by anneal at 600°C for 12 h at 120 mTorr in N_2. Polysilicon islands were formed and a 50 nm thick gate oxide was deposited by LPCVD using TEOS at 635°C or at 710°C with pressures of 100, 300, or 500 mTorr. Some wafers were then annealed at 635°C and 120 mTorr, for 72 h. Subsequently, a 500 nm polysilicon film was deposited and doped. Polysilicon gates were patterned followed by source / drain doping with As implant at 80 keV and 3×10^{15} cm^{-2}, for self-aligned gate structures. After activation at 600°C for 12 h in N_2, a 500 nm passivation oxide was deposited by LPCVD using

TEOS, contacts were opened and Al source, drain and gate electrodes were formed. A hydrogen plasma passivation of the grain boundaries was not performed.

3 Results and discussion

The structural differences of thermal oxides from TEOS deposited oxides can be studied using FTIR spectroscopy. Figure 1 (left) shows the IR transmission spectra of a thermally grown oxide at 950°C and two TEOS deposited oxides at 710°C and 635°C. All oxide thicknesses are from 14.1 nm to 14.6 nm. The TEOS oxides are more transparent by about 1%, and their transmission peak has a greater FWHM, compared to thermal oxides. These differences are attributed to the different growth temperatures, the different bond lengths and angles of the O–Si–O bridge and the granular form of the TEOS deposited oxides. The spectra of the two TEOS films are almost identical, indicating only minor differences in their chemical structure. Also, figure 1 (right) shows the IR transmission spectra of TEOS deposited oxides at 620°C and 635°C. The 620°C oxide spectrum has peaks corresponding to carbon bonded with oxygen or other carbon atoms. Carbon is incorporated because of partial dissociation of TEOS molecules in the gas phase due to the low temperature [3]. Another possible source of carbon contaminants is pyrolysis byproducts trapped inside the film [4]. Increasing the deposition temperature we offer enough thermal energy to fully break the organic part of the TEOS molecule and to reduce the time that pyrolysis byproducts occupy active areas of the growing film.

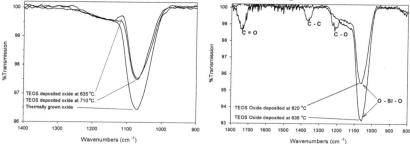

Figure 1. FTIR transmission spectra of a thermally grown oxide at 950°C and of two TEOS deposited oxides at 710°C and 635°C; the thicknesses of all films are between 14.1 and 14.6 nm (left). Transmission spectra of two TEOS deposited oxide films of similar thickness; the oxide deposited at 620°C is carbon contaminated while the oxide deposited at 635°C is not (right).

Polysilicon TFTs with gate oxides deposited using TEOS at 710°C or at 635°C and at 100, 300 or 500 mTorr exhibited on/off current ratios of 5×10^6 to 10^7. The leakage currents were of the order of 10^{-12} A for $V_{DS} = 0.1$ V or 10^{-11} A for $V_{DS} = 1$ V. The electron mobilities were 10 to 15 cm^2/V·sec and the threshold voltage was 5 to 13.5 V, as we have previously reported for TFTs fabricated at 710°C [5]; V_t increases, suggesting increasing oxide and interface trap charge densities, with increasing oxide deposition pressure, while the spread of the values of both the mobility and the threshold voltage across a wafer decrease, indicating progressively better uniformity of the oxide characteristics, with increasing pressure. Similar threshold voltages and slightly lower mobilities of 7 to 12 cm^2/V·sec were measured for TFTs fabricated at 635°C; the additional 72 h anneal in some devices resulted in a mobility increase of about 10%. The large V_t values indicate high oxide trapped charges; these are expected because of the granular structure and the possible carbon contamination of the TEOS-deposited oxides which results in more oxide traps compared to thermally grown oxides. The mobility is reasonable for a grain size of 0.3 μm, but lower than that resulting from the use of thermal oxides due to an expected rougher interface with higher trap density. However, V_t can be reduced by V_t-adjust implant and hydrogen plasma passivation, not done for these TFTs; the passivation would also improve the mobility and the subthreshold swing. All TFTs, fabricated at either 635°C or 710°C, had subthreshold swings of 1.34 to 1.44 V/dec.

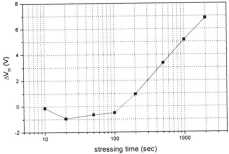

Figure 2. Threshold voltage degradation against stressing time for electrical stressing at V_{GS} = -30 V.

Although these parameters are inferior to those of TFTs currently realized with advanced fabrication methods, their evolution with electrical stressing can provide useful understanding. TFTs were stressed at various conditions and the degradation of their parameters and characteristics was studied. Figure 2 shows the threshold voltage degradation with stressing time for stressing at V_{GS} = - 30 V (V_{DS}= 0 V) for a typical device with 8 μm gate length, fabricated at 710°C. It is observed that after a short initial interval, for which V_{th} exhibits stability or some improvement, the threshold voltage is degraded towards more positive values with a logarithmic time dependence. This behavior, also observed for V_{GS} stressing at - 10 V or - 20 V, is indicative of charge trapping in the oxide or at the interface [6]; the positive shift corresponds to trapping of injected electrons. However, as seen from figure 3 the subthreshold swing, which is affected by grain boundary or intragrain traps in the polysilicon active layer or by interface traps, also shows logarithmic degradation for stressing times above 100 sec. This suggests that a concomitant degradation of the polysilicon-oxide interface occurs, involving at least a part of the charge trapping. During V_{GS} stressing electrons are injected into the oxide, high energy electrons can create interface states and weak oxide areas susceptible to electron trapping [7].

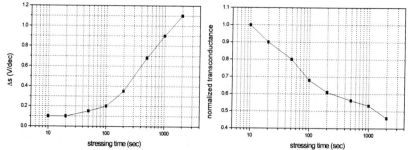

Figure 3. Subthreshold swing degradation (left) or transconductance degradation (right) against stressing time for electrical stressing at $V_{GS} = -30$ V.

Figure 3 also shows the degradation of the transconductance g_m, or the field effect electron mobility, with stressing time. The g_m degradation can be attributed to the creation of acceptor-like trap states [8] in the active area of the polysilicon film and at the interface. We believe that the stress-induced creation of acceptor-like traps at the interface, which have negative charge when occupied by electrons, accounts for the simultaneous positive V_{th} shift and subthreshold swing degradation, while being consistent with a g_m degradation due to created near interfacial bulk acceptor traps.

4 Conclusions

Polysilicon TFTs having gate oxides deposited using TEOS were fabricated and characterized. The lowest temperature for depositing SiO_2 using TEOS is 635°C; below that value the oxide was found to be contaminated with carbon. The TFT degradation under gate bias stress was investigated and a mechanism based on the creation of acceptor-like interface and bulk traps was deduced as its cause.

References

1. A.C. Adams and C.D. Capio, J. Electrochem. Soc., **126**, 1042 (1979).
2. D. Davazoglou, V.E. Vamvakas and C. Vahlas, J. Electrochem. Soc., **145**, 1310 (1998).
3. D.R. Secrist and J.D. Mackenzie, J. Electrochem. Soc. **113**, 914 (1966).
4. J. Arndt and G. Wahl, Electrochem. Soc. Proceedings, **97 – 25**, 147 (1997).
5. D.N. Kouvatsos, V.E. Vamvakas and D. Davazoglou, TFT Technologies IV Symp. Proceedings, Electrochem. Soc. (1998).
6. Y. Tai, J. Tsai, H. Cheng and F. Su, Appl. Phys. Lett. **67**, 76 (1995).
7. D.N. Yaung, Y.K. Fang, K.C. Huang, C.Y. Chen, Y.J. Wang, .C. Hung, S.G. Wuu and M.S. Liang, Semicond. Sci. Technol. **15**, 888 (2000).
8. T.M. Brown and P. Migliorato, Appl. Phys. Lett. **76**, 1024 (2000).

SOLID INTERFACE STUDIES WITH APPLICATIONS IN MICROELECTRONICS

S. KENNOU[1,2], S. LADAS[1,2], A. SIOKOU[2], I.DONTAS[1,2] AND V. PAPAEFTHIMIOU[1,2]

[1] *Department of Chemical Engineering, University of Patras, 26500 Rion, Patras, Greece*
[2] *ICEHT / FORTH, P.O.B. 1414, 26500 Rion, Patras, Greece*
E-mail: kennou@iceht.forth.gr

Two selected examples of interface studies relevant to microelectronis, using X-ray and Ultra-violet photoelectron spectroscopies (XPS,UPS) and Work Function (WF) measurements are presented. The study of the early stages of growth of the Cu /6H-SiC(0001) interface under ultra-high vacuum (UHV) conditions at RT and its behavior upon annealing up to 900K showed that the Cu film grows in a layerwise fashion and exhibits a thermally stable Schottky barrier height of about 0.6 eV. Thin layered films of the OOCT-OPV5 oligomer were grown on Au in UHV and their valence band structure, work function and band bending were determined, from which the band lineup at the interface was obtained.

1 Introduction

Since interfaces play an important role in microelectronics, surface sensitive techniques are used to probe the early stages of their formation. X-ray photoelectron spectroscopy (XPS) can be used for band bending and Schottky barrier height determination, whereas ultra-violet photoelectron spectroscopy (UPS) provides information about interface electronic structure. Two selected examples of interfacial studies using XPS and UPS, the Cu/ 6H-SiC and the OOCT-OPV5/ Au interfaces, are presented. Metal/SiC contacts are important for high temperature microelectronic devices [1], therefore their electrical behavior and thermal stability has been studied for a large number of metals [2-6]. As Cu /SiC is one of the least studied systems, the early stages of the Cu/(0001)6H-SiC interface formation at RT and the influence of annealing up to 900K was investigated. On the other hand, thin polymeric films are used in devices, like FET and p-LED. The metal/organic interface controls the injection of electrons or holes, hence, it is attracting great attention in this field [7–12]. The OOCT-OPV5 oligomer, a structural model of PPV, one of the common photonic polymers in organic screens, was deposited on polycrystalline Au at RT, with an aim to investigate the alignment between the molecular orbitals of the organic and the Fermi level position of the metal.

2 Experimental

The measurements were performed in two different UHV chambers (base pressure 5×10^{-10} mbar), both equipped with a hemispherical electron energy analyzer, a twin-anode X-ray gun and a UV-lamp [6,13]. Photoelectrons were excited using the

MgK$_\alpha$ line at 1253.6 eV (XPS) or the HeI resonance line at 21.23 eV (UPS). Single crystal (0001) 6H-SiC n-type wafers, from Cree Research Inc., were used as substrates, for copper e-beam evaporation in UHV. The 99.999% pure polycrystalline Au foil was sputter-cleaned prior to the sublimation of the oligomer [(2,5-Bis(4-styryl)styryl) 1,4-dioctyloxybenzene] powder.

3 Results and Discussion

3.1 The Cu / (0001)6H-SiC Interface

Upon Cu deposition at RT, the WF exhibits a sharp initial decrease up to 0.1ML, which is not yet fully explained and then increases continuously up to a saturation value for a 15Å film of metallic Cu (fig 1). The small amount of copper (close to 1ML) required to reach WF saturation indicates layer-like film growth.

Figure 1. Work Function change upon Cu deposition and after annealing of the final 15Å Cu/SiC contact. **Figure 2.** The C1s XPS peak before and after annealing of the 15Å Cu/SiC contact

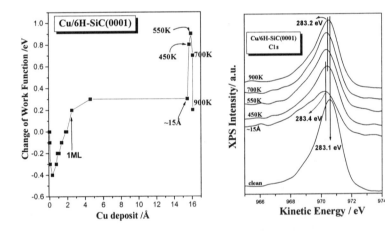

The binding energy upshift by 0.3 eV of the core level peak C1s from the SiC substrate upon 15Å Cu deposition (fig. 2) arises from band bending in the semiconductor. The Schottky barrier height, Φ_B, for the Cu/SiC interface was calculated from the XPS measurements using the method described by Waldrop and co-workers [3]. In brief, $\Phi_B = E_g - E_{fi}$, where $E_g = 3.03$ eV is the 6H-SiC band gap and E_{fi} is the interface Fermi level measured by XPS with respect to the top of the silicon carbide valence band; that is, $E_{fi} = E_{C1s} - (E_{C1s} - E_v)_0$, where the quantity in parentheses is the C1s to valence-band maximum binding energy difference for the clean SiC. This quantity was measured (281.0±0.1) eV. From the measured C1s binding energy of 283.4 eV after 15Å of Cu deposition, a Schottky barrier height of

(0.6±0.1) eV is obtained. This value indicates a nearly ohmic contact formation between copper and SiC . Upon gradual annealing up to 900K, the work function, after an initial - yet unexplained - increase, decreases nearly back to its original value (fig.1). Furthermore, annealing causes a small increase of the binding energy of the C1s peak (fig.2), corresponding to a small increase of the Schottky barrier height (0.7±0.1) eV and suggesting good thermal stability of the contact. This behavior indicates only a limited chemical interaction in the interface.

3.2 The OOCT-OPV5 / Au interface

From the sequence of UP spectra obtained after each oligomer evaporation step (fig.3), one can determine the WF of the Au substrate (5.2±0.05eV), the high BE

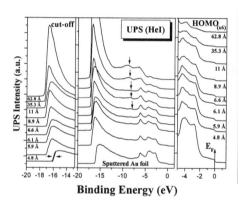
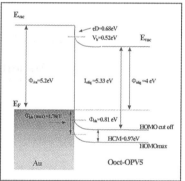

Figure 3. HeI UP spectra upon Ooct-OPV5 film growth. Details of the end regions of the full spectra (center) are shown magnified on both sides.

Figure 4. Energy level diagram for the Ooct-OPV5 interface constructed from combined XPS and UPS data.

cut-off shifts corresponding to a gradual lowering of the sample WF with increasing oligomer thickness (by up to ~1.2±0.05eV for ~11Å.) and the development of the characteristic features of the oligomer valence structure, with the final HOMO cut-off at −1.33 ±0.05 eV . This structure, which is complete at ~63Å , is very similar to that of the PPV [14].

Information about the orbital alignment of the oligomer relative to the Fermi edge and the vacuum level of the Au foil was obtained using the combined XPS and UPS data. The contribution of the band bending in the overlayer to the total WF change is determined from the total BE shift of the C1s peak and is V_b= 0.52±0.05 eV. The remaining 1.20eV-0.52eV=0.68±0.05 eV is the interface dipole. The hole injection barrier Φ_{bh} at the interface is determined from the 63 Å UP spectrum by subtracting the band bending from the the HOMO cut off position (1.33 eV). Thus: Φ_{bh} = 1.33eV-0.52eV=0.81eV. The HOMO maximum position was determined at 2.3 eV, yielding: Φ_{bh}(max) = 2.3eV-0.52eV=1.78eV. The electron injection barrier Φ_{be} cannot be determined here since the HOMO-LUMO optical gap of the oligomer is not known.

The resulting orbital alignment of the Ooct-OPV5 relative to the Fermi and Vacuum level of Au is shown in fig. 4. The interface dipole may result from a chemisorption process at the interface, in which the first monolayer is more strongly bonded than the following layers. The inert nature of Au, however, does not justify strong bonding with the oligomer. The electronic dipole could be the result of an interfacial state involved in a mechanism that works as a buffer at the interfacial charge exchange between the metal and the organic layer [15].

References

1. H.Morkoc, S.Strite, G.B.Gao, M.E.Lin, B.Sverdlov, M.Burns, J.Appl.Phys **76(3)** (1994) 1363.
2. V.M.Bermudez, J.Appl.Phys. **63(10)** (1988) 4951.
3. J.R.Waldrop, R.W.Grant, Y.C.Wang, R.F.Davis, J.Appl.Phys. **72(10)** (1992) 4757.
4. N.Lundberg, M.Oestling, Appl.Phys.Lett. **63(22)** (1993) 3069.
5. S.Kennou, J.Appl.Phys. **78(1)** (1995) 587.
6. S.Kennou, A.Siokou, I.Dontas, S.Ladas, Diamond and Related Materials **6** (1997) 1424.
7. H. Ishii, K. Sugiyama, E. Ito, K. Seki, Adv. Mater. **11 (8)** (1999) 605.
8. R. Schlaf, C. D. Merritt, L. A. Crisafulli, Z. H. Kafafi, J. Appl. Phys. **86 (10)** (1999) 5678.
9. N. Johansson, F. Cacialli, K. Z. Xing, G. Beamson, D. T. Clark, R. H. Friend, W. R. Salaneck, Synth. Metals **92** (1998) 207.
10. Y. Park, V.-E. Choong, B. R. Hsieh, C. W. Tang, T. Wehrmeister, K. Mullen, Y. Gao, J. Vac.Sci. Technol. A **15 (5)** (1997) 2574.
11. R. Schlaf, P. G. Schroeder, M. W. Nelson, B.A. Parkinson, P. A. Lee, K. W. Nebesny, N R. Armstrong, J. Appl. Phys. **86 (3)** (1999) 1499.
12. R. Schlaf, P. G. Schroeder, M. W. Nelson, B.A. Parkinson, C. D. Merritt, L. A. Crisafulli, H. Murata, Z. H. Kafafi, Surf. Sci. **450** (2000) 142.
13. S.Zafeiratos, S. Kennou, Surf. Sci. **443** (1999) 238.
14. W. R. Salaneck, S. Stafstrom, J-L. Bredas, Conjugated Polymer Surfaces and Interfaces, Cambridge University Press, (1996).
15. L. J. Brillson, Surf. Sci. Rep. **2** (1982) 123.

A COMPARISON BETWEEN POINT DEFECT INJECTING PROCESSES IN SILICON USING EXTENDED DEFECTS AND DOPANT MARKER LAYERS AS POINT DEFECT DETECTORS

D. SKARLATOS, D. TSOUKALAS and C.TSAMIS
Institute of Microelectronics, NCSR 'Demokritos', 15310 Aghia Paraskevi, Greece
E-mail:dskar@imel.demokritos.gr

M. OMRI, L.F.GILES and A. CLAVERIE
CEMES/CNRS, BP 4347, 31055 Toulouse Cedex, France

J.STOEMENOS
Department of Physics, Aristotle University of Thessaloniki, 54006 Thessaloniki, Greece

In this work we use dislocation loops and boron - doped δ-layers to monitor the interstitial injection during common and nitrous oxidation of silicon at low temperatures (850–950 °C). The interstitials captured by the loops are measured using Transmission Electron Microscopy. The number of Si atoms released after oxynitridation was calculated from the difference between the total number of atoms stored in the loops for oxidizing and inert ambient. We found that this number is larger compared with the same dry oxygen oxidation conditions. This result is also confirmed by measuring the diffusivity enhancement of boron δ - layers during oxidation under both ambients.

1 Introduction

It is well known that silicon oxidation acts as a source of excess silicon interstitials, which perturbates the equilibrium concentration of point defects and affects the dopant diffusion under the oxidized regions. Dopants as Boron and Phosphorus, which mainly diffuse via an interstitial-mediated mechanism, exhibit an enhancement in their diffusivity (OED phenomenon), while Antimony, which diffuses via a vacancy-mediated mechanism, exhibits a reduction of its diffusivity due to vacancy undersaturation (ORD phenomenon). The use of N_2O for the formation of thin gate oxides presents several advantages as higher immunity to interface trap generation as well as suppression of boron diffusion from the polysilicon gate into the MOS device channel area that makes it attractive for gate dielectric. In this work we investigate the effect of oxidation in nitrous ambient on point defect generation and compare it to common oxidation case.

2 Dislocation loops study

Our experiments have been launched with Si^+ implantation in CZ (100) p-type Si wafers at 50 keV and dose 2×10^{15} cm^{-2} through a thin sacrificial oxide (20 nm). After annealing the wafers at 900 °C for 10 min in N_2 a layer of dislocation loops was formed 110 nm below the surface. Subsequently, the wafers were cut into pairs of companion samples. One of the samples was oxidized in 100% O_2 or N_2O ambient with gas flow 3.5 SLM, while the other was annealed in N_2 for the same conditions. The thermal treatments were carried out for various temperatures (850°-950 °C) and times (up to 2 h). The net number of Si atoms released during common or N_2O oxidation was calculated by subtracting the total amount of interstitial atoms stored in the loops under N_2 conditions from their corresponding number but under oxidation conditions. The results are shown in Figure 1. We observe that the amount of the injected interstitials is larger (of the order of 30-50 %) in the oxynitridation case compared to the dry oxidation case despite that the oxidation rate in N_2O ambient is slower compared to common dry oxidation [1].We attribute this to the formation of a nitrogen-rich layer at the Si/SiO$_2$ interface during nitrous oxidation, which reduces the interstitial loss mechanisms[2].

3 Boron δ- layer study

A 50 nm thick 1×10^{18} cm^{-3} boron –doped δ-layer was grown in a 6-inch wafer followed by a 360 nm intrinsic silicon cap. Samples were annealed in an inert (N$_2$) ambient at 900°C for 30, 60 and 120 min, while an equal number of samples were oxidized in O$_2$ and N$_2$O ambient respectively at the same temperature and times. The boron profiles were determined by secondary ion mass spectrometry (SIMS) and the average diffusivity enhancement for the layer was extracted by comparison of our experimental results to simulations performed with the process simulator SILVACO SSUPREM4 using as fitting parameter the recombination velocity at the Si/SiO$_2$ interface using for the other parameters the default SSUPREM4 values. The results of the simulations are shown in Figure 2. The values of the recombination velocities obtained are:

$$K_{ox(O2)} = 0.23 \text{ cm sec}^{-1}$$

$$K_{ox(N2O)} = 0.11 \text{ cm sec}^{-1}$$

We observe that the surface recombination velocity during oxidation in dry O_2 is 100% higher than the one observed during nitrous oxidation, indicating the influence of the nitrogen – rich layer formed at the SiO$_2$/Si interface during N$_2$O oxidation.

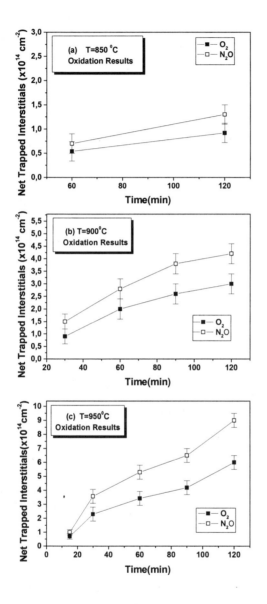

Figure 1: Interstitials injected under common dry and nitrous oxidation

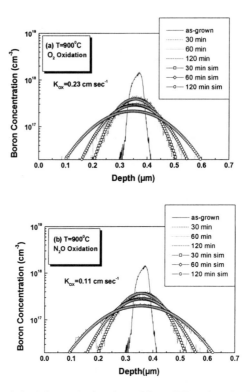

Figure 2: SIMS data and simulation results from boron δ-layer OED study under O_2 (a) and N_2O (b) oxidation conditions at 900°C using as fitting parameter the surface recombination velocity.

4 Conclusions

We found that the nitrogen interfacial layer in the case of nitrous oxidation is responsible for an enhanced interstitial injection in comparison to common dry oxidation

References

1. D.Skarlatos, M.Omri, A.Claverie, D.Tsoukalas, J. Electrochem. Soc.,**146**(6), 2276 (1999).
2. N. Koyama, T. Edoh, H. Fukuda and S. Nomura, J. Appl. Phys. **79**, 1464 (1996).

RAPID THERMAL ANNEALING OF ARSENIC IMPLANTED SILICON FOR THE FORMATION OF ULTRA SHALLOW N^+P JUNCTIONS

N.GEORGOULAS, D. GIRGINOUDI, A. MITSINAKIS, M. KOTSANI AND A. THANAILAKIS
Department of Electrical and Computer Engineering, Democritus University of Thrace, 67100 Xanthi, Greece

Ultra shallow n^+p junction formation by lamp based rapid thermal annealing of As implanted Si has been investigated. The effect of different heating rates, i.e. 70, 160 and 200°C/s, on electrical activation, sheet resistance, junction depth and leakage current has been studied. The results obtained have shown that RTA at a high temperature of 1050°C for 10sec, using a high heating rate of 200°C/s, leads to the formation of shallow n^+p junctions with low sheet resistance and leakage current, as well as to complete activation of the implanted As.

1 Introduction

Submicron ULSI technologies impose stringent requirements for dopant distributions and demand that the junctions should be extremely shallow and heavily doped. It has been shown that RTA leads to a high dopant activation and implantation damage recovery at high temperatures, while at the same time the transient enhanced diffusion of dopant resulting from the interaction with point defects is effectively reduced, due to its short annealing cycle [1-2]. However, there have been only few reports concerning the effect of heating and cooling rates, but the RTA cycle refers to the processing temperature and time [3]. Recently, the heating and cooling rates as a separate part of the overall RTA cycle have been studied and it was found that the entire annealing cycle influences the final activation [4]. In this work, a detailed study of ultra shallow n^+p junctions formed by lamp based RTA of As ion implanted p-Si, has been carried out in order to find the effect of annealing cycle, temperature, time and heating rate on sheet resistance, electrical activation and leakage current.

2 Experimental Procedure

Czochralski-grown p-type Si wafers of <100> orientation with a resistivity of 2-8 Ωcm have been used. The wafers were implanted with 70keV As at a constant dose of 1×10^{15}cm^{-2}. The RTA was carried out, using a tungsten-halogen lamp system, in a nitrogen ambient. Different RTA conditions of temperature (900, 950 and 1050°C), heating rate (70, 160 and 200°C/s) and time (10s or 20s) have been used. The sheet resistance $R_□$ of the annealed wafers was measured using the four-point probe technique. The As atomic concentration profiles have been measured using secondary ion mass spectrometry (SIMS), whereas the electrochemical C-V profiling technique was used to determine the carrier concentration profiles. Mesa diodes of 0.785×10^{-2} cm^2 in area were fabricated on these processed wafers and they were evaluated using I-V and C-V measurements.

3 Results and Discussion

The As atomic concentration profiles and the carrier concentration (electrically active As atoms) profiles of n^+p junctions for different RTA conditions of temperature and processing time, as well as heating rate, are shown in Fig.1.

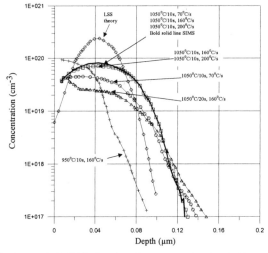

Figure 1. The As atomic (solid line) and carrier concentration (symbols) profiles of n^+p junctions for different RTA conditions.

At 1050 °C/10s the As atomic (SIMS) profiles (solid line) do not change measurably by increasing the heating rate from 70°C/s to 200°C/s. They are quite abrupt and a shallow junction depth of 0.127μm is obtained, at the As atomic concentration of 10^{17}cm^{-3}. The maximum As atomic concentration is 8×10^{19}cm^{-3}. It is also shown that the respective carrier concentration profiles almost follow the atomic profiles at the peak and at the tail regions. As the heating rate increases from 70°C/s to 200°C/s, the carrier concentration profiles and the As atomic profiles almost coincide, indicating that a high fraction of As atoms is in substitutional sites (electrically active), and this was found to be ≈95%. The junction depths, at the carrier concentration of 10^{17}cm^{-3}, are 0.127μm for 160°C/s and 200°C/s and

0.138µm for 70°C/s. When the annealing time increases from 10s to 20s a deeper electrical junction of 0.148µm at the carrier concentration of $10^{17} cm^{-3}$ is obtained.

However, at 950°C/10s the carrier concentration profiles are quite different from those at 1050°C /10s. A high electrical activity over the depth range 0-R_p is observed, indicating that a large fraction of As atoms is in substitutional sites, whereas the electrical activity falls rapidly at the deeper region. At the high As implantation dose of $1 \times 10^{15} cm^{-2}$ an amorphous Si layer is formed. During annealing, the movement of the a-Si/c-Si interface during epitaxial regrowth incorporates As atoms on lattice sites. However, the activation of As atoms deeper than the a-Si/c-Si interface requires annealing temperatures higher than 950 °C.

The values of R_s for different annealing conditions are shown in Table 1. As the temperature increases from 900 to 1050°C the R_s decreases about half an order of magnitude and this is attributed to the increase of both the electrical activation of As atoms and the electrical junction depth. However, a significant increase of R_s (from 45Ω/sqr to 80Ω/sqr) occurs at 1050°C as the time increases from 10 to 20s, due to the out diffusion of As atoms. The values of R_s for different heating rates also show that the higher heating rate results in the highest As activation. The lower R_s of 40 Ω/sqr was observed at 1050°C/10s, 200°C/s. At high heating rates the defect clustering is avoided and the damage annihilation is favored, whereas the low heating rates favor point defect clustering and retard defect recovery.

Table 1. Sheet resistance R_s, ideality factor η and leakage current I_r (at –2V) of n^+p junctions for different annealing conditions.

Temperature/time	Heating rate	R_s	η	I_r(-2V)
900°C/20s	160°C/s	230Ω/sqr	3.50	7.9×10^{-7} A
950°C/10s	160°C/s	120 Ω/sqr	1.80	3.6×10^{-8} A
950°C/20s		140 Ω/sqr	1.75	1.9×10^{-8} A
1050°C/10s	70°C/s	55 Ω/sqr	1.60	7.8×10^{-9} A
	160°C/s	45 Ω/sqr	1.55	6.0×10^{-9} A
	200°C/s	40 Ω/sqr	1.45	3.2×10^{-9} A
1050°C/20s	160°C/s	80 Ω/sqr	1.75	9.0×10^{-9} A

Figure 2 shows the logI-V characteristics of n^+p junctions for different RTA conditions, where it can be seen that as the annealing temperature increases, a significant improvement of logI-V characteristics is observed, the log I-V being more abrupt and the leakage currents lower. The experimental results, such as ideality factor η and leakage current I_r (at –2V), of n^+p junctions for different annealing conditions are summarised in Table 1. In all cases, the ideality factor is larger than 1.4. The leakage current I_r decreases by more than two orders of magnitude as the temperature increases from 900 to 1050°C/10s. Furthermore, an increase in heating rates also leads to a decrease in I_r by a factor greater than two. The residual defects within or near the junction depletion region degrade the electrical junction quality by providing generation-recombination centres resulting in increased currents. At temperatures 900°C and 950°C, the damage recovery level

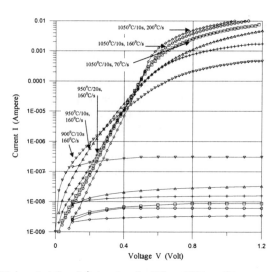

Figure 2. LogI-V characteristics of n^+p junctions for different RTA conditions.

as well as the dopant activation level, was low and, furthermore, the end-of-range defects were not completely recovered. Thus, the electrically activated arsenic concentration was low, and large number of generation-recombination centres led to high reverse leakage, as well as forward, currents. At the annealing temperature of 1050°C/10s and high heating rates 160°C/s-200°C/s, the increased damage annihilation level and dopant activation level contribute to better junction characteristics, demonstrated by the lower values of reverse and forward currents and sheet resistance.

References

1. R.B. Fair, J.J. Wortman and O.W. Holland, *J. Electrochem. Soc.* **132**, p.1962 (1985).
2. Y. Kim, H.Z. Massoud, and R.B. Fair, *J. Electron. Mater.* **18**, p.143 (1989).
3. K. Keyers, V. Lauer, and P. Balk, in proceedings of the *First International Rapid Thermal Processing Conference*, edited by R. B. Fair and B. Lojek, (RTP, Scottsdale, Az, 1993), p.135.
4. Aditya Agarwal, Anthony T. Fiory, Hans-Joachim L. Gossmann, Conor S. Rafferty and Peter Frisella, *Material Science in Semiconductor Processing* **1**, (December 1998) p.237.

SIMULATION OF THE FORMATION AND CHARACTERIZATION OF ROUGHNESS IN PHOTORESISTS

G.P.PATSIS, V.CONSTANTOUDIS, AND E.GOGOLIDES

Instisute of Microelectronics, NCSR "Demokritos", 15310, Aghia Paraskevi
GREECE
E-mail : evgog@imel.demokritos.gr

The aim of this work is to obtain a better understanding of the roughness formation in a lithographic process. To this end, a molecular type simulator of surface and line-edge (SR and LER) roughness for the lithography of a negative-tone chemically amplified epoxy resist is first briefly described. Then the problem of the quantitative characterization of the roughness is addressed. Two parameters, the root-mean square deviation (rms) and the fractal dimension D_f, are computed after each process step for different exposure doses and concentrations of the photoacid generator (PAG). It is found that for constant PAG content, the rms of both SR and LER shows a maximum at low doses, whereas the D_f is reduced as the exposure dose decreases. At high doses, the increase in PAG content is followed by a slight increase in D_f and a decrease in rms. Furthermore, it is shown that in all cases the development process reduces the D_f and increases the rms roughness. Hence, these roughness parameters seem to be not so irrelevant since changes in material properties and process conditions have opposite effects on them. Finally, a first comparison of the simulation results with experimental data is presented.

The continuing race to shrink the circuit elements of microelectronic devices has captured the interest of the scientific and technical world during last decades. One of the key elements in this race is the *lithographic process*, which for this decade is facing the demands of sub100 nm range of characteristic dimension. For such small dimensions, the *roughness* of lithographic materials (also at the nm scale) places a significant limitation to the quality of the printed lines and the CD control. Therefore, the performance and reliability of microelectronic devices necessitates a good understanding of the roughness formation [1]. In order to illuminate the dependence of roughness on material properties and process conditions it is worthwhile constructing a molecular type simulator that follows the different steps of the lithographic process, and then determining and computing reliable descriptors of the roughness of profiles or surfaces produced by the simulator. In this way we can provide a first order prediction of the optimum choices of material properties and process conditions involved in lithographic process.

Such a molecular type simulator of surface and line-edge roughness (SR and LER) for the lithography of a negative tone chemically amplified epoxy resist (EPR) has been developed [2]. A 2D square or a 3D cubic lattice, which is filled with the polymer chains by a self-avoiding random process, simulates the resist system. Here, as a first approximation we restrict ourselves to the 2D square lattice. The photoacid generator (PAG) is randomly distributed in the lattice according to its content in the actual material, and the successive steps of the lithographic procedure including reaction, acid diffusion, crosslink formation and development are incorporated in our simulator. A more detailed description of the simulator has been given elsewhere [2,3]. Actually, it can follow the appearance of SR and LER after each process step and provide us with the final profiles of the fabricated structure for different material and process effects.

After producing the line and surface profiles of the resist, the problem of the quantitative characterization of their roughness arises. There are several parameters which can characterize roughness. The most known and used is the *rms roughness*, which is the root-mean-square deviation of the profile coordinates from their average position. Rms roughness provides the important vertical magnitude of the roughness, but gives no spatial information. The spatial complexity of a profile or a surface can be quantified by the *fractal dimension* (D_F), which is related to the rate at which their fluctuations are reducing as we are examining smaller and smaller scales [4,5]. D_F for a rough profile is a non-integral number between 1 and 2 - the dimensions for a perfectly straight line with no roughness and an infinitely rough profile respectively. Therefore, the higher the D_F, the more complex the profile is.

The fractal dimension of a profile can be calculated by several algorithms, which are equivalent when applied to theoretical profiles. However, when the profile morphology is approximated with a finite set of data points then many algorithms fail to predict the correct D_F. Dubuc et al in [6] have proposed the *variation method* for estimating the D_F in such cases, and showed the superior performance of this method compared to other common algorithms. We applied this method to the calculation of D_F for the profiles produced from the simulator and indicated its equivalence to the cluster dimension used in [1]. In the following, using our simulator we examine the dependence of the roughness parameters, rms and D_F, on the exposure dose, the photoacid generator concentration and the development process.

As regards the dependence on exposure dose, we restrict ourselves to the case where PAG concentration equals to 5%. Similar results are obtained for other PAG contents. In figure 1, we show the rms roughness (1a) and the fractal dimension (1b) for SR and LER as a function of exposure dose as well as the contrast curve. Our results show that the rms roughness parameter goes through a peaked maximum at very low doses, and then quickly drops to small values at useful doses in agreement with [1]. In fact, the maximum roughness from crosslink positions is situated at the onset of the contrast curve. On the contrary, at low doses fractal dimension follows the rise in the contrast curve and then is almost stabilized at useful doses. Furthermore, development process seems to increase rms of both SR and LER, whereas the opposite holds for D_F.

At small doses, the polymer is not crosslinked and thus roughness and surface complexity are small leading to small rms and D_F. At high doses, the polymer is crosslinked, and not attacked by the developer (rms small). However, crosslinking gives surface complexity and so high D_F. At intermediate doses, low crosslinkikg leads to attack of the surface by the developer and thus to large rms.

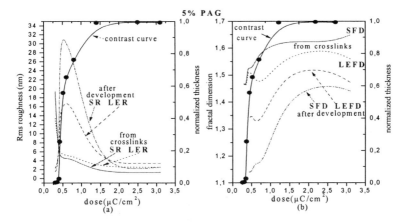

Figure 1: The dependence of the rms roughness (a) and the fractal dimension (b) for surface and line profiles before and after development on the exposure dose. The PAG content was set equal to 5%. Notice the opposite behavior of two roughness parameters.

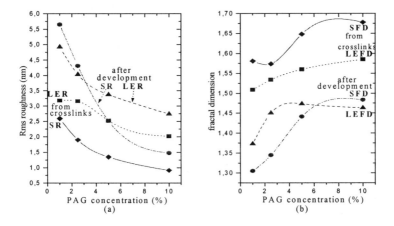

Figure 2: Rms roughness (a) and fractal dimension (b) vs PAG concentration for high exposure doses. Notice again the opposite behavior of rms and fractal dimension: as the PAG content increases, the former is reduced whereas the later is slightly increased.

Figure 2 shows the effects of PAG content on rms roughness (5a) and fractal dimension (5b) for profiles produced from our simulator before and after development process at high doses. One can observe that the increase in PAG content is followed by a decrease in the rms and an increase in fractal dimension.

For this crosslinking system, high PAG content induces high crosslinking and reduces acid diffusion length. This has two effects: On one hand, profiles cannot show large rough fluctuations, thus rms roughness is reduced. On the other hand, the acid catalyzed reactions take place in smaller areas, increasing the local crosslink density and thus triggering the development of more important small-scale structure in profiles, which are made more complex. The increase in complexity is quantified by the increment of fractal dimension observed in figure 2b.

Finally, figure 3 shows the experimentally measured rms values of SR for EPR with 1% PAG content at various prebake and PEB conditions. In all cases rms roughness goes to a maximum at small doses. On the same figure we also present the SR values from our simulations from the crosslink positions and for 1% PAG. Notice the good quantitative agreement with experimental data. Moreover, we show the simulated SR values after development for 5% and 1% PAG content. While the 5% PAG results show good quantitative agreement with the data, the 1% PAG SR values are much higher. This is a hint that our simplified development model is too aggressive for mildly crosslinked systems. The comparison of our simulator predictions for the fractal behavior with the experimental findings is now in progress.

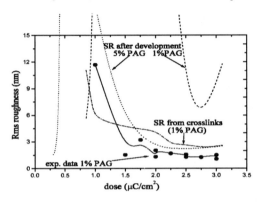

Figure 3:Comparison of simulations with experimental data for the rms roughness parameter.

References

1. D.He and F.Cerrina, J. Vac. Sci. Techmol. B **16**(6) (1998) pp. 3748-3751.
2. G.P.Patsis and N.Glezos, Microelec. Eng. **46** (1999), pp. 359-368.
3. G.P.Patsis, A.Tserepi, E.S.Valamontes, I.Raptis, N.Glezos, and E.Gogolides, J. Vac. Sci. Technol. B, Nov/Dec. 2000.
4. L.Spanos and E.A.Irene , J.Vac.Sci.Technol. A **12**(5) (1994) pp. 2646-2652.
5. L.Lai and E.A.Irene , J.Vac.Sci.Technol. B **17**(1) (1999) pp. 33-39.
6. B.Dubuc, J.F.Quiniou, C.Roques-Carmes, C.Tricot, and S.W.Zucker, Phys. Rev. A, **39** (1989) pp.1500-1512.

F_2 LASER (157 nm) LITHOGRAPHY: MATERIALS AND PROCESSES

E. TEGOU, E. GOGOLIDES, P. ARGITIS, C. D. DIAKOUMAKOS AND A. TSEREPI
Institute of Microelectronics, NCSR "Demokritos", Ag. Paraskevi, Attiki, 153 10 Greece
E-mail: E.Tegou@imel.demokritos.gr

A.C. CEFALAS AND E. SARANTOPOULOU
National Hellenic Research Foundation, Theoretical and Physical Chemistry Institute, 48, Vass. Constantinou Avenue, Athens, 116 35 Greece

J. CASHMORE AND P. GRUNEWALD
Exitech Limited, Hanborough Park, Long Hanborough, Oxford OX8 8LH, England

F_2 laser (157 nm) lithographic materials and processes are examined. Modified acrylic copolymers with tailored etch resistance enhancement are employed as both negative tone and positive tone single layer resists, while siloxanes are used for bilayer schemes. VUV absorbance spectra are studied for film thickness optimization. First results from preliminary exposures at the Exitech Ltd 157 nm prototype microstepper are presented. Aqueous base developed negative resist (ADNR) and pure poly dimethyl siloxane (PDMS) show high resolution potential.

1 Introduction

Lithography using the F_2 excimer laser at 157 nm has recently emerged as the industry – preferred technology for the post - 193 nm era. Critical issues regarding the successful implementation of 157 nm lithography for fabrication of devices with critical dimensions below 100 nm have already been presented by various research groups [1,2,3]. Among these issues, one with very demanding material challenges is the photoresist selection. As with transition to shorter wavelengths in the past, existing photoresists (initially developed for longer wavelengths) are too opaque at 157 nm laser light for practical use. The very high absorbance coefficients of the usual organic polymeric materials limit their use to a maximum thickness of 40-90 nm. Nevertheless, it is expected that by partial fluorination, a decrease in the absorbance value will be obtained [1]. Besides the transparency requirement, the possibly different photochemistry at this wavelength and its effect on imaging possibilities and outgassing effects is a very important subject [4]. Finally, single layer resists must also exhibit good etch resistance, thickness uniformity and developer solubility.

In this paper, two different lithographic approaches are examined. First, modified acrylic copolymers are employed as both negative tone and positive tone single layer resists. Second, siloxanes are used for bilayer schemes. Spectroscopic studies of acrylic polymers and siloxanes are used as a vehicle for a first screening of candidate materials [3]. Results from preliminary exposures at the Exitech Ltd 157 nm prototype microstepper are presented.

2 Experimental

First, poly (hydroxy ethyl methacrylate) (PHEMA), and other hydroxyl functionalised methacrylate copolymers synthesized and developed at IMEL, are examined as single layer resists. These polymers can exhibit both positive (as non-chemically amplified resists, non-CARs) and negative (as chemically amplified resists, CARs) lithographic behavior depending on their formulation. The copolymes consist of up to 4 monomers each, where the HEMA monomer is always one of them. Apart from HEMA, various monomers are used in the formulation of the copolymers, such as cyclohexylmethacrylate (CHMA), isobornylmethacrylate (IBMA), tert-butyl methacrylate, lauryl methacrylate, methacrylic acid and acrylic acid (AA). HEMA is primarily responsible for the aqueous base development of the copolymers. The modification of the chemical composition of PHEMA eliminates the swelling effect and enhances the etch resistance without any addition of etch resistance promoters [5]. The photoacid generator (PAG) used in the resist formulations is a triphenylsulfonium salt in 3% w/w on polymer.

The present work is mostly focused on the material which brings the code name ADNR (Aqueous Developed Negative Resist). The resin is a copolymer [poly (hydroxy ethyl methacrylate)-*co*-(cyclo hexyl methacrylate)-*co*-(iso bornyl methacrylate)-*co*-(acrylic acid), PHECIMA] (see Scheme 1). PHECIMA is obtained as a white solid, and a solution of 20% w/w in ethyl lactate is prepared. PAG is added and the solution is filtered through 0.20 μm filters. Typical process parameters are the following: (i) Prebake (PB): 160°C, 1 min (ii) exposure (iii) Post exposure bake (PEB): 90°C, 2 min (iv) Wet development in diluted AZ 726MIF (H_2O:AZ, 100:1) for 50 s.

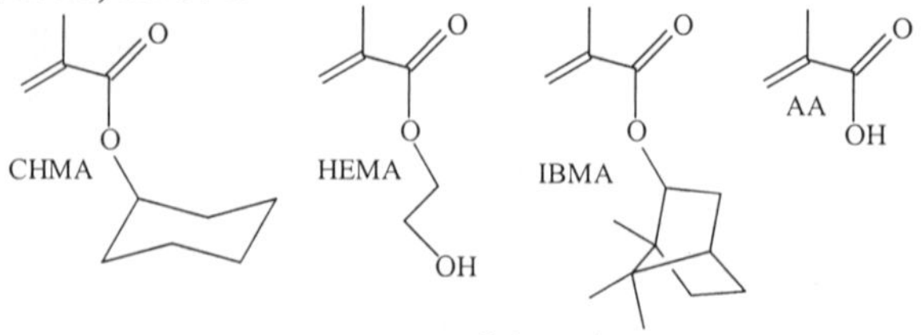

Scheme 1

Second, organosilicon materials are tested as bilayer resists. Based on data from 193 nm and e-beam exposures, pure poly dimenthyl siloxane (PDMS) has been selected as a high resolution material [6]. For PDMS, negative tone imaging is possible. The process parameters include: (i) PB: 90°C, 1 min (ii) PEB: 90°C, 1 min (iii) exposure (iv) Wet development in methyl isobutyl ketone for 90 s (v) Dry development (Reactive ion etching or high density plasma) in 2 steps including a break through etch step.

3 Results and Discussion

Spectroscopic studies of acrylic polymers, phenolic polymers and siloxanes have been used as a vehicle for a first screening of candidate materials [3]. Fig. 1 (a) presents the VUV spectrum of PHEMA in comparison to an epoxy novolac and poly methyl methacrylate (PMMA). The absorbance spectra of both PMMA and PHEMA show a high absorbance coefficient at 157 nm. The hydroxyl group of PHEMA does not add significantly to the absorbance. Due to their high absorbance coefficient (~6 μm^{-1}), acrylic polymers can be imaged at a thickness of ~100 nm.

The absorbance spectra of two different siloxane-based polymers of pure PDMS, and PDMS copolymerised with 5% diphenyl siloxane are indicated in Fig. 1(b). Both polymers seem to respond the same way below 180 nm, and to have the right absorbance coefficient of 2-4 μm^{-1} at 157 nm. Therefore, they can be imaged as negative tone resists at a thickness of ~120 nm for bilayer processes.

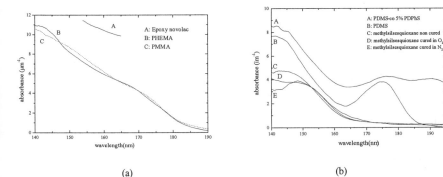

(a)   (b)
Figure 1 (a) Absorbance spectra in the VUV of an epoxy novolac (A), PHEMA (B) and PMMA (C). (b) Absorbance spectra in the VUV of PDMS-copolymer 5% PDPhS (A), PDMS (B), methylsilsesquioxane non-cured (C), methylsilsesquioxane cured in O_2 (D) and methylsilsesquioxane cured in N_2 (E).

For siloxanes and acrylic polymers, preliminary exposures have been carried out at the Exitech Ltd 157 nm prototype microstepper. The modified acrylic polymers exhibit both positive and negative lithographic behavior. Particularly for PHEMA it was noticed that positive behavior switches easily to negative at high exposure doses. In addition, the sensitivity at 157 nm is higher compared to 193 nm.

Among the negative tone single layer resists tested, ADNR shows high resolution potential since 0.5 μm L/S are easily resolved with no previous formulation optimization. Similarly PDMS can resolve 0.35 μm L/S which is close to the exposure tool's limitations. Representative results are presented in Figures 2

(a) and 2(b). For PDMS, the patterns were 250% overdeveloped due to a non-optimized etching process. The exposure dose is measured in arbitrary units.

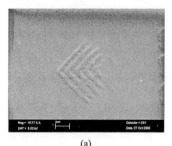 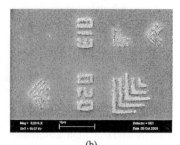

(a) (b)
Figure 2. (a) 0.5 μm L/S of ADNR single layer resist (exposure dose: 2.8 a.u.), (b) 1.0 μm L/S, 0.5 μm L/S, 0.35 μm L/S of PDMS on top of hard baked novolak. (exposure dose: 0.55 a.u.).

4 Conclusions

The high absorbance coefficient limits the use of acrylic copolymers to a thickness of ~100 nm, while organosilicon polymers can be imaged at the usual thickness of 120 nm as bilayer resists. First results for test materials, ADNR and PDMS, from 157 nm exposures are promising. High resolution experiments are needed in the near future.

5 Acknowledgments

This work was funded by ESPRIT project No 33562 ("Photoresist development for 193 nm photolithography and below" RESIST 193-157).

References

1. R.R. Kunz, T.M. Bloomstein, D.E. Hardy, R.B. Goodman, D.K. Downs, J. Vac. Sci. Technol. B **17**(6) (1999) pp. 3267-3272.
2. K. Patterson, M. Somervell, C.G. Willson, Solid State Technology **43** (3) (2000) pp. 41-8
3. A. C. Cefalas, E. Sarantopoulou, E. Gogolides, P. Argitis, Microelec. Eng. **53** (2000) pp. 123-126
4. A.C. Cefalas, E. Sarantopoulou, P.Argitis, E.Gogolides, Appl. Phys. A, **695** (1999) pp. 929-33
5. C. D. Diakoumakos, I.Raptis, A. Tserepi, P. Argitis, MNE 2000 Conference, Microelec. Eng. (in press)
6. A. Tserepi, E.S. Valamontes, E. Tegou, I. Raptis, E. Gogolides, MNE 2000 Conference, Microelec. Eng. (in press)

FABRICATION OF FINE COPPER LINES ON SILICON SUBSTRATES PATTERNED WITH AZ 5214™ PHOTORESIST VIA SELECTIVE CHEMICAL VAPOR DEPOSITION

D. DAVAZOGLOU,
NCSR "Demokritos", Institute of Microelectronics, POB 60228, 153 10 Agia Paraskevi, Greece.

S. VIDAL AND A. GLEIZES
CIRIMAT - UMR 5085, INPT/CNRS, Ecole Nationale Superieure de Chimie, 118 route de Narbonne, 31077 Toulouse cedex 4, France.

Copper features with dimensions down to 0.5 μm were fabricated on silicon substrates by selective chemical vapor deposition. For the fabrication oxidized (100) silicon substrates were used, covered with a film grown by LPCVD at 0.1 Torr and 550 °C, from $W(CO)_6$ decomposition. These substrates were subsequently covered with AZ 5214™ photosensitive polymer, which has been developed as both positive and negative tone resist. Copper was then chemically vapor deposited on the patterned substrates by 1, 5-cyclooctadiene Cu(I) hexafluoroacetylacetonate decomposition, at 1 Torr and temperatures of 110 and 140 °C. A vertical, cold-wall reactor was used, equipped with a UV lamp permitting photon-assisted deposition. Under UV illumination, copper was deposited on resist covered and uncovered parts of the substrate. In absence of illumination, the metal was selectively grown on the tungsten film only at relatively slow rates (1 and 3.5 nm/min at 110 and 140 °C respectively). Copper films had a granular form with a grain size increasing with temperature (150 and 550 nm at 110 and 140°C respectively). After depositions the resist was removed in an oxygen plasma leading to the formation of fine copper features.

1 Introduction

Copper was first introduced in the metallization of electronic circuits integrated on silicon substrates, near the end of nineties [1]. This metal exhibits lower resistivity than aluminium and is less sensitive to electromigration [2], so its use in the metallization of electronic circuits permits the fabrication of faster and more reliable devices. Furthermore, the use of copper, together with that of dielectrics with low dielectric constant, permits the reduction of the total number of metal levels while maintaining a low RC constant, in devices with dimensions below 0.35 μm [3].

Before the full introduction of copper in the manufacturing of integrated circuits, many problems have to be solved. The most important of them are: the high diffusivity of copper into silicon and the various dielectrics used in manufacturing, its poor adhesion on these materials and the difficulty to be patterned with traditional, plasma based, etching methods. In order to respond to these issues, other materials have to be introduced in the metallization that serve as diffusion barriers and as seed layers for copper. Two voids have been explored aiming in the solution of the problem of patterning. One involves the deposition of a blanket copper film over an etched surface followed by a chemical-mechanical polishing step (damascene process [4]) and the other consists in the growth of this metal over selected areas of the substrate [2].

In this work we report on the selective chemical vapor deposition (SCVD) of copper by 1, 5 cyclooctadiene Cu (I) hexafluoracetylacetonate (COD-Cu-hfac) decomposition, on silicon substrates covered with a commercial photo sensitive resist (AZ 5214^{TM}) and patterned, following various industrial processes. After SCVD of the metallic film, the remaining resist can be removed, leading so to the formation of various features on the substrates.

2 Fabrication and discussion

The details of fabrication of copper features and simple devices via SCVD have been described previously [5] so, only a brief description is given here. To fabricate MOS capacitors high quality oxides, 40 nm thick, were grown on 3 in. (100) silicon substrates at 950 °C in dry oxygen. These oxides were covered by polycrystalline silicon, 500 nm thick, on top of which a tungsten film, 50 nm thick was grown by LPCVD at 0.1 Torr and 550 °C from $W(CO)_6$ decomposition [6]. Some samples were also processed for which tungsten was deposited directly on the oxide. AZ 5214^{TM} was then applied and developed, in a first time, as a negative tone resist. Copper was subsequently chemically vapor deposited on the patterned substrates in a vertical, cold wall, stainless steel MOCVD reactor, using COD-Cu-hfac as molecular precursor, at a pressure of 1 Torr and a temperature of 110 °C, under UV illumination [7]. At these conditions copper was deposited smoothly on resist covered and exposed parts of the substrate. It was shown [5] that the polymer melts during growth and diffuses below the copper film near the edges of the various features. An attempt to remove the polymer after copper growth, in an ultrasonically agitated acetone bath, was proven ineffective, because the polymer was hardened due to the long heating (180 min) and because the copper film was deposited uniformly over the substrate topography, inhibiting the acetone to diffuse below it dissolving the polymer. In absence of UV illumination selective CVD was obtained; the copper was grown on the exposed parts of the substrate only. In this case also the polymer was difficult to remove in ultrasonically agitated acetone however, in some cases this proved to be feasible. The capacitance-voltage (C-V) characteristics of the MOS capacities thus formed were measured and showed that the electrical properties of the Si/SiO_2 interface were not affected by the processing. This was an indication that tungsten acts effectively against copper diffusion.

After this first series of experiments, a decrease of minimum features was attempted. For these experiments, the polymer was developed as a positive tone resist and in order to harden it and at the same time ensure its removal by plasma etching after copper deposition, after development it was treated by three different methods: a) Flood exposed with a deep UV lamp (Hg-Xe) for 5 sec and baked at

125 °C for 5 seconds. b) Flood exposed, with the same lamp, for 1 hour and baked for 15 min at 125 °C and c) Flood exposed for 1 hour and baked for 5 min at 140°C.

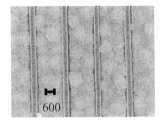 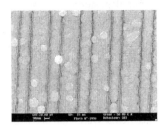

Figure 1. Copper lines formed at 140 °C on substrates patterned with photoresists treated with methods (a) (left) and (b) (right).

Copper was then chemically vapor deposited on the patterned substrates at the same as before conditions. In Fig. 1 copper films are shown grown by SCVD, at 140 °C on patterned substrates for which the resist was treated with the method a) and b) described above. It is seen that the copper film has a granular form and contours smoothly the polymer. Moreover, practically no copper nucleation occurs on the resist. For samples for which the resist was treated with method c) similar results were obtained. The average grain size of the copper film increases rapidly with temperature (150 and 550 nm at 110 and 140°C respectively).

Selective deposition of copper is a complex process which depends on the nature of the precursor, the surface, the reactor type and geometry, operating conditions, surface preparation, deposition time and other factors [2, 8, 9]. So, the investigation of the exact mechanisms resulting in selective growth is a very difficult and laborious matter. Within the frame of this work, mostly oriented towards the technological aspects of SCVD, it can be speculated that deposition is catalyzed by negative electric charges. This speculation is made in view of older

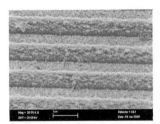 

Figure 2. Copper lines formed by SCVD after resist removal. Some polymer "scum" is shown between lines.

results relative to selective copper deposition from the same precursor on tungsten, in presence of SiO_2 [9]

After deposition the resist was removed in a reactive ion etcher (RIE) in an oxygen plasma. In Fig. 2 copper lines deposited at 110°C with widths below 1 μm and a thickness of 170 nm, are shown after resist removal. From such measurements the growth rate was estimated to be 1 and 3.5 nm/min for deposition temperatures of 110 and 140 °C. The layer of tungsten below the copper lines can be easily removed with plasma etching, using SF_6 chemistry. From four point probe measurements taken on samples before and after the two plasma treatments, it was found that the resistivity of the copper films was not affected. It must be noted however, that these measurements were not taken on the copper lines but on substrate areas blanket covered by copper.

3 Conclusions

SCVD of copper by COD-Cu-hfac decomposition on silicon substrates covered with tungsten and patterned with AZ 5214™ when this last is developed either as negative or positive tone resist, has been demonstrated. It is possible with this method to fabricate thin copper lines of width below one micron. The process is reproducible and robust, if one takes into account that sometimes SCVD took place several weeks after tungsten deposition and AZ5214™ patterning.

References

1. D. Edelstein, J. Heidenreich, R. Goldbatt, W. Cote, C. Uzoh, N. Lusting, P. Roper, T. McDevitt, W. Motsiff, A. Simon, J. Ducovic, R. Wachnick, H. Rathor, R. Schulz, L. Su, S. Luce and J. Slattery, Tech. Digest IEEE Int. Electron Devices Mtg., (1998) p. 773.
2. T. Kodas and M. Hampten-Smith, "The Chemistry of Metal CVD", VCH, Weinheim, 1994.
3. T. Seidel and B. Zhao in "Advanced Metallization for future ULSI", K. Tu, J. Mayer, J. Poate and L. Chen, EditorsMaterials Research Society (1996), p. 3.
4. P. Andricacos, The Electrochem. Soc., "Interface", Spring 1999 p. 32.
5. S. Vidal, A. Gleizes and D. Davazoglou, Microelec. Eng. (In press)
6. D. Davazoglou, A. Moutsakis, V. Valamontes, V. Psycharis and D. Tsamakis, J. of the Electrochem. Soc., **144** (1987) p. 595.
7. S. Vidal, Thesis, INP Toulouse 1999.
8. A. Jain, K. Chi, M. Hampten-Smith, T. Kodas, J. Farr and M. Paffet, J. Mater. Res. Vol. 7 (1992) p. 261.
9. A. Jain, J. Farkas, K-M Chi, M. Hampten-Smith and T. Kodas, Appl. Phys. Lett., **62** (1992) p. 2662.

INVESTIGATION OF THE NITRIDATION OF Al_2O_3 (0001) SUBSTRATES BY A NITROGEN RADIO FREQUENCY PLASMA SOURCE

S. MIKROULIS, V. CIMALLA, A. KOSTOPOULOS, G.CONSTANDINIDIS, G. DRAKAKIS, M. ZERVOS, M. CENGHER AND A. GEORGAKILAS

Foundation forResearch and Technology Hellas (FORTH), Institute of Electronic Structure and Laser(IESL), Microelectronics Research Group, P.O.BOX 1527, Heraklion, Crete, Greece
E-mail: spirosm@physics.uoc.gr

> The purpose of this study is to investigate the exact mechanism of the nitridation of c-plane Al_2O_3 and the role of this treatment in the properties of GaN epitaxial layers grown by radio frequency plasma assisted molecular beam epitaxy. Different nitridation conditions were used with high and low substrate temperatures and nitridated samples were studied by reflected high energy electron diffraction (RHEED), Auger electron spectroscopy (AES), and atomic force microscopy (AFM). The sapphire nitridation temperature was found to control the polarity of GaN epitaxial layers grown on a GaN nucleation layer. High temperature resulted to strong nitridation effect and Ga-face material, while low temperature resulted to weak nitridation and N-face GaN. However, Ga-face material was also grown when an AlN nucleation layer was used on substrate nitridated at low temperature. Finally AlGaN/GaN HFET devices were realized on Ga-face structures completely grown by MBE, using an AlN nucleation layer on a low temperature nitridated Al_2O_3.

1 Introduction

Group III nitrides are today a material system of great interest for their applications in short wavelength photonic devices [1] as well as for high power and high frequency electronics[2,3]. In the absence of suitable GaN (AlN) substrates the most commonly used material for heteroepitaxy is c-plane sapphire (Al_2O_3), despite its large lattice (12.6%) and thermal mismatch. However, the growth of thin AlN [4,5] or GaN [6] layers at low temperatures and a nitridation step [7,8] prior to depositing the GaN active layer have been proved to improve significantly the heterostructures in terms of both structural and optoelectronic properties. The purpose of the nitridation treatment is to convert the surface layers of the Al_2O_3 (0001) substrate to AlN by the exchange of surface oxygen with nitrogen, thus providing an improved surface for subsequent epitaxy.

In this article, we present a detailed study of nitridated Al_2O_3 (0001) by a radio frequency plasma source and the influence of nitridation conditions on the GaN layers grown by radio frequency plasma assisted molecular beam epitaxy. It was found that the nitridation temperature controls the polarity of the overgrown GaN films. Ga-face material appeared to grow through the nucleation and expansion of "Ga-face" AlN grains on the Al_2O_3 surface.

2 Experimental

(0001) sapphire substrates were degreased using organic solvents, rinsed with deionised water and blown dry with nitrogen gas before their loading in the MBE system for outgassing: Some samples were prepeared using wet chemical etching with a solution of HF/H_2O (1:4) instead of degreasing and characterized by fourier transform infrared spectroscopy (FTIR). Nitridation experiments were accomplished in a standard MBE growth chamber of a RIBER 32P two-module system, equipped with an Oxford Applied Research HD25 radio frequency (13.56MHz) nitrogen plasma source operating mostly at 400 and 500W. The substrate temperature during nitridation was 750 and 200°C for a nitrogen flux of 0.2-1.5sccm, equivalent to 0.6-4.2×10^{-5}Torr pressure in the growth chamber. Nitridation duration was 100min. The nitridation was monitored in-situ with reflected high energy electron diffraction (RHEED) using an electron beam of 12keV and recorded with a high resolution digital camera. Some samples were ex-situ characterized by Auger electron spectroscopy (AES) using an electron beam of 4.5keV and by atomic force microscopy (AFM) with a Digital Instruments Nanoscope IIIa operating in a contact mode. In most cases GaN layers were also overgrown using different nucleation layers; 15nm GaN and 20nm AlN at 400°C and in some cases $Al_{0.12}Ga_{0.88}N/GaN/Al_{0.12}Ga_{0.88}N$ HFET structures. On these structures, 2 and 4μm gate length transistor devices were fabricated using Ni/Au Schottky and Ti/Al ohmic contacts and their dc I-V characteristics were determined.

3 Results and discussion

Spectral analysis of the plasma source was accomplished for different powers and nitrogen fluxes. The N II line of ionic nitrogen at 648.2 nm was also observed superimposed on the molecular band of the rf plasma source, only at injected powers higher than 100W for a gas flow of 0.56 sccm. This increase is related to the increase of the population of ionic nitrogen in the RF plasma. According to H.P. Summers [9] the onset of this line might be due to the increase of the electron density in the plasma.

FTIR measurements on sapphire substrates prepared with chemical etching in a HF/H_2O solution showed a significant reduction of the reflectance intensity (10%) which may be due to crystal degradation. The most important result was the appearance of an absorption peak at approximately 832cm^{-1} attributed to Oxygen vacancies [10]. These vacancies might assist the nitridation process, as it was revealed by AES measurements which will be described later.

The deflection distance of the RHEED diffraction rods was used as a direct observation of the variation of the in-plane lattice constant of (0001) Al_2O_3 during nitridation (fig.1). AlN when grown directly on Al_2O_3 has a lattice constant 13.2% larger, so this would be the reduction in the distance of diffraction rods if a fully relaxed AlN layer was created on the substrate by nitridation. The analysis of RHEED observations during nitridation indicated that we had a partially relaxed AlN on Al_2O_3 independently of the nitridation temperature. This relaxation was in the order of 6-9% (fig.1). The spotty RHEED patterns during HT nitridation were

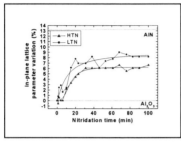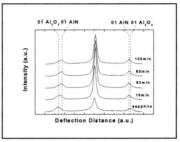

Figure 1. RHEED observations of AlN formed during nitridation. The right picture shows the reduction of the deflection distance between first order rods of (0001) sapphire towards a partially relaxed (0001) AlN, and the left graph the amount of this relaxation.

indicative of the 3-dimensional character of the surface while the streaky RHEED patterns observed during (LT) nitridation showed a 2-dimensional surface

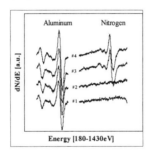

Figure 2. AES measurements. The N(KLL) peak was only observed for HT nitridation (samples #3, #4) and was more intense in the sample with HF/H_2O preparation (sample #3)

morphology and these results were confirmed with AFM.

AES was used as a more straightforward method to estimate the extent of the surface nitridation. Four samples were investigated, with different preparation methods and substrate temperatures (fig.2). We concluded that LT nitridation should be limited to a surface atomic plane and become unstable against re-oxidation. On the other hand HT nitridation should extent to several monolayers, probably in the form of 3-D islands and remain substantial during the sample's transfer from the MBE to the AES system.

RHEED surface reconstruction observations as well as KOH selective etching combined with AFM measurements were used for an investigation of the polarity of GaN overlayers grown with high and low nitridation temperatures using GaN nucleation layers. It was found that the nitridation temperature controls the polarity of the GaN epitaxial layers[11]. N-face material was grown for LT nitridation while Ga-face for HT nitridation. The use of an AlN nucleation layer, instead of GaN, produced a Ga-face material as already has been shown in the literature [12]. A

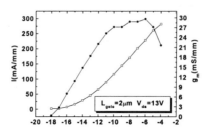

Figure 3. DC characteristics of AlGaN/GaN devices grown on (0001) sapphire

comparision of the properties of N-face and Ga-face GaN are presented elsewhere[11].
Finally $Al_{0.12}Ga_{0.88}N/GaN/Al_{0.12}Ga_{0.88}N$ HFET structures were grown on LT nitridated sapphire, using an AlN nucleation layer (Ga-face material). Transistors with 2 and 4μm gate length and 250μm width were fabricated. These devices could withstand drain-source voltages above 50V for a 50mA current and exhibited a maximum transconductance of 30mS/mm as shown in fig. 3.

4 Acknowledgements

This work has been supported by GSRT, Hellenic Ministry of Development through the PENED project no.: 99EΔ 320

References

1. S. Nakamura and G.Fasal, The blue Laser Diode (Springer, Berlin, 1997)
2. S. N. Mohammad et al., Proc. IEEE 83, 1306 (1995)
3. M. A. Khan et al. Electron Lett. 32 357 (1996)
4. I. Akasaki et al. J. Cryst. Growth 98, 209 (1989)
5. J. N. Kuznia et al. J. Appl. Phys. 73, 4700 (1993)
6. S. Nakamura et al. Jpn. J. Appl. Phys., Part 2 30, L1705 (1991)
7. S. Keller et al. Appl. Phys. Lett. 68, 1525 (1996)
8. N. Gradjean et al. Appl. Phys. Lett. 69, 2071 (1996)
9. H.P. Summers Mon. Not. R. Astron. Soc. 169 (1974) p. 663
10. M. D. Mc Cluskey et al., J.Appl.. Phys., V87, n8, 3593, (2000)
11. A. Kostopoulos, S.Mikroulis et al., Proc. of MMN 2000, Athens 20-22 Nov.
12. F. Widmann et al., J. Appl. Phys., 85, 1550, (1999)

SIMULATION OF SI AND SIO$_2$ FEATURE ETCHING IN FLUOROCARBON PLASMAS

GEORGE KOKKORIS

Institute of Microelectronics, NCSR "Demokritos", Aghia Paraskevi, Attiki, GREECE 15310, and Dept. of Chem. Eng., Nat Tech Univ of Athens, Zographou Campus, Attiki, GREECE 15780
E-mail: gkok@imel.demokritos.gr, gkok@central.ntua.gr

EVANGELOS GOGOLIDES

Institute of Microelectronics, NCSR "Demokritos", Aghia Paraskevi, Attiki, GREECE 15310
E-mail: evgog@imel.demokritos.gr

ANDREAS G. BOUDOUVIS

Dept of Chem. Eng., Nat Tech Univ of Athens, Zographou Campus, Attiki, GREECE 15780
E-mail: boudouvi@chemeng.ntua.gr

A surface model for open area etching of Si and SiO$_2$ is coupled with a model to calculate the local values of etching rate on each elementary surface of the structure being etched. Ion energy and composition, and the ratios –R- of neutral to ion flux (Fluorine atoms or carbon containing radicals) are used as independent variables, while local etching yields and rates are the dependent variables. The local etching model (essentially a local flux calculation model) includes shadowing effects of ions / neutrals and reemission, while charging effects are simulated only by an increased ion angular spread. Aspect ratio dependent (ARDE) and independent (ARIE) etching as well as transition from etching to deposition are predicted and studied as a function of plasma phase composition.

1 Introduction

Etching of Si and SiO$_2$ structures in fluorocarbon plasmas is a frequently encountered process in microelectronics fabrication (i.g. etching of contact holes). Etching rates of Si and SiO$_2$ features have been observed to depend on aspect ratio (AR) of the feature (depth/width of the feature). This scaling of etching rates to AR is described by the term Aspect Ratio Dependent Etching (ARDE). Several manifestations of ARDE during Si and SiO$_2$ structure etching have been reported. Reactive Ion Etching lag (*RIE lag*) is frequently encountered: etching rate is lower in narrower trenches and smaller diameter holes. Sometimes, intense RIE lag can lead to etching stop and *transition from etching to deposition* occurs. *Inverse RIE lag* has also been reported: etching rate is higher in narrower trenches and smaller diameter holes. All the above mentioned effects can cause several problems during etching of structures. For example, during etching of contact holes with different diameter, etching can stop in smaller diameter holes. The goal to achieve is Aspect Ratio Independent Etching (ARIE).

The effects of surface reactions and near surface ion and neutral transport on ARDE are studied. This effort requires: a) a *surface model for open area etching*, which takes into account surface chemistry and defines the quantitative effect of several parameters (i.e. ion energy, composition and flux and neutral flux) on etching rate. b) a *model to calculate the local values* of these parameters on each elementary surface of the structure being etched. c) coupling of model (a) with (b) to calculate the etching rate at each surface of the structure.

2 The surface model

An adsorption-type model [1] accounting for etching and deposition in fluorocarbon plasmas during Si and SiO_2 open area etching is used to describe the surface physical and chemical processes. In this phenomenological model, fluorine atoms, fluorocarbon radicals, and polymer (produced on the surface) are considered bonded to surface atoms.

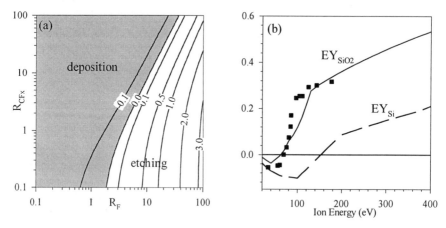

Figure 1. (a) Si etching yield vs R_F, R_{CFx}. Ion energy is 100eV and ion composition is 10% CF_3^+, 85% CF_2^+, 5% CF^+. (b) Si and SiO_2 etching yield vs ion energy. $R_F=7$, $R_{CFx}=10$, ion composition is similar to 1a. Experimental data [2] (squares) show the etching yield of a SiO_2 open area in a CF_4 plasma.

The basic equations of the model are F, CF_x, and Polymer site balances [1]. The independent variables for the model are the *ratios*, R_F *and* R_{CFx}, *of neutral species (F atoms and fluorocarbon radicals) fluxes to ion flux*, ion energy and ion composition. If etching rate is to be calculated, ion flux value is needed. Besides yield (Si atoms/ion) and rate, effective sticking coefficients, S_E, of neutral species are outputs of the surface model. S_E of a species represents the net loss (if $S_E>0$) or creation (if $S_E<0$) of this species during surface reactions.

In Fig.1a Si etching yield contours as a function of ratios R_F and R_{CFx} are shown. To move from deposition to etching region R_F increase of R_{CFx} decrease is needed. Generally, as R_F increases or R_{CFx} decreases, etching yield increases. In Fig. 1b Si and SiO_2 etching yields as a function of ion energy are shown. For low values of ion energy, deposition of a fluorocarbon film occurs, for higher values transition to a fluorocarbon suppression regime happens, and for even higher values oxide ion-enhanced etching regime is entered [2]. The experimental data in Fig. 1b are used only for qualitative comparison, since the absolute values of neutral densities needed for the quantitative comparison were not measured in [2].

3 The local flux calculator

The model used for the local value calculation of ion and neutral flux (flux calculator) on each elementary surface of a structure takes into account shadowing of ion and neutral flux and re-emission of neutral flux in trenches and holes [3]. Charging effects are not considered, but are simulated by an increased ion angular spread. This flux calculation model is modular. One can add different phenomena, different flux distributions, different structures, different surface models.

The total flux of a species (ion or neutral) at an elementary surface s of the structure is given by the equation:

$$j(s) = j_{direct}(s) + \int_0^{S_{end}} g(s,s')[1 - S_E(s')]j(s')ds' \quad (1)$$

$g(s, s')$ is a function which depends on the re-emission mechanism and the geometry of the structure being etched. j_{direct} is the flux coming directly from the plasma and expresses the effect of shadowing on the flux. The integral of eq. 1 stands for the flux coming at s from all other elementary surfaces s' by reemission.

In Fig. 2a total flux at the trench bottom center as a function of the trench AR for different values of S_E is shown. Flux is normalized to the flux at an open area. For small values of S_E the total flux is close to the flux to an open area. When S_E increases, the flux decreases. For $S_E=1$, the total flux is equal to direct flux.

4 Coupling of the surface model with the flux calculator

The calculation of local etching rates requires the coupling of the surface model with the flux calculator. The key point of the coupling algorithm is the simultaneous calculation of the effective sticking coefficient (by the surface model) and the total flux for every species (by a system of integral equations such as eq. 1).

In Fig. 2b the normalized etching rates of Si and SiO_2 as a function of trench AR for various gas phase conditions are shown. Conditions are expressed by the ratios of neutral to ion fluxes reaching the trench top, ($R_{F,0}$, $R_{CFx,0}$). Gas phase conditions define whether RIE lag, inverse RIE lag, or ARIE occurs, as well as the exact position of the transition from etching to deposition.

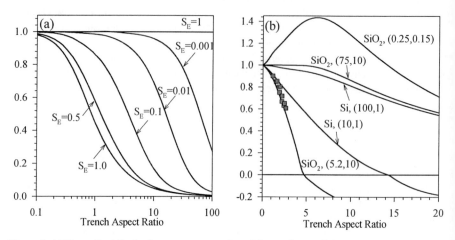

Figure 2. (a) Normalized (to the flux at an open area) total flux at the trench bottom center vs trench AR for different S_E. Notice the importance of S_E in total flux calculation. (b) Normalized Si and SiO_2 etching rates (to the etching rates at an open area) vs trench AR for different couples $(R_{F,0}, R_{CFx,0})$. Ion energy and composition is the same as in Fig. 1a. Experimental data [4] refer to SiO_2 trench etching in a C_2F_4 plasma.

5 Conclusions

The effects of surface reactions and near surface ion and neutral transport on ARDE of Si and SiO_2 features were studied through the coupling of a surface model accounting for etching and deposition with a local flux calculator inside features. The focal point of the coupling algorithm is the effective sticking coefficient calculation. Depending on the gas phase conditions, transition from etching to deposition, RIE lag, inverse RIE lag, and ARIE can be observed.

References

1. E. Gogolides, P. Vauvert, G. Kokkoris, G. Turban, and A. G. Boudouvis, *J. Appl. Phys.* **88**, (2000) pp. 5570-5584.
2. G. S. Oehrlein, Y. Zhang, D. Vender, and M. Haverlag, *J. Vac. Sci. Technol. A* **12**, (1994) pp. 323-332.
3. G. Kokkoris, E. Gogolides, and A. G. Boudouvis, presented in MNE 2000.
4. O. Joubert, G. S. Oehrlein, and Y. Zhang, *J. Vac. Sci. Technol. A* **12**, (1994) pp. 658-664.

EPITAXIAL ErSi$_2$ ON STRAINED AND RELAXED Si$_{1-x}$Ge$_x$

G. APOSTOLOPOULOS, N. BOUKOS, P. PAPANDREOPOULOS, A. TRAVLOS

Institute of Material Science, NCSR "Demokritos", 153 10 Ag. Paraskevi, Greece

We study for the first time the epitaxial growth of ErSi$_2$ on strained and relaxed Si$_{1-x}$Ge$_x$ substrates, as well as its structural properties. Epitaxy of ErSi$_2$ is achieved at a temperature of 550°C after the reaction of a very thin Er/Si template layer. ErSi$_2$ grows in the tetragonal phase, which has been previously observed for epitaxial growth on Si at higher temperatures. ErSi$_2$ layers grown on relaxed Si$_{1-x}$Ge$_x$ exhibit a higher crystal quality, due to the reduction of the lattice mismatch with increasing Ge content.

1 Introduction

Silicide-semiconductor epitaxial heterostructures are very interesting from the technological and the scientific point of view[1,2]. This type of structures may be implemented in the realization of various devices like metal base transistors, resonant tunneling diodes and transistors. Furthermore, they play a major role in the detection of infrared radiation. From the pure scientific point of view, the silicide-semiconductor heterostructure represents an ideal metal-semiconductor junction, with well-defined crystallographic orientations and quality of interface. Especially the rare-earth silicides are particularly important, due to their good epitaxial properties on silicon and their low Schottky barriers. The combination of silicides and Si$_{1-x}$Ge$_x$/Si heterostructures opens new possibilities for applications, and it may also serve as a means for making metal contacts to silicon-germanium.

In the present contribution we investigate for the first time the growth of erbium silicide, one of the most important rare-earth silicides, on Si$_{1-x}$Ge$_x$/Si(001) substrates. Two different kinds of substrate have been studied: (a) growth on strained Si$_{1-x}$Ge$_x$(001) with a 20% Ge content, and (b) growth on relaxed Si$_{1-x}$Ge$_x$(001) with a Ge content of 25%, which in turn is grown on the step graded Si$_{1-x}$Ge$_x$ buffer with Ge content ranging from 6% to 25%. Both situations have fundamental importance in most devices developed within the Si$_{1-x}$Ge$_x$ material system. Epitaxy of ErSi$_2$ is achieved at a temperature of 550°C after the reaction of a starting thin Er/Si bilayer. ErSi$_2$ grows epitaxially on Si$_{1-x}$Ge$_x$ in its tetragonal phase, which is not observed in the bulk form of the silicide. The tetragonal phase has been previously observed on Si substrates[3], but at higher substrate temperatures. The ErSi$_2$ films grown on relaxed Si$_{1-x}$Ge$_x$ present a higher crystal quality, which is attributed to the reduction of the lattice mismatch.

2 Experimental

The samples are prepared in an UHV chamber on Si(001) substrates by evaporating of the starting materials Si, Ge and Er from e-gun sources. The substrate temperature is measured by a thermocouple situated at the backside of the substrate, while the fluxes of the evaporating materials are determined by means of quartz crystal balances. After standard chemical cleaning the substrates are inserted into the vacuum chamber and heated at 900°C for 10 min under a low Si flux in order to desorb the native oxide. A 1000 Å Si buffer is then grown at 680°C. Strained $Si_{1-x}Ge_x$ layers with x=0.2 and a thickness of 500 Å are deposited at a temperature of 550°C. For the realization of relaxed $Si_{1-x}Ge_x$ layers, a 5000 Å step graded buffer is first grown with Ge content ranging from x=0.06 to 0.25, and substrate temperatures from 700°C down to 500°C. A 500 Å relaxed $Si_{1-x}Ge_x$ layer with x=0.25 is grown on top of the step graded buffer at a temperature of 500°C. After the preparation of the $Si_{1-x}Ge_x$ substrates, a thin Si layer (20 Å) followed by an Er layer (15 Å) are deposited at a low temperature (350°C) to avoid the segregation of Ge and its mixing with Er. After 10 min annealing at 550°C a starting layer of $ErSi_2$ is formed. Growth of the silicide continues with coevaporation of Er and Si at a stoichiometric ratio. The layers are characterized by x-ray diffraction (XRD) and cross-section transmission electron microscopy (TEM).

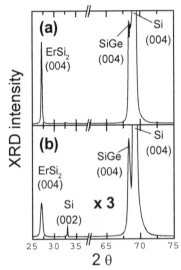

Figure 1. XRD results from $ErSi_2$ grown on (a) relaxed and (b) strained $Si_{1-x}Ge_x$

3 Results and Discussion

Figure 1 shows θ–2θ XRD scans for two characteristic samples. Apart from the two adjacent peaks from the (004) planes of Si and $Si_{1-x}Ge_x$, another peak is observed at lower angles which corresponds to the (004) peak of the tetragonal $ErSi_2$. Thus, the silicide grows epitaxially with the c-axis parallel to the [001] direction of $Si_{1-x}Ge_x$(001). The peak from the $ErSi_2$ grown on the relaxed $Si_{1-x}Ge_x$(001) substrate appears sharper and with a higher intensity, indicating a higher epitaxial quality of the silicide in this case. $ErSi_2$ has a lattice constant c=13.26 Å, as estimated from the XRD data, on both strained and relaxed $Si_{1-x}Ge_x$,

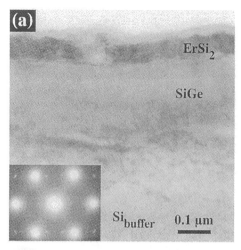

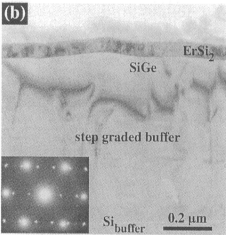

Figure 2. Bright field TEM images of ErSi$_2$ films on (a) strained (b) relaxed SiGe

despite of the 1% reduction of its lattice mismatch from the strained to the relaxed substrate. This shows that the silicide grows in the relaxed state. Figures 2a and 2b show cross section TEM bright field images of ErSi$_2$ films grown on strained and relaxed Si$_{1-x}$Ge$_x$ substrates, while the corresponding selected area electron diffraction (SAED) patterns are shown in the inserts. The electron beam direction is parallel to the [011]$_{SiGe}$ direction, as may be confirmed by the diffraction patterns. The ErSi$_2$ layers are continuous with large epitaxial regions. Indexing of the diffraction images showed that the crystal structure of the erbium silicide is indeed tetragonal, with the following orientation with respect to the SiGe substrate:

(001)ErSi$_2$ // (001)SiGe
[100]ErSi$_2$ // [110]SiGe
[010]ErSi$_2$ // [$\bar{1}$10] SiGe

and lattice constants

$$a = b = 3.95 \text{ Å}, c = 13.26 \text{ Å}$$

The above crystallographic orientations as well as the growth of the tetragonal and not the orthorombic phase are confirmed by an analysis of diffraction patterns at the [001]$_{SiGe}$ zone axis. It must be noted that the tetragonal phase has been observed also in the case of erbium silicide growth on Si(001), but at higher temperatures (~ 800 °C)[3].

As may be seen in Fig. 2, the epitaxially grown regions of ErSi$_2$ are separated by small crystallites of another orientation. The density of these crystallites is higher for the silicides grown on strained SiGe. Furthermore, by comparing the two bright field images, it is clear that the silicide layer grown on the relaxed SiGe substrate has a less crystal defects and a smoother surface. This is attributed to the smaller

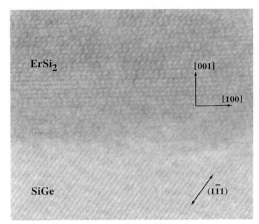

Figure 3. HREM image of an $ErSi_2$/ relaxed SiGe structure.

lattice mismatch (100)$ErSi_2$-(110)SiGe due the increase of the surface lattice constant of the relaxed SiGe in comparison to that of the strained SiGe.

In Fig. 2b one can also see the threading dislocations which are formed during growth of the graded SiGe buffer. While the dislocation density is high inside the buffer layer, it is significantly reduced inside the relaxed SiGe layer of constant Ge content.

Figure 3 shows a high resolution TEM image of the $ErSi_2$/relaxed SiGe interface, parallel to the [011]$_{SiGe}$ axis. The various crystal planes and directions are marked for clarity. The good epitaxial growth of $ErSi_2$ on the relaxed SiGe substrate is apparent, while the quality of the interface is reasonable.

4 Conclusion

We realized for the first time the epitaxial growth of tetragonal erbium silicide on strained and relaxed $Si_{1-x}Ge_x$. The silicide layers were prepared be coevaporation of Er and Si after the reaction of a starting thin Er/Si bilayer. $ErSi_2$ grows in the tetragonal phase. The crystal quality of the silicides grown on relaxed $Si_{1-x}Ge_x$ substrates exhibit a higher crystal quality due to the reduced lattice mismatch.

Acknowledgement

The current contibution is supported by the PENED99 research project "Development of metal/SiGe/Si heterostructures for the detection of IR radiation"

References

1. C. Schaeffer and M. Rodewald, J. of Cryst. Growth **165** (1996) 61
2. B. I. Boyanov, P.T. Goeller, D.E. Sayers, and R.J. Nemanich, J. Appl. Phys. **86** (1999) 1355
3. G. Kaltsas, A. Travlos, A.G. Nassiopoulos, N. Frangis, and J. Van Landuyt, Appl. Surf. Science **102** (1996) 151
4. K. Kugimiya, Y. Hirofuji, and N. Matsuo, Jap. J. Appl. Phys. **24** (1985) 564

DEVELOPMENT OF A NEW LOW ENERGY ELECTRON BEAM LITHOGRAPHY SIMULATION TOOL

D.VELESSIOTIS, X.ZIANNI, N.GLEZOS

Institute of Microelectronics, NCSR "Demokritos", P.Grigoriou, 15310, Aghia Paraskevi, Attiki, Greece
E-mail: dveles@imel.demokritos.gr

K.N.TROHIDOU

Institute of Material Science, NCSR "Demokritos", P.Grigoriou, 15310, Aghia Paraskevi, Attiki, Greece

An electron beam lithography simulator for low energy beam is developed and presented in this paper. The elastic scattering cross section is evaluated using the Partial Wave Expansion Method (PWEM). The cross section is evaluated by a numerical approach and introduced into the main Monte-Carlo program as a look-up table in order to reduce simulation time. The results are compared to the Born approximation in the case of PMMA and a polyoxometalate resist. The Point spread function (PSF) is evaluated by the Monte Carlo simulation technique and introduced intro the SELID simulator in order to obtain the final developed profiles. A study of the energy deposition versus depth is also carried out. It is deduced that the PWEM is necessary for energies below 500eV and high-Z substrates.

1 Introduction

Monte Carlo simulators have already been used widely in the study of electron beam lithography (EBL). The critical feature in a Monte Carlo sequence is the exact evaluation of the differential scattering cross-section. For high beam energies, the traditional approach is to use the Born approximation in order to calculate the scattering cross section. However, this approach should be revised in the case of electron beams with energies lower than 2keV. A method that can be applied in such case is the Partial Wave Expansion Method (PWEM). In this paper, the Born approximation is compared to PWEM, for low energies and two materials of different atomic number Z.

2 Monte Carlo Simulator

In the simulator, electrons are considered to be quantum particles undergoing wave scattering [3][4]. If an electron beam is introduced in a material, which is the case in EBL, each electron should be represented by an incoming wave function. In PWEM, this wave fuction is considered as a superposition of partial waves, each corresponding to a different value of the angular momentum. Outgoing wave is altered due to scattering. The differential elastic scattering cross section $d\sigma/d\Omega$ is expressed through a set of potential dependent phase shifts, which can be calculated by solving numerically the Dirac equation for the Thomas – Fermi potential. The calculated cross sections are used in order to form a look up table for the main Monte - Carlo code. Other authors [2] include this step in the form of an empirical relation for the total elastic Mott scattering [5]. The process described in the present paper is more accurate.

In the main simulator, the energy loss is modeled through the modified Bethe equation and the standard geometry is used in order to describe electron trajectories [6]. The point-spread function (PSF) can be found and used as an input to SELID (®™ Sigma C Inc.) EBL simulator, so that the final developed profiles can be obtained. The dependence of the energy deposition from the depth can also be calculated.

3 Results and discussion

First the differential cross section is evaluated in the energy range 100eV-2000eV and compared to that obtained by the Born approximation. Typical results are demonstrated in figure 1 for 500 and 1000eV. For low-Z materials, such as PMMA, the difference is small and it is limited to large angle scattering, especially for higher energies (1keV). In contrary, the differences are significant in the case of high-Z materials such as resist substrates or metal containing resists, as shown in the case of a polyoxometalate resist used in DUV lithography [7]. It is obvious that the Born approximation is not successful even for small angle scattering events.

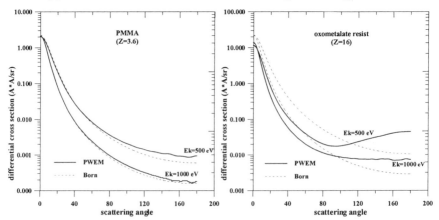

Figure 1: Differential cross section for PMMA (left) and for a polyoxometalate resist (right) for energies 500eV and 1000eV

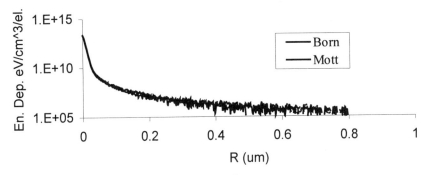

Figure 2: PSF comparison of the Born approximation to the PWEM for a 50nm film of oxometalate resist using beam energy of 1keV

The PSF has been evaluated by the Monte-Carlo code for both resists and films of 50nm thickness over Si. The e-beam energies are in the range of 100-2000eV. It is interesting to investigate the possibility that the differences in the differential cross sections appear in the PSF. In figure 2, the PWEM is compared to the Born approximation in the case of a 50nm polyoxometalate resist film over Si. Compared to figure 1 the Monte-Carlo results present less significant differences. The PSF is used as an input to the simulation tool SELID [8] in order to obtain the final developed profiles of the resist films.

The same conclusion can be reached by the study of the energy deposition versus depth. As seen in figure 3, the energy deposition for a 50nm film of PMMA is smooth -and so lithographically appropriate- for energies above 1keV. In the case of the oxometalate resist, the energy of 2keV should be used in order to achieve optimum results for lithography (figure 3).

4 Conclusions

In this paper a low energy electron beam lithography simulator is presented. Results for the range 100eV-2000eV has been obtained and incorporated into the SELID code in order to obtain the final developed resist profile. It is demonstrated that PWEM provides an accurate physical description of the scattering of the electron, especially for energies lower than 500eV therefore it is necessary. For energies higher than 500eV, electron scattering events are adequately described by the existing semiclassical models.

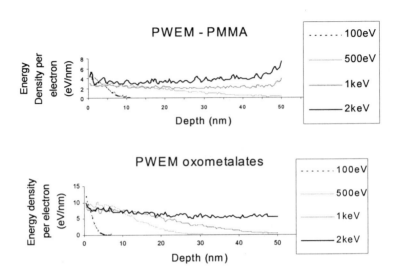

Figure 3: Energy deposition per electron versus depth for various beam energies in 50nm films of PMMA (up) and oxometalates (bottom)

References

1. P.A.Peterson, Z.J.Razimski, S.A.Schwalm and P.E.Russel, J.Vac.Sci. Technol. B **10(6)** (1992) 3088
2. S.H.Kim, Y.M.Ham, W.Lee and K.Chun, Microel. Eng. **41/42** (1998) 179
3. S.Ichimura and R. Shimizu, Surf. Sci. **112** (1981) 386
4. Z.Czyzewski, D.O'Neill, A.Roming and D.Joy, J.Appl.Phys. **68(7)** (1990) 3066
5. R.Browning, T.Z.Li, J.Ye, R.W.Pease, Z.Czyzewski and D.C.Joy, J.Appl.Phys. **76(4)** (1994) 2016
6. W.Williamson,Jr. and G.C.Duncan, Am.J.Phys. **54(3)** (1986) 262
7. P.Argitis, R.A.Srinivas, J.C.Carls and A.Heller, J.Electrochem.Soc. **139** (1992) 2889
8. A.Rosenbusch, N.Glezos, M.Kalus and I.Raptis, 16th Annual BACUS Symposium on Photomask Technology and Management, SPIE Proceedings (1996) 435

CMOS Devices and Devices Based on Compound Semiconductors

ADVANCED SOI DEVICE ARCHITECTURES FOR CMOS ULSI

F. BALESTRA (invited)
Laboratoire de Physique des Composants à Semiconducteurs (CNRS/INPG),
ENSERG, 23 Avenue des Martyrs
BP 257, 38016 Grenoble, France.
E-mail : balestra@enserg.fr

In this paper, a review of the electrical properties and the performance of SOI MOSFETs realized with various device architectures (single gate, double gate with volume inversion, ultra-thin film, ground plane, DTMOS, etc.) is given. The subthreshold operation and the special SOI electrical and thermal floating body effects are addressed. The short channel and hot carrier effects are studied for gate lengths down to deep sub-0.1μm. The impact of the SOI material and the transistor architecture on device reliability is also outlined.

1 Introduction

A number of advantages, suitable for many applications, are obtained with the SOI structure which allows to push back the technological and physical limits intrinsic to the bulk Si structure [1-3]. In view of the huge amount of work recently carried-out by many companies, it seems that the SOI CMOS technology is really taking off.
The aim of this paper is to give a review of the main advantages, drawbacks, performance and physical mechanisms of deep submicron SOI MOSFET. In this respect, the basic electrical properties (subthreshold swing, drain current), the special SOI phenomena (kink, parasitic bipolar, self heating), and the critical behaviors for advanced devices (short channel and hot carrier effects) will be thoroughly discussed as a function of device architecture.

2 Subthreshold swing of the SOI MOSFET

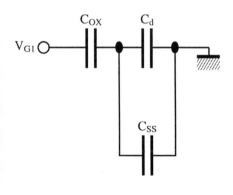

Fig. 1. Equivalent circuit in weak inversion of a bulk MOSFET.

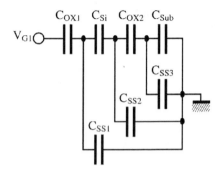

Fig. 2. Equivalent circuit in weak inversion of a fully depleted SOI MOSFET.

A prominent advantage of fully depleted SOI MOSFETs formed on good quality thin film (with SIMOX, WB or Smart Cut) is due to the possible improvement of the subthreshold swing as compared to partially depleted Si film or bulk silicon transistors [4]. The equivalent circuits in weak inversion of a bulk-like MOSFET (bulk or partially depleted SOI transistor) and a fully depleted SOI MOSFET [5] are shown in Figs. 1 and 2, respectively. For a fully depleted device (Fig. 2), the depletion capacitance C_d (=$dQ_d/d\Phi_s$, Q_d being the depletion charge and Φ_s the potential at the Si/SiO$_2$ interface) is suppressed because the depletion charge is limited by the thickness of the Si film and thereby does not vary with the gate voltage or the surface potential. C_d is replaced by a series of capacitances.

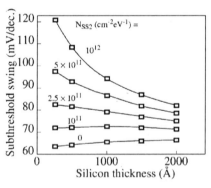

Fig. 3. Simulation of the subthreshold swing as a function of the Si film thickness for various back interface state densities (tox1=27 nm, tox2=350nm, 10^{15}cm^{-3} Si film doping, 10^{15}cm^{-3} substrate doping).

If the fully depleted SOI MOSFET is fabricated with a thick buried oxide and/or a low doping density in the silicon substrate (under the buried oxide) leading to small buried oxide and substrate (depletion) capacitances, the subthreshold swing S can be substantially lower than that observed in bulk devices. For small interface state densities at the various Si/SiO$_2$ interfaces, which is usually the case for present technologies, the swing can reach the minimum theoretical limit of about 60 mV/dec at 300 K (Fig. 3) [4,5]. This offers the opportunity to achieve both a low threshold voltage and a small leakage current. These fully depleted SOI devices are very interesting for high performance low voltage-low power integrated circuits.

3 Kink and transient effects

The buried insulator of the SOI structure leads to floating body effects. Indeed, contrary to the case of bulk silicon device, there is usually no substrate (or body) contact to determine the potential of the active Si layer and to collect free carriers induced by various mechanisms (for instance carriers created by impact ionization). When a body contact is available, the efficiency of this contact depends on the thickness of the Si film and on device architecture, owing to possible series resistance effect and/or potential barrier between the active silicon layer and the body contact.

For partially depleted SOI transistors, a kink effect is obtained at high Vd. For a fully depleted SOI MOSFET realized on ultra-thin films, the depletion capacitance Cd vanishes which suppresses the shift of the threshold voltage, the excess drain current and, thereby, the kink effect [6].

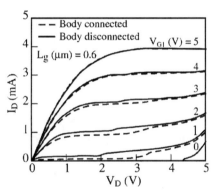

Fig. 4. Output Id(Vd) characteristics of n-channel thin film SIMOX MOSFET showing influence of body connection.

However, it has been shown that a moderate kink effect can still be obtained for fully depleted SOI MOSFETs with intermediate Si film thickness (Fig. 4) [7]. This effect can be modulated by connecting the body terminal (Fig. 4), if available. This kink is due to the substantial barrier height of the source/thin film diode in moderately fully depleted transistors.

On the other hand, parasitic current transients can also be suppressed in fully depleted SOI MOSFETs where the source potential barrier is small and the threshold voltage is not sensitive to a small amount of excess holes.

4 Latch and Breakdown Phenomena

Another floating body effect is the parasitic bipolar transistor (PBT) action which can be observed for SOI MOSFETs in a certain voltage range. This PBT induces latch and premature breakdown as compared to bulk Si devices [8]. As the kink effect, this parasitic phenomenon is triggered by the impact ionization current which leads to an internal forward bias of the (floating) base/emitter (thin Si film/source) junction of the PBT. A large collector (drain) current is thus created giving a substantial enhancement of the drain current. This PBT action portends power-consumption problems in SOI CMOS, hysteresis in the subthreshold characteristics and eventually the loss of gate control.

Breakdown and latch phenomena have been reported for both N- and P-channel enhancement- and depletion-mode devices [9-12]. For P-channel transistors, these effects are attenuated by the lower impact ionization rate of holes. On the other hand, a body contact can reduce or suppress this phenomenon.

5 Noise

Flicker noise in microelectronics devices, observed at low frequency with a variation inversely proportional to the frequency (1/f), is very harmful for analog or RF applications. It is attributed to carrier mobility or carrier number fluctuations. In MOS transistors, the main fluctuations are due to dynamic exchange of channel carriers with interface or oxide traps (carrier number fluctuations).

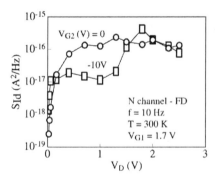

Fig. 5. Experimental drain current noise as a function of drain voltage, for various back gate biases (N-channel thin film (80nm) enhancement-mode SOI MOSFET); a kink effect and an excess noise is obtained when the device is partially depleted (Vg2=-10V).

In SOI MOSFETs, supplementary fluctuations are obtained owing to floating body effects. For a thin film fully depleted transistor, a conventional drain current noise (bulk-like) is obtained as a function of Vd (Fig. 5 with Vg=0). However, when a negative bias is applied to the substrate (back gate), the device becomes partially depleted (accumulation of the back interface) and a kink effect is obtained leading to an excess noise around the kink (Fig. 5 with Vg2=-10V) [13]. This excess noise can be attributed to the dynamic trapping of carriers created by impact ionization in the Si layer or at the interfaces of the SOI structure as well as the impact of shot noise at the source/body junction.

Another floating body effect, the parasitic bipolar transistor action which can be observed in both partially and fully depleted SOI MOS transistors, also leads to a peculiar low frequency noise behavior. The noise measurements are shown in Fig. 6. For small drain current, below the threshold of the parasitic bipolar transistor, similar noise magnitudes are obtained for the various V_d values. However, for large I_d, the PBT regime is reached for a sufficiently high drain bias ($V_d \geq 3.3V$) leading to a significant increase of the noise. Similar results for the drain current noise are obtained in the case of forward- (Vd=3.35V) and reverse (Vd=3.3V) -V_g-scan latch [14].

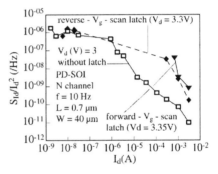

Fig. 6. Normalized drain current noise vs. Drain current for N-channel partially-depleted SOI MOSFET (tsi=100nm, tox1=4.5nm, tox2=80nm); an excess noise is observed in the presence of both forward- and reverse-Vg-scan latch.

This excess noise is attributed to supplementary fluctuations associated with the impact ionization (generation) and carrier capture in the SOI layer (recombination). Note also that the impact-ionization current leads to a biasing of the base of the parasitic bipolar transistor, and further fluctuations can be caused by the PBT carrier transport between emitter (source) and collector (drain).

On the other hand, an interesting behavior in the low frequency noise has been observed in volume inversion with a double gate operation (see below). In this case, an important part of the carrier transport is carried out in the Si volume far from the interfaces, inducing a reduction of the dynamic exchange of carriers with the oxide traps and a screening of the oxide charge fluctuations, thus enabling a significant reduction of the noise of the MOS transistors.

6 Self-Heating Effects

SOI MOSFETs suffer from self-heating effects conveyed by the low thermal conductivity of the buried oxide. At high power levels, one observes the onset of negative output conductance in the saturation region [15]. This behavior is mainly attributed to the reduction of the mobility with increasing channel temperature by self-heating. However, other device parameters (threshold voltage, saturation velocity, etc.), have to be taken into account for accurate modeling. Self-heating also leads to an increase of the interconnect temperature which is critical for electromigration considerations.

The temperature rise is proportional to the power, and is much larger in a SOI device than in a bulk Si transistor. As the silicon layer is thinner, the channel temperature substantially increases. The channel temperature is also raised with increasing the buried oxide thickness and the channel-metal contact separation [16]. The reduction of this important parasitic effect in SOI technology may require the device structure as well as the film and buried oxide thicknesses be optimized. Fortunately, lower self-heating effects are obtained under dynamic operation. For thin buried oxides, the self-heating phenomena are also substantially reduced (Fig.7).

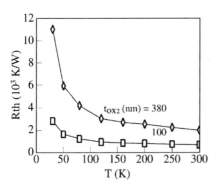

Fig. 7. Simulated variations of the thermal resistance versus temperature with various buried oxide thicknesses for SIMOX MOSFETs.

7 Short-Channel Effects

Short channel effects can become harmful in the deep submicron range. In this respect, two phenomena have to be optimized in order to obtain a reliable device and circuit operation. The charge sharing effect is due to the increased influence of the depletion region at the source and drain junctions with scaling down the MOSFETs. This leads to a reduction of the depletion charge controlled by the gate and thereby a decrease of the threshold voltage of the transistor which can induce substantial leakage currents. The drain induced barrier lowering (DIBL) is due to the electrostatic influence of the drain potential on the source/Si film barrier height at high Vd. This phenomenon has been shown to jeopardize deep submicron device operation with, in particular, a significant drain leakage current.

Fig. 8. Experimental variations of DIBL as a function of Si layer thickness for various gate lengths.

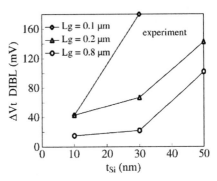

SOI structures offer unique options for the reduction of short channel effects. However, a careful adjustment of the SOI parameters is necessary to improve the performance of deep submicron devices. For instance, Fig. 8 shows typical DIBL experimental results oberved for various gate lengths. A substantial decrease of this short channel effect is obtained with reducing the Si layer thickness, especially for 0.1µm devices [17].

8 Hot-Carrier Effects

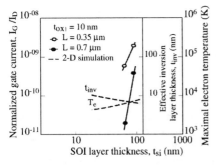

Fig. 9. Variation of the experimental normalized gate current as a function of Si film thickness compared with simulated inversion layer thickness and maximum electron temperature.

Hot carrier effects (HCE) are created by the large lateral electric field observed at high Vd close to the drain/Si film junction inducing high energy carriers in the conduction channel. These electrons or holes give the substrate current by impact ionization and the gate current for carriers which can overcome the Si/SiO_2 energy barrier, as well as photon emission. The high energy carriers can also be injected in the gate oxide creating interface or oxide defects (positive or negative fixed charges and interface states). A substantial reduction of the normalized gate current is found for decreasing the Si film thickness. 2-D simulations (including the energy balance equation) show that the inversion layer thickness increases with reducing the film thickness in fully depleted SOI transistors, which results in a decrease in the maximum electron temperature (Fig. 9) [18].

On the other hand, photon emission measurements have shown that a number of hot carriers are also created in the parasitic bipolar regime for small gate biases ($0 \leq Vg \leq Vt$) and high Vd [19]. Substantial degradation have been pointed out in this gate voltage range. Indeed, the worst case degradation of fully depleted SOI nMOSFET is often obtained in the PBT regime in off state operation.

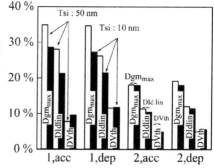

Fig. 10. Degradation of 0.15 μm SIMOX nMOS electrical properties after 50 000s of electrical aging @ Vd = 3.0 V and Vg= 0 V, dark bar: Tsi = 10 nm and white bar: Tsi = 50 nm.

For thick films (50 nm), the degradation is enhanced as compared to ultra-thin films (Fig. 10). The electrical properties at the front interface with a back interface accumulation (1,Acc) or with a back interface depletion (1,Dep), and the electrical properties at the back interface with a front interface accumulation (2,Acc) or a front interface depletion (2,Dep) are investigated in order to determine the impact of interface coupling effects. The degradations of the transconductance and the drain current controlled by the front gate (1,Acc or 1,Dep) are substantially lower for a device realized on a 10 nm silicon layer. A small increase of the threshold voltage degradation is observed for a 10nm Si film. The back interface degradation (2,Acc or 2,Dep) of 10nm and 50nm Si films are similar (with a low reduction of drain current degradation for a 10nm Si layer). This behavior can be explained by the reduction of carrier energy in thinner films which leads to a decrease of the front interface degradation and a similar aging of the back interface whatever the Si-film thickness is, due to the enhancement of the coupling effect. These very interesting results show that ultra-thin Si films lead to an improvement of long term device reliability [20].

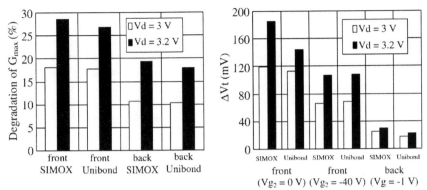

Fig. 11. Comparison of transconductance and threshold voltage degradation of SIMOX and UNIBOND MOSFETs.

It can also be mentioned that the buried oxide degradation in UNIBOND materials (Smart Cut® process) are lower than that observed in SIMOX devices owing to the different buried oxide formation (Fig. 11) [21].

9 Ultimate device architectures

9.1 Ground Plane :

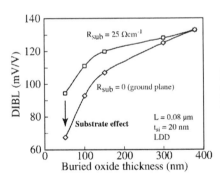

Fig. 12. DIBL in traditional and ground plane SOI MOSFETs (Lg=0.08µm) as a function of buried oxide thickness.

Using the SOI flexibility, a ground plane (highly doped or metallic) can be realized under the buried oxide. This leads to interesting behaviors in the sub-0.1µm range. For instance, the DIBL is shown in Fig. 12 for traditional and ground plane (GP) SOI MOSFETs. This short channel effect is substantially reduced in the case of a GP structure, in particular for thin buried oxides.

9.2 Low k buried insulator :

Fig. 13 shows the variations of the DIBL as a function of the Si film thickness for various device architectures. The DIBL decreases for a low k buried insulator and a ground plane as compared to the case of conventional SOI transistors whatever the silicon layer thickness is.

Fig. 13. DIBL in traditional, low k buried insulator and ground plane SOI MOSFETs (Lg=0.08µm) vs. Si layer thickness.

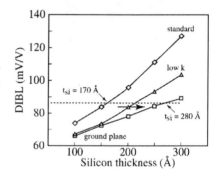

9.3 DTMOS :

The Dynamic-Threshold MOSFET realized on SOI is well-known for its high performance at very low voltage (down to around 0.5V). The gate is connected to the body of the device inducing a high threshold voltage at low Vg (low leakage current) and a low threshold voltage at high Vg (high driving current). Fig. 14 also demonstrates that this interesting device leads to a reduction of short channel effects compared with traditional and body tied SOI MOSFETs [22].

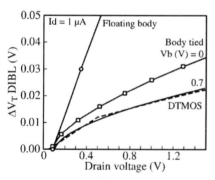

Fig. 14. Threshold voltage shift due to the DIBL effect vs. Vd for floating body, body tied and DTMOS devices.

9.4 Double gate MOSFET with volume inversion :

The inversion channel induced by the gate of a MOSFET is located at the Si/SiO$_2$ interface with a typical length of a few nanometers. The double-gate control of a SOI MOSFET allows forcing the whole silicon film (interface layers and volume) in strong inversion and gives rise to the "volume inversion (VI)" concept [23]. The fact that the current drive of the VI-MOSFET is governed by two gates and carriers are no longer confined at one interface presents remarkable advantages: enhancement of the number of minority carriers, increase in carrier mobility and velocity due to reduced influence of scattering associated with oxide charges and surface roughness, increase in drain current and transconductance, ideal subthreshold slope [23].

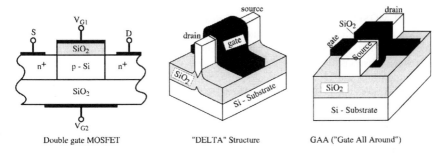

Fig. 15. Various SOI structures using the concept of volume inversion.

Various SOI structures (double-gate [24], DELTA [25], GAA (Gate-All-Around) [26] have been proposed in order to take advantage of this original feature (Fig. 15). Fig. 16 shows an example of the performance improvement during volume inversion operation as compared to a conventional single-gate SOI operation [26] : the GAA device exhibits a transconductance up to 3 times larger.

Fig. 16. Transconductance in a conventional SOI MOSFET, a double-gate transistor without volume inversion, and a GAA device with volume inversion (W/L = 3µm / 3µm, Vd=100mV).

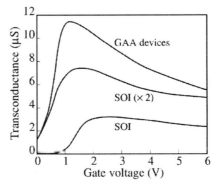

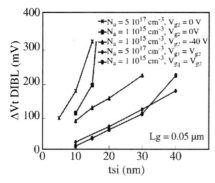

Fig. 17. DIBL effect vs. silicon film thickness (tox2=380nm, tox1=3nm) for 0.05µm single gate SOI MOSFETs with high doping (Na=5×10^{17}cm^{-3}, Vg2=0V), low doping (Na=10^{15}cm^{-3}, Vg2=0V), back channel accumulation (Na=10^{15}cm^{-3}, Vg2=-40V), and for double gate SOI MOSFETs (Na=10^{15}cm^{-3} and Na=5×10^{17}cm^{-3} with Vg1=Vg2).

The VI-MOSFET seems also to be an ideal device for alleviating short-channel effects in ultimate ultra-short channel MOS transistors. A substantial reduction of the DIBL (Fig. 17) and charge sharing phenomena in volume inversion operation has been observed [27]. These advantages are obtained together with very large driving current and very small leakage currents for drain bias up to 1.5V in a wide Si film thickness range. For 0.05µm SOI-MOSFETs, the short channel effects (DIBL and charge sharing) are also reduced with decreasing the silicon film thickness and doping (down to 5nm) whatever the architecture is (single gate and double gate (Fig. 17)). Besides, a reduced sensitivity on the silicon film thickness and doping is observed for the double gate devices which is very interesting for the optimization of their electrical properties.

It is also worth noticing that the electron temperature is reduced with decreasing the Si film doping which is very interesting for long term device reliability. The lowest electron temperature has been obtained in the case of a double gate SOI MOSFET.

This transistor seems to be the best candidate for the ultimate integration of silicon and could be proposed for devices down to at least 20nm gate length without the need of an alternative gate dielectrics.

References

1. J.P. Colinge, *Silicon-On-Insulator technology: materials to VLSI*. Kluwer Academic Publishers, 1991.
2. S. Cristoloveanu and S.S. Li, *Electrical characterization of Silicon-On-Insulator materials and devices*. Kluwer Academic Publishers, 1995.
3. F. Balestra, *SOI devices*, Wiley Encyclopedia of Electrical and Electronics Engineering, 1999.

4. J.P. Colinge, Subthreshold slope of thin-film SOI MOSFET's, *IEEE Electron Dev. Lett.*, EDL-7, p. 244, 1986.
5. F. Balestra, M. Benachir, J. Brini, and G. Ghibaudo, Analytical models of subthreshold swing and threshold voltage for thin- and ultra-thin-film SOI MOSFET's, *IEEE Trans. Electron Dev.*, ED-37, p. 2303, 1990;
6. J.P. Colinge, Reduction of kink effect in thin film SOI MOSFETs, *IEEE Electron Dev. Lett.*, EDL-9, p. 97, 1988.
7. F. Balestra, T. Matsumoto, M. Tsuno, H. Nakabayashi, Y. Inoue, and M. Koyanagi, "Moderate kink effect in fully depleted thin-film SOI MOSFETs, *Electronics Letters*, 31, p. 326, 1995.
8. C.E.D. Chen, M. Matloubian, R. Sundaresan, B.Y. Mao, C.C. Wei, and G.P. Pollack, Single-transistor latch in SOI MOSFET's, *IEEE Electron Dev. Lett.*, EDL-9, p. 636, 1988.
9. M. Yoshimi, M. Takahashi, T. Wada, K. Kato, S. Kambayashi, M. Kemmoshi, and K. Natori, Analysis of the drain breakdown mechanism in ultra-thin-film SOI MOSFET's, *IEEE Trans. Electron Dev.*, ED-37, p. 2015, 1990.
10. J.Y. Choi and J.G. Fossum, Analysis and control of floating body effects in fully depleted SOI MOSFET's, *IEEE Trans. Electron Dev.*, ED-38, p. 1384, 1991.
11. J. Gautier and A.J. Auberton-Hervé, A latch phenomenon in buried N-body SOI NMOSFET's, *IEEE Electron Dev. Lett.*, EDL-12, p. 372, 1991.
12. F. Balestra, J. Jomaah, G. Ghibaudo, O. Faynot, A.J. Auberton-Hervé, and B. Giffard, Analysis of the latch and breakdown phenomena in N and P channel thin film SOI MOSFET's as a function of temperature, *IEEE Trans. Electron Dev.*, ED-41, p. 109, 1994.
13. J. Jomaah, F. Balestra, and G. Ghibaudo, Low-frequency noise in SOI MOSFET's from room to liquid helium temperature : experimental and numerical simulation results, *Proc. of ESSDERC'93*, p. 111, 1993.
14. J. Jomaah, D. Dixkens, J.L Pelloie, C. Raynaud, and F. Balestra, Impact of latch phenomenon on low frequency noise in SOI MOSFETs, *Proc. ESSDERC'96* (Les Editions Frontières), p. 87, 1996.
15. L.J. McDaid, S. Hall, P.H. Mellor, and W. Eccleston, Physical origin of negative differential resistance in SOI transistors, *Electronics Letters*, 25, p. 827, 1989.
16. L.T. Su, K.E. Goodson, D. A. Antoniadis, M.I. Flik, and J.E. Chung, Measurement and modeling of self-heating effects in SOI NMOSFETs, *IEDM Tech. Dig.*, 1992, p. 357.
17. F. Balestra, Performance and physical mechanisms in deep submicron SOI MOSFETs, Electron Technology, 32, pp. 50-62, 1999.
18. Y. Ohmura, An improved analytical solution of energy balance equation for short-channel SOI MOSFET's and transverse-field-induced carrier heating, *IEEE Trans. Electron Dev.*, ED-42, p. 301, 1995.
19. S.H. Renn, J.L. Pelloie, and F. Balestra, Photon emission in deep submicron N-channel SOI MOSFETs, *Electronics Lett.*, 33, p. 1093, 1997.

20. O. Potavin, S. Haendler, J. Jomaah, F. Balestra, C. Raynaud, Reliability of ultra thin film deep submicron SOI nMOSFETs, Proc. ULIS'2000, Grenoble, January 2000, pp. 43-46.
21. S. H. Renn, C. Raynaud, J. L. Pelloie and F. Balestra, "A Thorough Investigation of the Degradation Induced by Hot-Carrier Injection in Deep Submicron N- and P-Channel Partially- and Fully-Depleted Unibond and SIMOX MOSFETs", *IEEE Trans. Electron Devices*, 45, p. 2146, 1998.
22. T. Ernst, D. Munteanu, S. Cristoloveanu, J.L. Pelloie, O. Faynot, C. Raynaud, Detailed analysis of short channel SOI DT-MOSFET, Proc. ESSDERC'99, p. 380.
23. F. Balestra, S. Cristoloveanu, M. Benachir, J. Brini, and T. Elewa, Double-gate silicon-on-insulator transistor with volume inversion: a new device with greatly enhanced performance, *IEEE Electron Dev. Lett.*, EDL-8, p. 410, 1987.
24. K. Suzuki, T. Tanaka, T. Tosaka, H. Horie, and Y. Arimoto, Scaling theory for double-gate SOI MOSFETs, *IEEE Trans. Electron Devices*, ED-40, p. 2326, 1993.
25. D. Hisamoto, T. Kaga, Y. Kawamoto, and E. Takeda, A fully depleted lean-channel transistor (DELTA) - A novel vertical ultrathin SOI MOSFET, *IEEE Electron Dev. Lett.*, 11, p. 36, 1990.
26. J.P. Colinge, M.H. Gao, A.R. Rodriguez, and C. Claeys, Silicon-on-insulator gate-all-around device, *IEDM Tech. Dig.*, 1990, p. 595.
27. E. Rauly, O. Potavin, F. Balestra, C. Raynaud, On the subthreshold swing and short channel effects in single and double gate deep submicron SOI MOSFETs, Solid-State Electronics, vol. 43, p. 2033, 1999.

RECENT DEVELOPMENTS AND RELIABILITY OF POLYCRYSTALLINE SILICON THIN FILM TRANSISTORS

C. A. DIMITRIADIS AND J. STOEMENOS

Department of Physics, University of Thessaloniki, 54006 Thessaloniki, Greece
E-mail: cdimitri@skiathos.physics.auth.gr

F. V. FARMAKIS, J. BRINI AND G. KAMARINOS

LPCS/ENSERG, 23 Rue des Martyrs, 38016 Grenoble Cedex 1, France
E-mail: kamarino@enserg.fr

The structural and electrical properties of excimer laser annealed polycrystalline silicon thin film transistors (polysilicon TFTs) are investigated in relation to the laser energy density. The TFT performance parameters are correlated with the structural properties of the polysilicon films and their electrically active defects, the basic variable being the laser energy density. Simple offset gated TFTs are investigated in relation to the intrinsic offset length and the polysilicon quality. It is demonstrated that the leakage current is completely suppressed without sacrificing the on-current when the polysilicon is of high quality and the offset length is 2 µm. Hot-carrier experiments indicate that the position of the grain boundaries in the channel with respect to the drain end may lead to a different behavior of the device degradation.

1 Introduction

Polycrystalline silicon thin-film transistors (polysilicon TFTs) have been extensively investigated for applications in large-area electronics, especially for liquid crystal displays (LCDs). Among the various techniques available to prepare polysilicon for fabrication of TFTs is solid phase crystallization (SPC) of as-deposited amorphous silicon films at relatively low crystallization temperatures [1]. However, such polysilicon films have a high density of intra-grain defects that degrade the TFT performance [2]. Excimer laser annealing (ELA) has proposed as an alternative low-temperature processing method for improving the polysilicon quality [3]. Further improvement of the material quality and of the scaling down in fabricating sub-micron devices, will enable TFT devices to compete silicon-on-insulator devices in the near feature [3].

In this paper, we examine the use of different laser energy densities to improve the quality of SPC polysilicon films. The structural properties of the polysilicon films are correlated with the electrical characteristics of the TFTs. In addition, offset-gated TFT structures were used to reduce the leakage current. The effect of the offset length on the leakage current was studied in relation to the polysilicon quality. Finally, hot-carrier phenomena were investigated in order to clarify the role of the grain boundaries on the device reliability.

2 Experiments

Polysilicon TFTs were fabricated on fused quartz glass substrates covered by 200 nm thick SiO_2. First, amorphous silicon film (50 nm thick) was deposited by low pressure chemical vapor deposition at 425°C and pressure 1.1 Torr and then crystallized by furnace annealing at 600°C for 24 h in nitrogen ambient. Then, the SPC specimens were irradiated at room temperature by KrF excimer laser with varying laser energy density (LED). A 120 nm thick SiO_2 gate oxide was deposited by electron cyclotron resonance plasma enhanced chemical vapor deposition at 100°C. A standard n-channel self-aligned and offset-gated NMOS process was used to fabricate devices of gate width W=10 μm and gate length L=10 μm. The offset length was ΔL=0.5, 1, 1.5 and 2 μm. Details for the device fabrication processes are presented elsewhere [4, 5]. The structure of the polysilicon films was studied by transmission electron microscopy (TEM) analysis. The transfer characteristics were measured using a HP4155 semiconductor parameter analyser.

3 Results and Discussion

3.1 Effect of ELA on the Structural and Electrical Properties of Polysilicon TFTs

The structural parameters (mean grain size, roughness in % of the film thickness and in-grain defect density) obtained from TEM analysis are summarized in Table 1. The as-deposited SPC film has smooth surface and large grains (about 2500 nm) with very high in-grain defect density. After laser irradiation at 260 mJ/cm^2, the grain size remains unchanged and the in-grain defect density decreases remaining still high. Due to laser irradiation, mass transfer towards the grain boundaries occurs resulting in a surface roughness of about 45% of the film thickness. When the LED increases to 320 mJ/cm^2, the quality of the SPC polysilicon material is improved showing the lowest in-grain defect density although the mean grain size remains about 2500 nm. However, at this energy density the surface roughness increases to about 112%. A further increase in energy density to 340 mJ/cm^2 results in 80% melting of the film area. Small grains with mean diameter of about 100 nm coexist with larger grains, revealing superlateral growth.

The performance parameters of the effective electron mobility and threshold voltage, extracted from the drain current I_D in the linear region, are also shown in Table 1. The device parameters are improved as the laser energy density increases until the silicon film is completely melted. It was found that the TFT performance parameters are strongly correlated with the gap state density distribution derived from low-frequency noise measurements [4].

Table 1. Structural parameters of SPC polysilicon films prepared by excimer laser annealing at various laser energy densities (LEDs).

LED (mJ/cm^2)	Grain Size (nm)	Roughness (%)	In-grain defect density	Mobility (cm^2/Vs)	Threshold Voltage (V)
0	2500	0	Very High	50.2	8
220	-	-	-	44.7	7.4
240	-	-	-	65.4	8.2
250	-	-	-	72.8	5
260	2500	45	High	81.9	3.3
280	2500	45	High	86.6	3.3
320	2500	112	Low	137	4.8
340	100	-	Low	113.8	5

3.2 Offset-Gated Polysilicon TFTs

The effect of the in-grain defect density on the on-state and off-state currents of polysilicon TFTs with various offset lengths is shown in Fig. 1. In the TFT

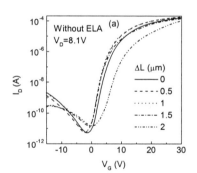
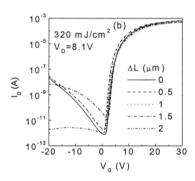

Figure 1. I_D-V_G characteristics of offset-gated SPC polysilicon TFTs with different offset lengths ΔL.

fabricated on polysilicon with high in-grain defect density, for $\Delta L>1$ μm the on-current is reduced due to the increased series resistance of the intrinsic region, while the leakage current initially increases and passes through a maximum before decreasing at high $|\text{-}V_G|$ values (Fig. 1a). However, in the device of the best polysilicon quality (lower in-grain defect density) and for $\Delta L=2$ μm, the leakage current is suppressed without sacrificing the on-current (Fig. 1b). One reason for improvement of the on-current can be the implantation damage of the drain region

[5]. Annealing for dopant activation at 300°C is not sufficient to anneal out the defects induced by ion implantation. Using an offset region of small parasitic resistance, the damaged interface between offset and drain regions is apart from the channel area under the gate-electrode, resulting in a final improvement of the on-state characteristics. The complete suppression of the leakage current without sacrificing the on-current for the best polysilicon quality and the peculiar behavior of the leakage current for the lower quality polysilicon are not understood yet and further work is required to clarify the leakage current mechanisms.

3.3 Hot-Carrier Effects

Hot-carrier effects have been investigated in self-aligned n-channel TFTs fabricated on large grain (2500 nm) polysilicon films. The bias-stress conditions for maximum device degradation (V_D=12 V and V_G=6 V) have been determined by photon emission measurements. Under these stress conditions, devices with identical transfer characteristics show either a negligible or a strong degradation with reduced electron field-effect mobility. Theoretical simulations, performed by employing the simulation program "Silvaco" which is based on a continuous trap model, indicate that the different behavior in the device degradation is due to the non-uniform structure of the polysilicon material [6]. More specifically, it is proposed that the degradation of large grain polysilicon TFTs occur mainly at the grain boundaries which are located near the drain junction. Numerical simulations have shown that more grain boundary band tail states are created by hot-carriers generated by the enhanced electric field near the drain [6]. Therefore, the position of the grain boundaries in the channel with respect to the drain junction may lead to a different behavior of the device degradation.

References

1. Voutsas T. and Hatalis M., J. Appl. Phys. **76** (1994) p. 777.
2. Choi K.-Y and Han M.-K., J. Appl. Phys. **80** (1996) p. 1883.
3. Kuriyama H. et al., Jpn J. Appl. Phys. **30** (1991) p. 3700.
4. Angelis C. T., Dimitriadis C. A., Miyasaka M., Farmakis F. V., Kamarinos G. and Brini J., J. Appl. Phys. **86** (1999) p. 4600.
5. Dimitriadis C. A. and Miyasaka M., IEEE Electron Dev. Lett. **21** (2000), Dec.
6. Dimitriadis C. A., Kinura M., Miyasaka M., Inoue S., Farmakis F. V., Brini J. and Kamarinos G., Solid State Electron. **44** (2000) p. 2045.

DIFFERENT TYPES OF SINGLE CRYSTALLINE GALLIUM NITRIDE THIN FILMS GROWN DIRECTLY ON VICINAL (001) GALLIUM ARSENIDE SUBSTRATES

A. GEORGAKILAS, K. AMIMER, M. ANDROULIDAKI, K. TSAGARAKI
*Microelectronics Research Group (MRG), FORTH / IESL, P.O. Box 1527, 711 10, and University of Crete / Physics Department Heraklion-Crete, Greece
E-mail: alexandr@physics.uoc.gr*

B. PECZ, L. TOTH
Research Institute for Technical Physics and Materials Science, H-1525 Budapest, P.O. Box 49, Hungary

AND M. CALAMIOTOU
Physics Department, University of Athens, 15784 Zografos, Athens, Greece

The approach of direct growth of GaN on vicinal (001) GaAs substrates, by RF-plasma source molecular beam epitaxy, has been investigated. It has been found possible to grow GaN thin films with several different kinds of crystal structure: polycrystalline hexagonal, single-crystalline hexagonal with (0001) or (-1012) orientation and single-crystalline cubic with epitaxial relationship to the GaAs substrate. The GaN crystal structure was controlled by the GaAs surface nitridation, the annealing of an initial low temperature GaN buffer layer and the N/Ga flux ratio conditions. It was also found that cubic or inclined C-axis hexagonal (-1012) crystals could overgrow on initial polycrystalline or mixed phase GaN buffer layers.

1 Introduction

The realization of device quality GaN layers on GaAs substrates [1-7] is very interesting, since it could offer several unique advantages: (i) use of well developed GaAs substrates for GaN technology, (ii) monolithic integration of GaN and GaAs devices and (iii) exploitation of any superior physical or technological properties of the cubic GaN material that can be grown on (001) [1]. Most of the GaN-on-GaAs work has been concentrated on the growth of cubic material [1-5]. In the molecular beam epitaxy (MBE) method, an optimization of the GaAs 2x4 surface (by growing a GaAs buffer layer) always proceeded the GaN growth [1, 2].

We have been investigating the growth of GaN by radio-frequency (rf) nitrogen plasma source MBE (RFMBE) directly on vicinal (001) GaAs substrates, at significantly high growth rates, without using an As beam during GaAs oxide desorption or for the growth of a GaAs buffer layer.

2 Experimental

Thin films of 1.0-1.8μm GaN were grown by RFMBE [3-6] on vicinal (001) GaAs substrates, misoriented by 2° from [001] toward [100], using a two-step GaN growth. The GaAs surface was exposed to the N-plasma beam (nitridation) at substrate temperatures in the range of 420 to 630°C and an initial 10-15nm GaN buffer layer was grown. Buffer layer growth initiation at 540°C without previous GaAs nitridation was also examined. Finally, the continuation of the GaN growth at the final growth temperature, either without interruption or by employing growth interruption and annealing of the buffer layer up to 700°C, was investigated. A variety of techniques were used to analyze the structural and optoelectronic properties of the GaN thin films [3-6].

3 Results and discussion

3.1 Polycrystalline GaN

Polycrystalline GaN was grown whenever an initial low growth temperature (such as ~420°C) was used for the buffer layer, as well as when a high temperature was used for GaAs nitridation and GaN buffer layer growth. Fig. 1a shows a cross-sectional transmission electron microscopy (XTEM) micrograph for the interfacial region of a GaN/GaAs film grown at 630°C, after GaAs nitridation also at 630°C. The GaN film exhibited a columnar structure and it was polycrystalline, according to selected area electron diffraction (SAED) patterns as that shown in Fig. 1b, with a strong (0002) type texture. A high density of defects, mainly stacking faults, occurred in the layers. Voids also existed at the GaAs surface, suggesting that GaAs was consumed for formation of GaN during the first period of growth. This was, obviously, the result of three-dimensional GaN growth and no complete coverage of the entire GaAs surface.

3.2 Single crystal (-1012) GaN

Nitridation and buffer layer growth at 540°C, with growth continuation at ~600°C, without interruption for buffer layer's annealing, resulted to single crystalline hexagonal GaN [4, 6]. The [0001] axis of the GaN layer was not parallel to the surface normal but inclined from it by about 43°. A rather similar observation has been reported for HVPE grown GaN [7], but in that case the [0001] axis coincided perfectly with the [111] GaAs direction, which was not fulfilled in our case. The epitaxial relationship of GaN, with respect to GaAs, was (-1012) GaN // (001) GaAs and [11-20] GaN // [110] GaAs. The films were characterized by abrupt GaN/GaAs interface, without any evidence of damage on the GaAs surface. Extensive conventional and high resolution electron microscopy observations, revealed the initial nucleation of two different hexagonal columns at the buffer layer and the eventual overgrowth of one domain.

a   b

Figure 1. (a) XTEM micrograph and (b) SAED pattern, for {110} cross-section from the interfacial region of a 1μm hexagonal polycrystalline GaN film grown after GaAs nitridation at ~630°C. SAED spots from GaAs and rings from GaN are simultaneously visible.

3.3 Single crystal (0001) GaN

Single crystal hexagonal GaN with the [0001] axis parallel to the surface normal could be grown by using high temperature annealing (700°C) of a thin buffer layer grown similarly to the previous case [6]. The films consisted of thin columns of GaN, which were oriented in the same way, epitaxially to GaAs. The GaN layer itself was nearly defect free, but pores were typically observed between the columns. The results suggest that separate grains of hexagonal GaN were nucleated with the same orientation at the beginning of growth and grew. The epitaxial relationship could be described as (0001) GaN // (001) GaAs and [11-20] GaN // [110] GaAs. Evidence for significant tensile stress characterized these layers.

3.4 Cubic vicinal (001) GaN

Cubic GaN was grown without GaAs surface nitridation (GaAs was exposed simultaneously to the N and Ga beams) and under stoichiometric N/Ga flux ratio conditions [3-5]. The other details of the two-step GaN growth were similar to the case of (-1012) films. X-ray diffraction and SAED patterns at the interfacial GaN/GaAs region revealed that the cubic GaN followed the crystallographic orientation of the GaAs substrate. Figure 2 is a typical XTEM micrograph for the cubic GaN films and reveals a high density of stacking faults along (111) planes and a stepped surface morphology. The top GaN layers were single crystalline GaN but polycrystalline regions were also observed near the GaN/GaAs interface, suggesting that cubic GaN was overgrown on different initial material phases. Typical near band edge photoluminescence (PL) spectra at 15K and 300K from cubic GaN are given in Fig. 3. The 300K PL peak occurred at 3.216 eV with a linewidth of 58 meV, which - to our knowledge - is the best result achieved for MBE grown cubic

GaN. Causes for the appearance of hexagonal mixing in the cubic (001) layers have been also identified [5].

 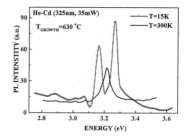

Figure 2. XTEM micrograph for {110} cross-section of a 1.8μm cubic GaN film.

Figure 3. Typical photoluminescence spectra at 15K and 300K for the cubic GaN/GaAs films

4 Conclusions

Two different orientations of single crystalline hexagonal GaN and cubic GaN can be grown by RFMBE directly on vicinal (001) GaAs. The final crystal structure appears to be the result of competition growth between different initial domains.

5 Acknowledgements

Research programs of the GSRT, Hellenic Ministry of Development and a research grant of University of Crete have supported this work. OTKA project No. T030447 is also acknowledged for financial support.

References

1. H. Okumura et al., J. Cryst. Growth **178** (1997) 113.
2. O. Brandt et al., Appl. Phys. Lett. **71** (1997) 473.
3. A. Georgakilas et al., Phys. Stat. Sol. (a) **176** (1999) 525.
4. K. Amimer et al., Appl. Phys. Lett. **76** (2000) 2580.
5. A. Georgakilas et al., presented in MBE-XI, J. Crystal Growth (in print).
6. A. Georgakilas et al., presented in EMRS 2000, Mater. Sci. Eng. B (in print).
7. A. Sakai et al., J. Cryst. Growth 183 (1998) 49.

EPITAXIAL Y_2O_3 ON Si (001) BY MBE FOR HIGH-k GATE DIELECTRIC APPLICATIONS

G. VELLIANITIS[(A)], A. DIMOULAS, A.TRAVLOS

Institute of Materials Science, National Center for Scientific Research "DEMOKRITOS", Athens, Greece
[(A)] *E-mail: geovel@ims.demokritos.gr*

Yttria (Y_2O_3) thin films were grown directly on Si (001) substrates by e-beam evaporation in an MBE chamber under UHV conditions. Based on x-ray diffraction data, it can be inferred that Y_2O_3 epilayers of high crystalline quality can be obtained at an optimum growth temperature ~450°C. At this temperature the heteroepitaxial relationship is $Y_2O_3(110)//Si(001)$ favoring the formation of potentially harmful complex microstructure, as can be seen by HRTEM. Although theory predicts a good thermodynamic stability for Y_2O_3 on Si, these materials react. At moderate growth temperature ~610 °C a YSi_2 phase appears. In addition, a non-uniform interfacial layer with thickness ranging between 5 to 15 Å is observed.

1 Introduction

Metal oxides (MOs) have rich and superior properties (e.g. high dielectric constant k) which can be used to improve high frequency performance, density and functionality of integrated circuits. One of the challenges [(1)] is the replacement of SiO_2 gate dielectric by suitable high-k materials to enable the downscaling of transistors in the deep submicron range (50-100 nm) by year 2011. In order to optimally exploit the properties of metal oxides, direct epitaxy of high (crystalline) quality material on Si may be required which raises the issue of interface control to avoid oxidation of the underlying silicon substrate.

Yttrium oxide (Y_2O_3) is a good candidate (among other materials[(2)]) mainly due to its predicted thermodynamic compatibility with Si [(3)]. In addition, $\alpha(Y_2O_3)$ = 1.06 nm $\approx 2\alpha(Si)$ so that commensurate growth is possible. There are a few works in the literature which report epitaxial Y_2O_3 films on Si with good crystalline quality by employing several methods such as Ionized Cluster Beam (ICB)[(4)], oxidation in furnace[(5)] and e-beam evaporation[(4)-(7)]. In all these works the experiment was performed in an O_2 environment (by inserting gas into the chamber).

In this paper, Y_2O_3 thin films were grown directly on Si(001) substrates by MBE under UHV conditions. X-ray diffraction (XRD), high-resolution transmission electron microscopy (HRTEM) and energy dispersive x-ray analysis (EDX) were used for characterization of the microstructural quality of the Yttria films. This information is necessary to assess the suitability of Y_2O_3 as a high-k replacement of SiO_2 gates in future transistor devices.

2 Film Growth

Y_2O_3 thin films (40 nm) were grown on Si(001) in an UHV chamber (base pressure ~ 2×10^{-10} Torr), equipped with 3-pocket e-gun and high-temperature effusion cells. Prior to growth the substrate surface was heated up to 770^0C for 3 min under Si flux of ~1.5×10^{14} atoms/sec cm^2 to remove the oxide. At a temperature of 650^0C a Si buffer layer (20nm) was grown at a deposition rate of 0.3 Å/sec using e-beam evaporation. The Y_2O_3 epilayer was also grown by e-beam evaporation at various temperatures (280 – 610^0C) using a powdered target of Yttria. In an attempt to minimize the oxidation of the Si substrate, O_2 was not introduced into the chamber. During the growth the partial pressure of O_2 was ~10^{-8} Torr, monitored by a mass spectrometer.

3 Structural Characterization and Discussion

A. Description of the Data

The crystallographic properties of Y_2O_3 films and Y_2O_3/Si interfaces were investigated by x-ray diffraction and cross-sectional & plan-view HRTEM. The x-ray diffraction data for two of the samples are shown in Fig.1. The heteroepitaxial relationship between the epilayer and the substrate depends on the temperature of the growth. The results can be summarized as follows:

$Y_2O_3(111)//Si(001)$ at 280^0C

$Y_2O_3(111)$ and $Y_2O_3 (110)//Si(001)$ are mixed at 320^0C

$Y_2O_3(110)//Si(001)$ at 450^0C (optimum growth temperature).

At a growth temperature of 610^0C, peaks associated with YSi_2 hexagonal phase appear in the x-ray diffraction pattern. The presence of YSi_2 results in a degradation of the dielectric quality of the film (the film becomes conducting).

It should be noted here that $Y_2O_3(001)//Si(001)$ (cube-on-cube epitaxy)was not observed despite the fact that $\alpha(Y_2O_3) \approx 2\alpha(Si)$.

The film grown at 450^0C, which showed the best epitaxial crystalline quality, was further investigated by TEM. It can be inferred from TEM observation [8] (not shown here), that a film with a uniform thickness and a flat surface at least over a few microns is obtained It can be seen from Fig.2 that films of very good crystalline quality can be obtained, although a complex microstructure consisting of large domains and smaller (~20 nm) inclusions with perpendicular crystallographic orientations is present in the film. It should be noted that the planes imaged in Fig. 2 correspond to nominally forbidden (011) diffraction, which become allowed in the thin film due to stresses originating from heteroepitaxy and/or processing conditions. Cross-sectional TEM [8] indicate that the domains are aligned with respect to the Si substrate in such a way that Y_2O_3 [001] is parallel to either Si[110] or Si [1$\bar{1}$0]. In addition, an amorphous layer of thickness between 5 to 15 Å is present at the Y_2O_3/Si interface (see inset of fig 2). This layer, (in general Y-Si-O silicate), is attributed here to SiO_2. Despite the presence of this amorphous layer, Y_2O_3 grow with good crystalline quality, which could be explained by considering that SiO_2 is formed during the deposition of Y_2O_3, probably due to diffusion of oxygen that reaches the substrate and reacts with Si.

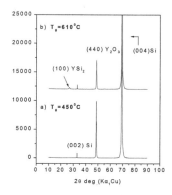
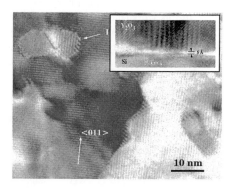

FIG. 1. X-ray θ-2θ scan a) for a sample grown at 450°C. The single strong diffraction peak (440) indicates good crystalline quality heteroepitaxial material such that $Y_2O_3(110)$//Si(001) b) for a sample grown at 610°C. In addition to Y_2O_3 (440), a weak YSi_2 (100) diffraction peak appears.

FIG 2. Plan view high resolution TEM image of sample grown at 450°C showing two different domains and 17 nm inclusion (marked by I) in one of them. Note that {011} family of planes changes orientation by 90° when crossing the boundary from one domain to the other. A cross-sectional image is shown in the inset. An amorphous layer is present at the interface.

Examination of the specimen by EDX in transmission mode yields a non-stoichiometric oxide of the type $Y_2O_{2.55}$ with a small variation across the sample area. This is to be expected since the growth was performed under an oxygen-deficient atmosphere.

B. Discussion and Results

The purpose of this work is to check the suitability of Yttria as a gate dielectric in future transistor devices. The complex microstructure present in the Y_2O_3 epilayer (fig 2) is expected to cause undesired fluctuations in the dielectric properties. In addition, the defective areas at the domain boundaries may promote Si and gate-metal diffusion. The formation of this potentially harmful microstructure is due to the particular heteroepitaxial orientation Y_2O_3 (110)//Si(001) adopted by the epilayer. In an effort to explain why Y_2O_3 grows with such a heteroepitaxial relationship on Si substrates we consider the following: the (001) planes of Y_2O_3 consists of only Y or O atoms. This means that in order to grow $Y_2O_3(001)$//Si(001), only Y or only O atoms should bond to Si atoms. However, in real experiment conditions, Y and O fall simultaneously on the Si substrate and it is reasonable to assume that both Y and O will bond to Si. This observation can be a good explanation as to why the Y_2O_3 (001) is not parallel to Si (001) plane. Furthermore, the (110) Y_2O_3 plane has the correct stoichiometry (there are 8 atoms of Y and 12 atoms of O belonging at the same unit cell of the (110) surface net) thus favoring the Y_2O_3 (110)//Si(001) heteroepitaxy.

In this work, the optimum growth temperature of Y_2O_3 (on the basis of x-ray data) was found to be 450^0C. At this temperature an epilayer with good heteroepitaxial quality is formed, in contrast to previous works[7], where growth temperatures as high as $700-800^0C$ were employed in order to achieve high quality heteroepitaxial growth. At moderate temperatures (~610^0C) YSi_2 phases start to appear due to reaction of the epilayer with the substrate, despite theoretical predictions for thermodynamic stability of Y_2O_3/Si interface. Finally, the formation of an amorphous layer could not be avoided, although O_2 partial pressure was kept at low levels.

These observations make it clear that the experiment conditions used do not offer a significant advantage. It should be noted though that all samples except the one grown at 610^0C were insulating, although detailed leakage current measurements on metal-insulator-semiconductor structures were not performed.

References

1. Semiconductor Industry Association, *The International Technology Roadmap for Semiconductors*, Front-End Processes p.p. 123-128, 1999 edition & 2000 update.
2. See e.g. Proceedings, MRS *High-k Gate Dielectric Workshop*, San Jose CA, June 1-2, 2000.
3. K. J. Hubbard and D.G. Schlom, J. Mat. Res. **11**, 2757 (1996)
4. S.C. Choi, M.H. Cho, S.W. Whangbo, C.N. Whang, S.B. Kang, S.I. Lee, and M.Y. Lee, Appl. Phys. Lett. **71**, 903 (1997).
5. K. Harada, H. Nakanishi, H. Itozaki, S. Yazu, Japan. J. Appl. Phys. **30**, 934 (1991).
6. A. Bardal, O. Eibl, Th. Matthee, G. Friedl, J. Wecker, J. Mater. Res., **8**, 2112 (1993).
7. H. Fukumoto, T. Imura, Y. Osaka, Appl. Phys. Lett., **55**, 360 (1989).
8. A. Dimoulas, A. Travlos, G. Vellianitis, N. Boukos, to be published, J. Appl. Phys., 2001

PERFORMANCE OF GaAs/AlGaAs LASER DIODES FABRICATED BY EPITAXIAL MATERIAL WITH SIGNIFICANTLY DIFFERENT NUMBERS OF QUANTUM WELLS

D. CENGHER, G. DELIGEORGIS, E. APERATHITIS, M. SFENDOURAKIS
FORTH/IESL and University of Crete, Physics Dpt., Microelectronics Research Group, P.O. Box 1527, 71110 Heraklion-Crete, Greece

G. HALKIAS
NCSR "Demokritos", Institute of Microelectronics, P.O. Box 60228, 15310 Agia Paraskevi, Greece

Z. HATZOPOULOS, A. GEORGAKILAS
*FORTH/IESL and University of Crete, Physics Dpt., Microelectronics Research Group, P.O. Box 1527, 71110 Heraklion-Crete, Greece
E-mail: alexandr@physics.uoc.gr*

In this work, we determined experimentally the effects of the number of quantum wells (QWs) on the parameters of laser diodes. Separate confinement heterostructures (SCH) with 2, 4 and 8 QWs and graded index-SCH structures with 2, 4, 8 and 16 QWs were investigated. The influence of the number of QWs on the threshold current density and differential quantum efficiency of the lasers was studied.

1 Introduction

The fabrication of monolithically integrated optoelectronic circuits (OEICs) [1], combining photonic circuits based on III-V optoelectronic devices [2] and Si electronic ICs, is a possible method to increase the speed and the bandwidth of the actual electronic circuits. The integration of the III-V and Si devices is an important point for the fabrication of OEICs and one possible method is to bond epitaxially grown III-V wafers with Si wafers [3]. Therefore, the simplest solution is to use the same III-V multi quantum well (MQW) structure to fabricate all the necessary types of devices: laser diodes (LDs), photodiodes (PDs) or modulators [4, 5].

In this work, we determined experimentally the effects of the number of quantum wells (QWs) on the parameters of LDs. This information is required for the optimum selection of the single III-V structure, since the increase of the number of QWs will increase the efficiency of PDs or modulators, but it will also increase the threshold current density of LDs. Such experimental results did not exist in the literature, although the development of GaAs/AlGaAs MQW LDs has reached significant maturity [6, 7].

2 Experimental

Molecular beam epitaxy (MBE) was used to grow separate confinement heterostructures (SCH) with 2, 4 and 8 QWs and graded index-SCH structures (GRINSCH) with 2, 4, 8 and 16 QWs. The QWs consisted of 10nm GaAs layers, spaced by 6nm of $Al_{0.20}Ga_{0.80}As$. These were embedded between two $Al_{0.26}Ga_{0.74}As$ layers, 0.29 μm thick, and two $Al_{0.45}Ga_{0.55}As$ cladding layers. Graded composition $Al_xGa_{1-x}As$ layers were included in the GRINSCH structures [5] in order to reduce the threshold current density and the series resistance of the LDs.

Ridge geometry laser devices, with 50 μm wide stripes and lengths (L) of 170 – 920 μm were fabricated and measured in pulsed conditions. The values of threshold current density and differential quantum efficiency of the LDs were calculated for all devices, using emitted power versus current characteristics.

3 Results and discussion

For both types of structures, the threshold current density (J_{th}) increased with the number of QWs. For each MQW structure, the logarithm of J_{th} was plotted versus the inverse device length (1/L), as shown for the GRINSCH LDs in figure 2, exhibiting the expected [6] linear behavior. From such plots, J_{th} values were determined for 1/L=0 and they were plotted in figure 2 for both types of structures.

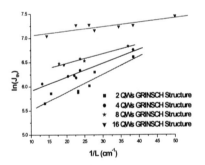
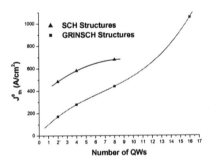

Figure 1. Threshold current density vs. inverse device length (1/L) for graded index separate confinement heterostructures with 2, 4, 8 and 16 QWs

Figure 2. Dependence of the threshold current density on the number of quantum wells for infinite LD device length.

Figure 2 shows that for both types of LD structures, the threshold current density increased slowly with the number of QWs for heterostructures with less than 8QWs. However, a fast increase of the threshold current density with the number of QWs occurred for samples grown with more than 8 QWs.

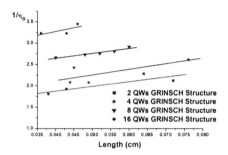

Figure 3. Inverse differential quantum efficiency vs. length of the device for graded index separate confinement heterostructures with 2, 4, 8 and 16 QWs

The inverse differential quantum efficiency ($1/n_D$) varied linearly with the LD length and the results for the GRINSCH LDs are shown in figure 3.
The coefficients of the fitting lines in the ($1/n_D$)-L plots are related to the absorption coefficient (α) and internal quantum efficiency (η_i) of the structures according to the formula [7]:

$$\frac{1}{\eta_d} = \frac{1}{\eta_i} + \frac{1}{\eta_i} \cdot \frac{\alpha}{\ln\left(\frac{1}{R}\right)} \cdot L$$

where R is the reflection coefficent of the mirrors and L the length of the device.

A comparison of the differential quantum efficiencies must take into account two parameters. One parameter is the absorption coefficient and the other one is the quantum internal efficiency. The absorption coefficient is similar for all the GRINSCH structures, respectively for all the SCH structures, so the differences in the differential quantum efficiency should be related to deviations of the internal quantum efficiency.

Figure 4 presents the dependence of the quantum internal efficiency on the number of QWs. The quantum internal efficiency decreased with the number of QWs for both types of structures and for the GRINSCH structures it could be fitted with an exponential decay curve (figure 4). The decrease of the quantum internal efficiency may be attributed to non-uniform distribution of carriers in the quantum wells along the multi quantum well structure. Increased leakage current may not be a significant factor, as the dependence of leakage current with the number of quantum wells generally considered to be weak.

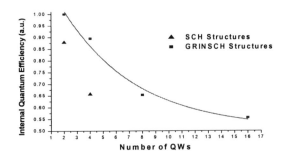

Figure 4. Dependence of the internal quantum efficiency on the number of quantum wells

4 Conclusions

The influence of the number of QWs in a GaAs/AlGaAs structure, on the threshold current density and differential quantum efficiency of LDs was studied. The threshold current density increased slowly with the number of QWs, for structures with 2 to 8 QWs and much faster for structures incorporating more than 8 QWs. The internal quantum efficiency decreased continuously with the number of QWs for both the SCH and GRINSCH structures.

5 Acknowledgements

This work has been supported by EU under the frame of ESPRIT 28998 project, BONTEC.

References

1. A. V. Krishnamoorthy, D. A. B. Miller, *IEEE J. Select. Topics Quantum Electron*, **2** (1996) pp. 55
2. W. J. Grande, J. E. Johnson, C. L. Tang, *A.P.L..*, **57 (24)** (1990) pp. 2537
3. Q-Y. Tong, G. Gha, R. Gafiteanu, U. Goesele, *J. Micromechanical Systems*, **3 (1)**, (1994) pp. 29.
4. D. Cengher at al, presented in EXMATEC 2000, Mater. Sci. Eng. B (in print)
5. Z. Hatzopoulos et al, presented in MBE-XI, J. Cryst. Growth (in print)
6. F. Yang, P. Blood, J.S. Roberts, *Appl. Phys. Lett.*, **66 (22)** (1995) pp. 2949
7. P. Bhattacharya, Semiconductor Electronic Devices (Prentice Hall, New Jersey), (1997)

MICROHARDNESS CHARACTERIZATION OF EPITAXIALLY GROWN GaN FILMS. EFFECT OF LIGHT ION IMPLANTATION

P. KAVOURAS, M. KATSIKINI, Ph. KOMNINOU, E. C. PALOURA,
J. G. ANTONOPOULOS, Th. KARAKOSTAS
Aristotle University of Thessaloniki, Department of Physics, 54006, Thessaloniki, Greece
E-mail: komnhnoy@auth.gr

> The microhardness value and surface microcrack propagation pattern of hexagonal and cubic GaN polytypes, grown with different deposition techniques, is studied. The implantation induced surface hardening is examined in the hexagonal symmetry films. The implantation species were N^+ and O^+. Additionally, the microhardness anisotropy with respect to the film's main crystallographic directions is revealed. Finally, the indentation induced surface microcrack initiation is studied. In the case of the cubic polytypes, the anisotropic surface microcrack propagation is monitored and correlated with the crystallographic directions. Additionally, the film's microhardness value is measured and deconvoluted from that of the Si substrate. The microhardness values and the microcrack propagation patterns between the two polytypes are compared.

1 Introduction

During the past decade the vast majority of research in GaN films is concentrated on their optoelectronic characteristics. On the contrary, research on the mechanical properties has not drawn equal attention. A reason is that the growth of large GaN monocrystals is still a problematic issue; hence classical methods of elastic moduli measurements are not applicable. As a result static indentation tests are applied, since the small dimensions of the examined specimens is not a problem. Due to the lattice mismatch and significant difference in thermal expansion coefficients between GaN films and the available substrates, the films are characterized by different mechanical properties. Consequently, it is becoming increasingly evident that research on the mechanical properties of GaN films will be needed for a more successful application of this semiconductor.

2 Experimental

Details on the growth conditions and characterization of the films have been published previously [1, 2]. The microhardness of the films was measured using a Knoop indenter (microhardness tester Anton Paar, MHT-10) in the range of 0.05 – 0.6 N for cubic GaN and 0.02-0.6 N for hexagonal GaN. The microhardness value H_K is calculated from equation

$$H_K = 14229 P d^{-2},$$

where **P** is the load and **d** is the length of the long diameter of the indentation. The Vickers indenter geometry was used in order to inspect the microcrack propagation, since this indenter geometry induces microcracks for lower applied loads than the Knoop indenter.

3 Results

3.1 Hexagonal symmetry GaN specimens

3.1.1 The as-grown specimen

The depth profile of the microhardness was obtained for each sample by using different values of applied load (P) of the indenter. The microhardness profile for the as-grown sample is shown in Fig. 1. In order to reduce the statistical error, a number of 10 indentations were averaged for each load. The large microhardness value for shallow indentations is an indication of the well-known Indentation Size Effect (ISE). We assign the value of microhardness at the plateau to that of the film. The bulk microhardness for the as-grown GaN sample is found equal to 14.6±0.3GPa, a value that is in agreement with hardness measurements by Nikolaev *et al.* who reported microhardness equal to 14GPa [3].

3.1.2 The implanted specimens

The microhardness depth profiles for the implanted samples are also shown in Fig.1. The three curves have similar shape but those of the implanted samples are shifted towards higher values of hardness. After implantation, the bulk microhardness of the implanted GaN samples is measured to be about 3 Gpa, higher than that of the as-grown samples. This difference is larger compared to the measurements' standard deviation.

Since the implanted epilayer is just 0.2 μm thick, its absolute value of microhardness cannot be obtained directly from the measurements of the implanted samples, although the information of the surface hardening is qualitatively present. In order to obtain the microhardness value of the implanted epilayer more accurately a deconvolution method, originally developed by Jönsson and Hogmark [4] for Vickers indenter and adapted to the geometrical configuration of Knoop indenter by Iost [5], is applied to the original microhardness data. Using the microhardness data of the substrate (the data of the as grown specimen) and the data of the composite microhardness (implanted specimens) the modeled microhardness value of the implanted epilayer is obtained. The implanted epilayer microhardness calculated with this method was found equal to 29 GPa for N^+ and O^+ implanted specimens for doses 5×10^{14} cm^{-2}.

3.1.3 Microhardness anisotropy

A series of Knoop measurements were performed in the as grown specimen, in order to monitor a possible anisotropy of microhardness values. For this reason the Knoop indenter was rotated with an interval of five degrees around an axis normal to the film's surface. The microhardness value exhibited a sinusoidal fluctuation with a period equal to 60° and amplitude of 2 GPa. The positions of the microhardness maxima are related to specific crystallographic orientations of the film and are achieved when the large diagonal of the indentation is parallel to the <11$\underline{2}$0> directions of GaN. Accordingly, the microhardness minima are obtained when the indentation diagonal is rotated by 30° from the <11$\underline{2}$0> direction, namely when it is parallel to the <10$\underline{1}$0> direction.

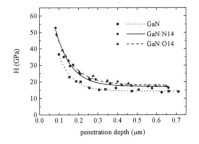

Figure 1. Microhardness depth-profiles of the as grown GaN (○) and the implanted samples GaN:N14 (●) and GaN:O14 (♦).

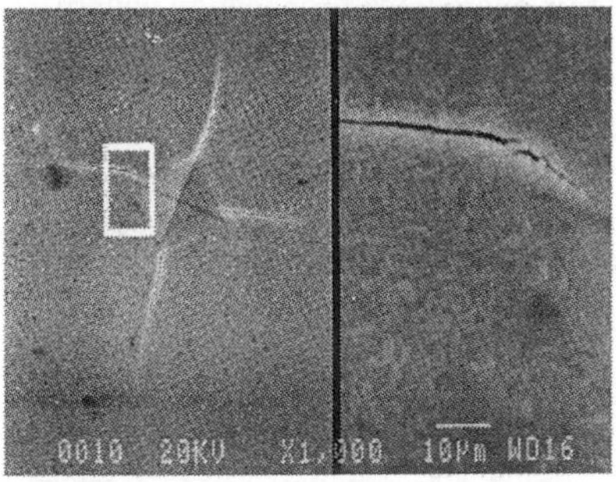

Figure 2. Point of microcrack bending due to its misorientation in respect to the <110> orientations.

3.2 Cubic symmetry GaN specimens

The microhardness of cubic symmetry GaN films grown on (001) Si was found using the same technique and deconvolution method mentioned previously. The microhardness value was found to be 11.6 GPa. The implication of these measurements was that indentations above a load of 0.6 N induced microcrack propagation, only for impressions parallel to the <110> directions. The microcrack propagation was found to have preferential directions with a series of Vickers indentations. These are parallel to the <110> directions. The microcracks are transmitted through the film following the characteristic tile-like surface morphology. As it is reported the tiles are oriented along the <110> directions [1].

4 Discussion

It is proposed that the main hardening mechanism due to implantation is related to the formation of nitrogen interstitials where the dislocations are pinned and thus prohibit the plastic deformation of the sample. The N interstitials are detected in the NEXAFS spectra where they introduce a transition at 1.4eV below the absorption edge [6]. The anisotropic value of microhardness is related to the different activation of the prismatic and pyramidal slip systems that takes place during the rotation of the indenter. The microcrack propagation pattern, in cubic GaN, could be attributed to the indentation induced dislocation accumulation to the borders of the tiles. In this way the borders become sources of microcrack initiation. However, cleavage fracture should be also considered as another possible mechanism for the generation of this microcrack propagation pattern

5 Acknowledgements

This work has been supported by the EU contract HPRN-CT-1999-00040 and the GSRT contract No 99ED 320.

References

1. Lei T., Moustakas T. D., Graham R. J., He and Berkowitz S. J., Epitaxial growth and characterization of zinc-blende gallium nitride on (001) silicon. *J. Appl. Phys.* **71** (1992) pp. 4933–4943.
2. Komninou Ph. Kehagias Th., Kioseoglou J., Sarigiannidou E., Mikroulis S., Tsagaraki K., Georgakilas A., Nouet G. Ruterana P., Dimitrakopulos G. P.,Karakostas Th., Structural defects of GaN films epitaxially grown on (0001) Al_2O_3 by MBE. In *XVI Conference of Solid State Physics* ed. by Efthymiou P. (Athens University).
3. Nikolaev V. I., Shpeizman V. V., Smirnov B. I., Determination of elastic moduli of GaN epitaxial layers by microindentation technique. *Phys. Sol. Stat.* **42** (2000) pp. 437–431.
4. Jönsson B. and Hogmark S., Hardness measurements of thin films. *Thin Solid Films* **114** (1984) pp. 257–269.
5. Iost A., Knoop hardness of thin coatings. *Scrip. Mater.* **39** (1998) pp. 231–238.
6. Kavouras P.,Komninou Ph., Katsikini M., Papaioannou V., Antonopoulos J., Karakostas Th., Anisotropic microhardness and crack propagation in epitaxially grown GaN films. *J. Phys. Cond. Mat.* **12** (2000) in print.

MULTIPLE QUANTUM WELL SOLAR CELLS UNDER AM1 AND CONCENTRATED SUNLIGHT

E. APERATHITIS, Z. HATZOPOULOS, M. KAYAMBAKI, V. FOUKARAKI
Microelectronics Research Group, Institute of Electronic Structure & Laser, Foundation for Research and Technology – Hellas, P.O. Box 1527, Heraklion 71110, Crete, Greece
E-mail: eaper@physics.uoc.gr

M. RUŽINSKÝ, V. ŠÁLY, P. SIROTNÝ
Slovak University of Technology, Faculty of Electrical Engineering and Information Technology Ilkovičova 3, SK-812 19 Bratislava, Slovak Republic

P. PANAYOTATOS
Rutgers, The State University of New Jersey, Department of Electrical and Computer Engineering, 94 Brett Rd., Piscataway, New Jersey 08854-8058, USA

GaAs/AlGaAs Multiple Quantum Well (MQW) solar cells, with different ratios of well to barrier number and thickness, have been examined with temperature, between -175C and 148C, and under AM1.5 illumination and 5x concentration conditions. The ideality factor of the diodes in the dark improved with temperature. The fabrication of 1cmx1cm MQW solar cells with different AlGaAs barrier thicknesses and illumination under 1.5 AM1 conditions showed that there is an improvement in the performance of the device as the thickness of the barrier layer is reduced. 1cmx1cm MQW solar cell with the optimum QW geometry (number of wells and barriers thickness) exhibited J_{sc}=56.8mA/cm^2 and V_{oc}=1.041V at 5x concentration when compared to J_{sc}=10.2mA/cm^2 and V_{oc}=1.009V at AM1.5 illumination conditions.

1 Introduction

III-V solar cells have been reported as exhibiting the greatest efficiencies under one sun illumination and under concentrated sunlight over any other solar cell [1]. Due to cost considerations however, they have been used mainly for space applications, whereas for terrestrial applications new design concepts with enhanced efficiency and the use of concentrators have been considered necessary [2].
One of the new solar cell designs which has attracted research interest is the Multiple Quantum Well (MQW) solar cell: a p/i/n solar cell structure with MQWs in the i-region [3]. In these cells, the carrier escape from the wells is primarily thermally activated, thus making their use at high temperature operations (such as under concentration conditions) more advantageous over the conventional cells without quantum wells [4]. Furthermore, the photocurrent in the MQW solar cells has been shown to consist of two major components: the thermionic emission, and the tunneling component. The contribution of each component to the total photocurrent depends greatly on the geometry of the quantum wells and the operating conditions of the cell. It has been found that the photocurrent is dominated by tunneling of carriers

at low temperatures, while thermal currents dominate at high temperatures [5]. These properties of MQW solar cells can also find applications in space missions where the cells have to operate under greatly different conditions of intensity and temperature. Even though there has been some controversy about the predicted operation of MQW solar cells, until now a GaAs/AlGaAs p/i/n MQW solar cell with an 6.25mm^2 area has been reported to have an 14% efficiency under AM1.5 illumination [3].

In this work, we report on results obtained from experiments concerning fundamental aspects of GaAs/AlGaAs MQW solar cells, in order to understand better the operating mechanisms in such low dimensional structures, as well as, for the first time, we report on the behaviour of these structures as 1cmx1cm MQW solar cells under AM1.5 illumination and 5x concentration conditions.

2 Experimental Details

The p/i/n MQW solar cell structures were grown by MBE on n$^+$GaAs substrates. The basic cell was p/i/n $Al_{0.36}Ga_{0.64}As$ structure, with p-type emitter and n-type base doping of $1.5 \times 10^{18} cm^{-3}$. GaAs wells were incorporated in the i-region which was 0.37μm thick. Details about the epitaxial structures can be found elsewhere [4]. Conventional p/n or p/i/n GaAs and $Al_{0.36}Ga_{0.64}As$ solar cells (without quantum wells) were also grown for comparison. Table 1 shows details about the i-region geometry of the solar cells examined in this work.

Table 1. Details of the i-region structure of p/i/n solar cells. Widths: L_W for the GaAs well, L_b for the $Al_{0.36}Ga_{0.64}As$ barrier, and L_i for the i-region.

SAMPLE	WELL PERIOD	L_w (nm)	L_b (nm)	L_i (μm)
GS64	40	5.4	3.6	0.37
GS65	23	5.4	10.8	0.37
GS63	p/i/n $Al_{0.36}Ga_{0.64}As$			037
GS62	p/i/n GaAs			0.37

The solar cells structures were processed using standard photolithographic techniques. Some of the solar cell structures were processed as photodiodes having ring geometry (520μm diameter window and 720μm diameter mesa) and tested with temperature under a halogen lamp illumination. AM1.5 illumination and 5x concentration conditions (ORIEL 6722 solar simulator) was also used for structures processed as solar cells having 1cmx1cm area, with top contact grid of 14 lines/cm and SiN_x as antireflecting coating after selectively removing the GaAs cap layer over the AlGaAs window layer between the lines of the top contact grid. All the processed structures had (Ge/Au)/Ni/Au for n-ohmic contacts and (Pt/Ti)/Pt/Au for p-ohmic contacts, both activated in a rapid annealing system.

3 Results and Discussion

3.1 MQW photodiodes in the dark

Dark current measurements with temperature, ranging from -175C to 148C, were performed on photodiodes with MQW geometries as seen in Table 1. Figure 1 depicts the behaviour of the MQW solar cell diodes at the two extreme temperatures, -175C and 148C, of the temperature range examined in this work, along with the two conventional solar cells consisting of the well material (GaAs) alone and the barrier material ($Al_{0.36}Ga_{0.64}As$) alone, respectively. The dark current of the MQW cells lied below the dark current of the conventional cell consisting of the well material alone and above the conventional cell which is composed of the barrier material alone. Furthermore, the spread in the I-V curves of the MQW cells observed at -175C almost disappears at elevated temperatures. The increased series resistance for these samples observed at high forward bias is partially due to the fact that these samples had not been specifically processed to minimize series resistance and partially due to defects associated with the quality of the grown material.

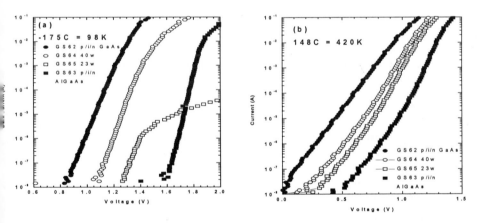

Figure 1. Dark I-V characteristics of p/i/n solar cells at two extreme temperatures: (a) -175C and (b) 148C. Details about the I-region geometries are given in the inserts.

Information about the quality of these diodes was extracted by fitting the linear part of the dark current curves by the exponential voltage dependence $I=I_{sat}exp(qV/nkT)$ at two different dark current levels. As seen in Fig.2(a) for low current levels (from

1×10^{-7}A up to 1×10^{-5}A) and at elevated temperatures, the ideality factors are very close to 2 for all diodes. At high current levels, from 1×10^{-5}A up to 1×10^{-3}A, Fig.2(b), there is a spread in the values of the ideality factors, with the diodes having thicker barriers showing improved diode quality as the temperature increases. It is apparent that there is a change of the quality of the diodes with temperature because of the increase role of defect recombination. The improvement of the ideality factor of these diodes with temperature suggests that recombination is due to these defects whose capture effectiveness is affected by temperature.

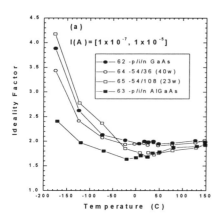
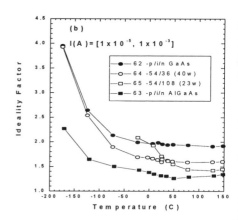

Figure 2. Temperature dependence of ideality factor of the p/i/n solar cells as extracted by fitting the I-V characteristics at two different current regions.

3.2 MQW solar cells under AM1.5 illumination

It has been shown, by fitting experimental results to theoretical calculation, that in such MQW structures there is a non-equal contribution of the two major components of the photocurrent, namely the thermionic emission/escape from the wells over the barriers and tunneling along the growth direction through the barriers [5]. It was found that tunneling is the dominant mechanism of carriers transport at low temperatures over thermionic emission, whereas the roles reverse at high temperatures where currents due to thermionic emission dominate their tunneling counterparts.

In order to see in a more profound way the effect of tunneling currents (effect of barrier thickness) on the total photocurrent, the MQW structures were fully processed as 1cmx1cm solar cells. A photograph of a processed 1cmx1cm solar cell can be seen in Fig.3. The top contact grid consisted of 14 lines/cm (50µm wide and 700µm apart) covering ~9% of the front 1cmx1cm surface of the cell. The semi-circular p-ohmic contact was the extension of the top contact grid for extracting the photogenerated carriers to the external probe. The n-ohmic contacts were made at the backside of the n⁺GaAs substrate. All solar cells had the same structure except for the geometry of the i-region and were processed identically. The performance of the cells was examined under AM1.5 illumination and 5x concentration conditions. The behaviour of the 1cm x 1cm solar cells under AM1.5 illumination has been tabulated in Table 3.

1 sun AM1.5	GS62 p/i/n GaAs	GS64 p/i(MQW)/n AlGaAs Barrier=3.6nm	GS65 p/i(MQW)/n AlGaAs Barrier=10.8nm
Isc (mA)	16	10.2	9.4
Voc (V)	0.981	1.009	1.004
FF	0.658	0.713	0.735
n (%)	10.33	7.34	6.93

Figure 3. Photograph of a typical 1cm x 1cm solar cell, with top contact grid. Circular test patterns are also seen.

Table 2. Performance of 1cm x 1cm p/i(=0.37µm)/n solar cells (of Table 1) under AM1.5.

The solar cell with the wider barriers (GS65) showed lower short-circuit current than the cell with thinner barriers (GS65) suggesting reduced tunneling through the barriers. However, the conventional GaAs p/i/n solar cell which had exactly the same structure like the MQW cells but without quantum wells exhibited higher overall performance. The reduced short circuit current of the MQW solar cells could not be compensated by the increased open circuit voltage, thus resulting in cells with poorer performance than the conventional cell.

Recent experimental results after testing GS64 under 5 suns revealed Isc=56.8mA and Voc=1.041V but with reduced fill factor resulting in lower efficiency than the efficiency obtained under 1 sun illumination. The reduced fill factor was due to resistance problems of the top contact grid.

Further work is in progress to examine in detail the behaviour of all the MQW solar cells processed in this work with temperature and under concentrated sunlight and using a denser top contact grid (20 lines/cm).

4 Conclusions

The behaviour of p/i/n $Al_{0.36}Ga_{0.64}As$ solar cells with $GaAs/Al_{0.36}Ga_{0.64}As$ MQWs in the i-region has been studied with temperature under illumination and compared to that of a p/i/n GaAs and AlGaAs solar cell. The dark currents of the MQW diodes were found to lie between those of conventional cells formed from the well material and from the barrier material alone. It was found that the increase in the dark current of the diodes with temperature was accompanied by an improvement of diode ideality factor.

The performance of 1cmx1cm MQW solar cells under AM1.5 illumination exhibited inferior performance than the conventional p/i/n GaAs solar cell without quantum wells. However, the performance of the MQW cell with thinner barriers was better than that with thicker barriers. Preliminary results of the MQW cell with thin barriers under 5x concentration conditions gave an increase in short circuit current by a factor of 5.6 and an increase of open circuit voltage by a factor of 1.03, but with reduced fill factor. More experiments are in progress to improve the top contact grid of the cells and thus the fill factor of the I-V curves under 5 suns.

It can be inferred from these results that for p/i/n GaAs/AlGaAs MQW structures of high quality material the relative contribution of tunneling and thermionic emission currents as a function of temperature can be determined by modeling and fitting and should provide guidance for the optimized design of MQW solar cells tailored for operation in specific temperature ranges and illumination levels. Inter-subband engineering of solar cells can play a vital part of solar cell design.

Acknowledgements

These results are part of a joint research project within the framework of a bilateral programme between Greece and Slovakia and have been financially supported by the GSRT (General Secretariat for research and Technology) of Ministry of Development Greece under project No. 15185/22-12-98 and Ministry of Education of the Slovak Republic under project No. Gr/Sl A90.

References

[1] A.W. Bett, et al., "III-V compounds for solar cells applications", *Appl. Phys.* **A69**, 1999, pp. 119-129.
[2] L. Kazmerski, "Photovoltaics: A review of cell and module technology", *Renewable and Sustainable Energy Reviews*, **1**, 1997, pp. 71-170.
[3] K.J.W. Barnham, et al., "Quantum well solar cells", *MRS Bulletin*, October 1993, pp. 51-55.
[4] E. Aperathitis, et al., "Effect of temperature on GaAs/AlGaAs MQW solar cells", *Mat. Sci. & Eng.* **B51**, 1998, pp. 85-89.
[5] A.C. Varonides, et al., "Significance of tunneling and thermionic emission currents in performance enhancement of MQW GaAs/AlGaAs solar cells", 2nd *World Conf. And Exhibition on PV Solar Energy Conversion*, Vienna, Austria, 1999, pp. 66-69.

PROCESSING WITH IN SITU DIAGNOSTIC TECHNIQUES FOR THE INTEGRATION OF GaAs-BASED OPTO-ELECTRONIC DEVICES BONDED ON Si C-MOS WAFERS

G. DELIGEORGIS, E. APERATHITIS,
D. CENGHER, Z. HATZOPOULOS and A. GEORGAKILAS
FORTH/IESL and Univ. Crete/Physics Dpt., Microelectronics Research Group, P.O. Box 1527, 71110 Heraklion, Crete, Greece, E-mail: eaper@physics.uoc.gr

Reactive Ion Etching (RIE) with laser interferometry has been demonstrated as a simple technology for the fabrication of GaAs devices bonded on Si C-MOS wafers for optical interconnect applications. Laser interferometry as end point control was used during the processing of the GaAs-based laser and photodetector optoelectronic devices. An optimized BCl_3 RIE process was developed for the fabrication of smooth and vertical facets for the formation of laser mirrors. The isolation of the individual GaAs optoelectronic devices was achieved in a mixture of BCl_3 and Cl_2 gases without affecting the bonding material.

1 Introduction

The fabrication of photonic integrated circuits by monolithic integration of active and passive optoelectronic devices is very interesting for applications in high-speed optical communications and optical computing [1]. A planar processing technology of photonic integrated circuits [2], with characteristics similar to the VLSI technology of electronic circuits, is ideally required for the industrial implementation of advanced optoelectronic chips. Such a technology has to consider not only the growth of a suitable epitaxial structure but also the processing of different types of integrated optoelectronic devices.
In this work we report on the processing technology, and particularly the use of Reactive ion etching (RIE) with in-situ diagnostic techniques, for the fabrication of different kinds of GaAs-based optoelectronic devices, such as edge emitting lasers and photodetectors, from the same single-growth epitaxial GaAs/AlGaAs wafer [3] which has been bonded on fully processed Si C-MOS wafer. The Si C-MOS wafer consisted of the laser driver and photodetector receiver integrated circuits. The MBE-grown GaAs/AlGaAs structure on the GaAs substrate had been successfully bonded on the Si C-MOS wafer by a low temperature wafer bonding technique [4]. The RIE system was a parallel plate reactor operating at 13.56MHz (VACUTEC AB), employing chlorine-based gases, and equipped with a Laser Interferometry (900nm) and CCD camera (Jobin Yvon - Sofie) for in-situ monitoring of the etching process.

2 Results and Discussion

2.1 Laser mirror fabrication

One of the most demanding process for the fabrication of etched mirrors on GaAs/AlGaAs structures by RIE is the process which will yield anisotropically etched walls and similar etch rates for GaAs and AlGaAs for obtaining smooth etched mirrors. The anisotropic and smooth etching of chlorine–based gases was examined both on GaAs and AlGaAs layers and on the structures used in this work. The use of pure Cl_2 gas (3 sccm, 5 mTorr, 75W) resulted in faster etching of GaAs over AlGaAs by 1.5 times, with rough sidewalls. Further increase in total pressure and power resulted not only in increased etch rate but also at the formation of HCl vapours in the chamber. The use of Boron Trichloride (BCl_3) as etching gas resulted in smooth, vertical sidewalls and no HCl vapours as by-product. The anisotropic etching obtained with a 50W power BCl_3 RIE recipe (10 mTorr total pressure and 10 sccm BCl_3) gave equal etch rates for GaAs and AlGaAs and produced smooth sidewalls. By increasing the BCl_3 flow from 0.2 sccm to 2 sccm and then to 10 sccm, the ratio of AlGaAs to GaAs etch rate was reduced from 1.35, to 1.20 and then to 1.05, respectively. Similar results were obtained by reducing the power to 25W. The etch rate was 23.5nm/min whereas the selectivity of the photoresist mask over the etched material was 1:50. No improvement was achieved by using mixtures of Cl_2 and BCl_3 gases. SEM photograph of a fully processed laser diode (LD) next to a photodetector (PD), both having the same structure, is seen in Fig.1 (a). The laser mirror facing the photodetector was formed using the optimized RIE conditions.

(a) (b)

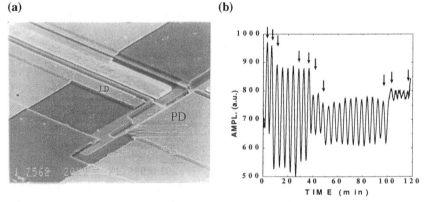

Fig.1. (a) SEM photograph of fully processed laser diode (LD) and photodetector (PD) and (b) Laser interferometry signal as a function of etching time for the laser mirror formation of (a). The arrows indicate heterointerfaces.

The laser interferometry with the CCD camera, with which the RIE system was equipped, was used as an in-situ method of etch rate determination as well as heterolayer identification, end point detection and observation of the quality of the surface morphology on the field. A typical example is shown in Fig.1(b), where in

Fig.3(a) the laser interferometry signal is plotted as a function of etching time during the optimized etching process for the formation of mirrors in the p/i/n laser structure seen in Fig.1(a). The layers of the structure [3] were identified and they are denoted by the arrows above the interferogram of Fig.1(b). The last five small peaks correspond to the GaAs substrate etching.

The signal shape is an indication of the interfaces quality of the MBE-grown structure and the constant etch rate during etching through the structure into the substrate or down to the buffer layer and consequently the very good endpoint detection control. Interferograms like the one shown in Fig.1(b) ensures the smooth surface morphology on the field during the whole etching process time. Any increased roughness of the field, which results in rough walls, due to micromasking arising from small sputtered particles and redeposition on the field [5], can readily be identified by the laser interferometry as a decrease in the reflected signal which in the worst case is reduced to zero.

2.2. Device Isolation

The electrical and optical isolation of the individual GaAs-based optoelectronic devices was performed by dry etching (RIE). The end point detection system was set to identify the removal of all the unmasked GaAs material, even at the areas between the photodetector and the laser which could be as close as 5µm, as seen in Fig.1(a). The isolation of the devices was performed in a BCl_3 and Cl_2 mixture (10sccm of BCl_3 and 1sccm of Cl_2) at 10mTorr total pressure and 25W power. The laser interferometry signal of this process is seen in Fig.2. The differences in the signal amplitude after around 800 seconds correspond to the removal of the GaAs/AlGaAs structure and the beginning of etching of the 2µ GaAs layer. The removal of all the unmasked GaAs material is clearly denoted in the interference signal at the end of the process. The bonding material is not attacked by the dry etchants used in this process. The etch rate was 185nm/min and the total thickness of the removed material was 5.3µm. Vertical and well-defined isolated geometries were obtained, with all the GaAs material removed even from places with quite different aspect ratios.

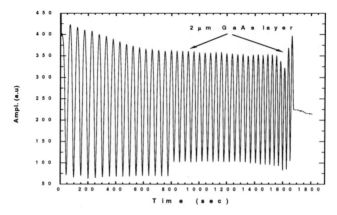

Fig.2 Laser interferometry signal during the removal of unmasked AlGaAs/GaAs material bonded on Si wafer.

4. Conclusions

Reactive Ion Etching with laser interferometry was used for precise end-point control during the processing of GaAs-based optoelectronic devices bonded on Si C-MOS wafer with SOG as the bonding media. An optimized BCl_3 RIE process was developed for the fabrication of smooth and vertical facets for the formation of laser mirrors. The isolation of the individual GaAs optoelectronic devices was achieved in a mixture of BCl_3 and Cl_2 gases without affecting the bonding material.

The precise real-time control of the etching of the laser diode heterostructure layers was demonstrated. This control was applied on two basic processing steps towards the integration of optoelectronic devices. Furthermore, this in-situ control gives also a unique capability for the engineering of complex processes for the fabrication of sophisticated photonic circuits, since it allows the selective removal of parts of the epitaxial heterostructure.

References

1. A. V. Krishnamoorthy, et al, IEEE J. Select. Topics Quantum Electron., vol. **2**, pp. 55-76, 1996.
2. W. J. Grande, et al, Appl. Phys. Lett., vol. **57**, no 24, pp. 2537-2539, 1990.
3. D. Cengher, et al, Mater. Sci. Eng. B (in print).
4. Q-Y. Tong, et al, J. Micromechanical Systems, vol. **3**, no 1, pp. 29-35, 1994.
5. L. Richeboeuf, Jobin Yvon – Sofie, laser interferometry simulation.
6. S.J. Pearton, Material Science & Engineering **B40** (1996) 101.

INVESTIGATION OF DIFFERENT Si(111) SUBSTRATE PREPARATION METHODS FOR THE GROWTH OF GaN BY RF PLASMA-ASSISTED MOLECULAR BEAM EPITAXY

M. Androulidaki[a], K. Amimer[a], K. Tsagaraki[a], M. Zervos,[a], G. Constantinidis[a], Z. Hatzopoulos[a], A. Georgakilas[a], F. Peiro[b], and A. Cornet[b]

[a] *FORTH/IESL and Univ. Crete/Physics Dept., Microelectronics Research Group, P.O. Box 1527, 71110 HERAKLION-CRETE, GREECE*

[b] *University of Barcelona, Electronics Department, Marti i Franques 1, 08028 Barcelona, Spain*
E-mail: pyrhnas@physics.uoc.gr

GaN films were grown on Si (111) substrates by nitrogen rf plasma source molecular beam epitaxy. Reflection high energy electron diffraction (RHEED), atomic force microscopy, transmission electron microscopy, infrared transmittance and photoluminescence results, characterize the properties of GaN films grown under different in-situ substrate preparation and growth initiation methods. Very smooth surfaces, exhibiting surface reconstruction in RHEED, were achieved by using an AlN buffer layer. However, these films produced weak photoluminescence, compared to that of the GaN layers directly deposited on a reconstructed 7x7 Si(111) surface.

1. Introduction

Gallium nitride (GaN) is of current interest as a wide band gap semiconductor for applications in ultraviolet-visible optoelectronic and high power-temperature and frequency electronics. Due to their large wafer size and low cost silicon substrates have many advantages over other materials, but most important is the possibility of GaN device integration with Si microelectronics[1,2].
Our experimental work aimed at the definition of a convenient Si(111) substrate preparation method and an efficient GaN growth initiation by radio-frequency (RF) plasma source MBE. A growth process compatible with GaN-Si monolithic integration requires the minimum thermal load of Si during the GaN deposition. The preparation of an atomically clean Si surface is a critical step for epitaxial growth and usually involves the chemical preparation of a passivating oxide and its desorption by heating to 900-1000C heating step inside the MBE chamber[2]. This investigation aimed to develop a growth procedure compatible with monolithic integration. Thus, in order to minimize the substrate heating requirement, the H-passivated Si(111) chemical preparation method was adopted and different ways for in-situ preparation of the Si surface and GaN growth initiation were compared.

2. Experiment

The GaN films were grown on Si(111) substrates, using an elemental Ga source and a nitrogen RF plasma source, operating at 400W. The growth temperature was 700C and the N_2 flux was 0.6 sccm. The GaN growth conditions remained constant for all samples. Five sets of films were grown on Si(111) surfaces with different in-situ preparation before the initiation of the epitaxial growth. (A) Hydrogen passivated and heated up to 400C, (B) 7x7 surface reconstruction after heating up to 700C, (C) √3 x √3 Ga reconstruction after Ga deposition on a 7x7 surface, (D) amorphous Al deposition on a 7x7 surface, annealing at 750C and deposition of AlN buffer and (E) 1x1 reconstruction after heating up to 800C.

3. Results and Discussion

AFM observations of the GaN surfaces yielded different values of rms roughness for each type of growth initiation condition. The roughness was approximately 30-50 nm for samples A,B,C,E (Fig.1a) while sample D, with the AlN buffer layer exhibited a roughness of only 8nm (Fig.1b)

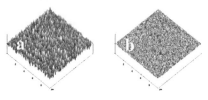

Fig. 1: AFM photographs of the surface of GaN films (a) 7x7 Si(111) surface and (b)with AlN buffer layer on Si(111) 7x7 surface.

All the samples have been examined in plane view and cross-section. We defined very well the columnar structure and the non-uniformity of "grains" size. Sample D, with an AlN buffer, gave us the best quality. The dislocation density was of order of 10^{10} cm^{-2}. For sample A and B, we observed a thin layer of Si_xN_y, between GaN and the Si substrate. In sample D we observed the AlN buffer layer. In Figs 2a,b, we present the TEM images of the samples A and D.

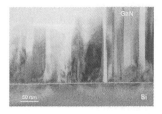 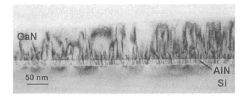

Fig. 2: TEM measurements of the GaN-on-Si. (a) sample A with Si_xN_y thin layer and (b) sample D with AlN thin layer, between the GaN and Si.

Infrared transmittance was used to study the optical phonon modes in the samples. For all samples we observed the characteristic transverse (TO) and longitudinal (LO) optical phonon frequencies of GaN at 560 cm^{-1} and 734 cm^{-1} respectively

[3,4,5]. In addition the Si-O chemical bond from the substrate appeared at 610 cm^{-1} and 1100 cm^{-1} [5,6]. For sample A, we observed at 872 cm^{-1} the stretching mode from Si-N bonds[5,6,7] in a thin layer of Si_xN_y. The shift of the peak position from the main Si-N mode at 840 cm^{-1}, indicates a non-stoichiometric Si_xN_y layer, grown

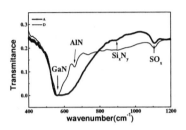 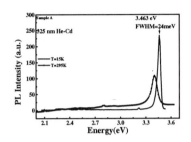

Fig. 3: IR spectra for two representative films of GaN-on-Si, grown by using different substrate preparation and growth initiation conditions.

Fig. 4: Low and room temperature PL

obviously between the Si substrate and the GaN epilayer[7]. The sample exhibited also a peak at 2960 cm^{-1} related to N-H bonds[7]. For sample B, the IR spectrum was the same as of sample A except that the Si-N peak appeared at 840 cm^{-1} and this is indicative of the formation of stoichiometric Si_3N_4 at the GaN/Si interface. For sample C, we observed a very weak absorption at 888 cm^{-1} and peaks at 1730 cm^{-1} and 2960 cm^{-1}. The peak at 888 cm^{-1} is probably related to oxygen-vacancy (V-O) complex[5] in the bulk Si and it is expected for Czochralski grown Si substrates. The peaks at 1730 cm^{-1} and 2960 cm^{-1} correspond to Ga-H and N-H bonds respectively [6,7]. The observation of both Ga-H and N-H bonds suggests the existence of both N and Ga-vacancies in this sample. In the samples with initial Al deposition on Si (D), we observed clearly the TO mode for the AlN layer at 656 cm^{-1}. This is the characteristic frequency for polycrystalline AlN with small size crystals. The Si-N bond disappeared in this case, indicating the efficient protection by the Al layer of the Si surface from reacting with the nitrogen beam. Finally, for sample E we observed a very weak peak at approximately 800 cm^{-1}, corresponding to amorphous SiC material. This is related to contamination of the Si surface by hydrocarbons, which were dissociated and reacted with Si during the high temperature heating. In Figs.3 we present the IR spectra from samples D, with AlN buffer.

15K PL spectra exhibited a main peak at 3.46eV(Fig4.) which may be assigned to the donor-bound exciton transition (DAP)[8]. The comparison of the PL intensity of the samples revealed correlation between the size of the "grains" and the substrate roughness. The sample exhibiting the stronger band-edge PL (linewidth of 24 meV and 60meV at low and room temperature) has the larger-surface grain size while the

sample with the flat surface (no granular morphology) exhibited a broad and weak PL.
This PL dependence on grain size is attributed to the radiative processes that occur prior to trapping, as carriers move into highly defective regions. We believe that the granular surface morphology is related to a growth process that proceeds through the nucleation and coalescence of three-dimensional islands on the substrate surface. The granular surface morphology and the increased roughness should be the results of the initial nucleation of a low density GaN islands on Si, which reach coalescence at large sizes. On the contrary, coalescence of a higher density of small islands should occur with the use of an AlN buffer layer, leading to a flat surface morphology. The defect structure in (0001) GaN films follows a columnar pattern with a high concentration of defects at boundaries of the columns and these columns are obviously related to the extent of the initial island grains. It is self-evident that if the grain size is small enough the trapping of carries, defects at the boundaries, happens so quickly that the luminescence is drastically reduced.

4.   Conclusions

Alternative methods for low temperature, in-situ preparation of Si (111) substrate surfaces for GAN MBE growths have been compared. The deposition of an Al layer on 7x7 Si(111) surface and the subsequent growth of an AlN bufer layer, led to the formation of Si_3N_4 at the GaN/Si interface and resulted in smooth GaN surface. However, the best photoluminescence results were obtained for GaN films, which exhibited a granular surface morphology and a disordered or contaminated GaN/Si interfacial region.

5.   References

[1] T. Lei, M. Fanciulli, Appl. Phys. Lett., **58**, 944 (1991)
[2] G.S. Higashi, R.S. Bercer, Y.J. Chabal and A.J. Becker, Appl. Phys. Lett., **58**, 165, (1991)
[3] A. S.Barker, Jr. and M. Ilegems , Phys. Rev. B, **7**, n2, 743 (1973)
[4] G. Wetzel, E. E. Haller, H. Amano, I. Akasaki, Appl. Phys. Lett., **68**, 2547, (1996)
[5] G. Mirjalili , T. J. Parker, S. Farjami Shayesteh, M. M. Bulbul, S. R. P. Smith, T. S. Cheng, C. T. Foxon, Phys. Rev. B, **57**, n8, 4656 (1998)
[6] T. Deguchi, D. Ichiryu, K. Toshikawa, K. Sekiguchi, T. Sota, R. Matsuo, T. Azuhata, M. Yamaguchi, S. Chichibu, S. Nakamura, J. Appl. Phys., **86**, n4, 1860, (1999)
[7] M.D. Mc Cluskey, J. Appl. Phys., **87**,n8, 3593, (2000)
[8] E. Godlewski, Appl. Phys. Letts, **69**, 2089, (1996)

MATERIAL PROPERTIES OF GaN FILMS WITH Ga- OR N- FACE POLARITY GROWN BY MBE ON Al_2O_3 (0001) SUBSTRATES UNDER DIFFERENT GROWTH CONDITIONS

A. KOSTOPOULOS, S. MIKROULIS, E. DIMAKIS, E.-M. PAVELESCU, M. ANDROULIDAKI K. TSAGARAKI, G. CONSTANTINIDIS, A. GEORGAKILAS

FORTH/IESL and Uuniversity of Crete, Physics Dpt., Microelectronics Research Group,
P.O.Box 1527, 71110 Heraklion-Crete, Greece
E-mail: kosto@physics.uoc.gr

PH. KOMNINOU, TH. KEHAGIAS, TH. KARAKOSTAS

Aristotle University of Thessaloniki, Physics Department, 54006 Thessaloniki, Greece

The marerial properties of GaN films grown on Al_2O_3 (0001) substrate nitridated at high and low temperatures were investigated. The effects of GaN or AlN nucleation layers and different III/V flux ratio on the polarity of the films were also investigated. Reflected high energy electron diffraction (RHEED) was used for in-situ monitoring of the surface structure and films polarity. The optoelectronic properties were determined by low and room temperature photoluminescence measurements. Finally a method combining etching in a KOH solution and atomic force microscopy (AFM) was developed to identify the polarity of the GaN films.

1 Introduction

In the past few years progress has been made in developing nitride materials both for optoelectronics devices in the blue spectral range [1,2] as well as for high-voltage, high-power and high-temperature electronic applications [3,4], power diodes, AlGaN/GaN high electron mobility transistor (HEMT).

Previous experimental work has established well that both GaN polarities may be grown on (0001) sapphire substrate, depending on the employed growth technique [6] (MBE or MOCVD) and the type of the buffer layer [7] (GaN or AlN). For MOCVD heterostructures with GaN buffer layer, Ga (Al)-face polarity is always observed with the two dimensional electron gas (2 DEG) located at the interface with AlGaN, whereas in MBE grown samples, N-face is also observed for which the electrons are not confined [5]

In this work, it was found that the nitridation temperature of sapphire controls the polarity and is one of the main factors controlling the overall material properties of the over grown GaN films.

2 Experimental

GaN thin films and heterostructures were grown on c-plane substrate [(0001) Al_2O_3] by radio-frequency (rf) plasma assisted molecular beam epitaxy (RFMBE). The growth rate was 0.35 µm/h and growth occurred under approximately stoichiometric V/III flux ratio.

N-face GaN was grown with low temperature nitridation while Ga-face GaN was grown with high temperature nitridation of sapphire, using GaN nucleation layers. However Ga-face material was also grown when an AlN nucleation layer was used on substrate nitridated at a low temperature. The GaN polarity was determined by reflected high energy electron diffraction (RHEED) observation of characteristic surface reconstructions and by comparing the reactivity of the GaN surface with a KOH solution. The surface morphology of the layers was observed by atomic force microscopy (AFM) before and after etching in a solution of KOH/H_2O (1/3.5 by weight) at room temperature for 30 minutes.

3 Results and discussion

Figure 1 compares the RHEED observations on GaN films of both polarities. N-face sample grown with low temperatures (T=200 ^0C) nitridation of sapphire exhibiting the characteristic 3x3 reconstruction (Fig.1a). On the contrary a 2x2 RHEED pattern of Ga-face material grown with high temperatures (T=800 ^0C) nitridation, is shown in (Fig.: 1b).

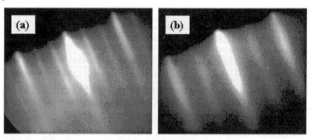

Figure 1. (a) N-face sample grown with low temperature nitridation : 3x3 reconstruction, (b) Ga-face sample grown with high temperature nitridation : 2x2 reconstruction.

The optoelectronic properties of GaN thin films grown on Al_2O_3 (0001) substrate were determined by low and room temperatures photoluminescence (PL) measurements. The 325 nm UV line of an He-Cd laser (35 mW) was used for the excitation. N-face 0.3 µm thin films GaN, exhibited strong PL at both 16K and 300K, while PL spectra of similar Ga-face films were broad as several orders of magnitude weaker in intensity. Figure 2 compares the PL spectra of N-face material at 300K and 16K respectively. The ratio of the two intensities has a state of the art value of 1/12.6 .

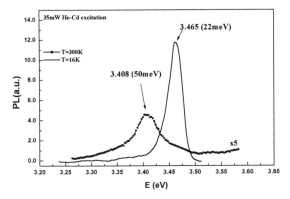

Figure 2. Photoluminescence spectra for 0.3 µm thick N-face GaN

Transmission electron microscopy (TEM) analysis indicated substantially lower defect densities and smoother GaN/Al_2O_3 interfaces for N-face material with significant micro-stuctural differences in the interfacial region. TEM also showed the presence of cubic phase micro-domains. This result suggests that excellent quality N-face GaN can be grown by RFMBE.

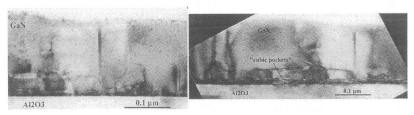

Figure 3. (a) N-face GaN [L.T nitridation Al_2O_3] with low defect density, (b) Cubic GaN inclusions exist near the GaN/ Al_2O_3 interface.

The surface morphology of both N-face and Ga-face materials depends on the value of the V/III flux ratio. Under Ga-rich conditions the samples have a lowest roughness for both polarities (Fig. 4). These results are characteristic of the RFMBE GaN growth. Etching in KOH solutions always resulted is significant increase of the surface roughness of the N-face material. On the contrary the KOH solution did not modify the surface morphology of the Ga face material [8].

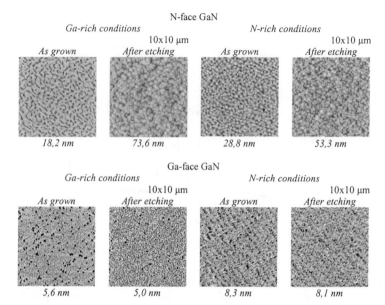

Figure 4. Atomic force microscopy analysis Ga-face and N-face GaN as grown and after etching in KOH : H$_2$O (1:3.5) at room temperatures. The number below each figure is the *Rms roughness*.

The different chemical reactivity of the surface of the two type of material is reasonable considering the different number of unbounded electrons of the atoms at the surface plane of a Ga- or an N-face material.

4 Acknowledgements

This work has been supported by GSRT, Hellinic Minestry of Development through PENED project no.: 99EΔ 320.

Reference

1. S. Nakamura et al., *Appl. Phys. Lett.* **70** (1997) pp. 1417.
2. H. Sakai et al., *J. Cryst. Growth* **189/190** (1998) pp. 831.
3. C. Nguyen et al., *Electron. Lett.* **34** (1998) pp. 309.
4. R. Gaska et al., *Appl. Phys. Lett.* **74** (1999) pp. 287.
5. O. Ambacher et al., *J. Appl. Phys.* **85** (1999) pp. 3222.
6. A. R. Smith et al., *Appl. Phys. Lett.* **72** (1998) pp. 2114.
7. M. J. Murphy et al., *J. Vac. Sci. Techol.* **B17** (1998) pp. 1252.
8. J. L. Weyher et al., *J. Cryst. Growth* **182** (1998) pp. 17.

THE INFLUENCE OF SILICON INTERSTITIAL CLUSTERING ON THE REVERSE SHORT CHANNEL EFFECT

C. TSAMIS AND D. TSOUKALAS

Institute of Microelectronics, NCSR "Demokritos"
15310 Aghia Paraskevi,, Athens,
Greece
E-Mail : C.Tsamis@imel.demokritos.gr

In this work we investigate the influence of controlled damage generated by Si^+ implantation in the S/D region on the extent of the Reverse Short Channel Effect (RSCE) of NMOS devices. The implantation that is performed prior to S/D formation is followed by either RTA or furnace annealing using different ramp rates. The influence of the ramp rates on the RSCE is then studied. The experimental data are analysed using combined process and device simulations by taking into account the dissolution kinetics of {113} defects.

1 Introduction

There is a general consensus on the explanation of the observed RSCE of NMOS devices. The phenomenon is attributed to the non-equilibrium diffusion of the channel boron atoms towards the Si/SiO_2 interface as a result of interstitial gradients.

On the other hand non-amorphizing implants in silicon are known to induce the formation of extended defects. The formation and dissolution kinetics of these defects has been linked to the phenomenon of Transient Enhanced Diffusion [1]. In a recent work, Cowern et al. [2] demonstrated that during the annealing of non-amorphizing implants two phases of dopant TED can be distinguished. An initial phase of ultrafast TED that lasts for a short time period (f.e. 100sec for 700 °C), where the supersaturation of interstitials is high due to the ripening of very small interstitial clusters precursors to the nucleation of {113} defects and a second long lasting phase. This phase of lower TED persists until the end of transient diffusion.

Within this work experiments were performed in order to study the influence of {113} defects on RSCE and to distinguish if possible the effect of ultrafast TED and of the much slower TED on threshold voltage roll-on.

2 Experiments

Two sets of experiment (A and B) were carried out. For experiment A, NMOS devices were fabricated with gate lengths from 1μm to 20 μm. A retrograde Boron channel profile was used. After polysilicon patterning, part of each wafer was implanted with a non-amorphizing Si^+ implant (10^{14} cm^{-2}, 50 keV). Then the wafers were split in two lots. One lot has seen a temperature anneal using a ramp-up rate of 15 °C/min (Fast Ramp) from 700 °C to 900 °C and the other lot a ramp rate of 1 °C/min (Slow Ramp) within the same temperature range. The fabrication process was followed for all the wafers by the same S/D implantation and annealing conditions. For the experiment B, the devices were fabricated in a similar manner until the Si implant. Subsequently the wafers were split in two lots. One lot has seen an RTA annealing at 700 °C for 100 sec, while the other lot was furnace-annealed at 700 °C for 100 min. For the subsequent formation of S/D regions the As implantation energy was chosen in such a way that the Si clusters that were formed during the annealling following the Si implant, are located inside the amorphous layer formed by As implant

On fig. 1a we present the obtained results of ΔV_T -defined as the threshold voltage difference between a 1 µm and a 20 µm device- for experiment A. It is clear that the Si^+ implanted devices that have seen the slow ramp rate (15 °C/min) exhibit a much more important threshold voltage roll-on than the corresponding devices with the fast ramp rate. For the non-implanted devices the roll-on value is similar for both ramp-rates.

On fig. 1b we show the results obtained from experiment B. We notice that the RTA-annealed devices exhibit a less voltage roll-on than the devices that were furnace annealed. This is in agreement in previous reported results. However, it is important to note that for the RTA-annealed devices presented here, the voltage roll-on is due only to the ultrafast TED phase.

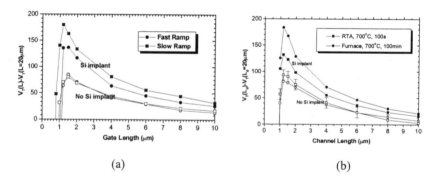

Figure 1. (a) Threshold voltage roll-on for Si-implanted and non-implanted devices as a function of the furnace ramp rate. (b) Threshold voltage roll-on for Si-implanted and non-implanted devices after RTA and furnace annealing at 700°C

3 Simulations and Discussion

It is now established that when silicon atoms are implanted in the substrate at relatively low doses and during the subsequent annealing the formation of point and extended defects (especially {311}) takes place. The interstitial supersaturation will depend on the dissolution kinetics of these defects. A first order model describing the kinetics of these defects was proposed by Rafferty et al. [3] and is presented below :

$$\frac{\partial C}{\partial t} = 4\pi \alpha a D_I I C - C \frac{D_I}{a^2} \exp\left(-\frac{E_b}{kT}\right)$$

where C and I are the concentration of clustered and free interstitial respectively.

The experimental results presented previously were successfully simulated using coupled process-device simulations. Three simulators were used : SUPREM IV, ALAMODE (a PDE solver) and Silvaco Tools (ATHENA and ATLAS). Another parameter of importance for the simulations is the surface recombination velocity of silicon interstitials, which determines the loss rates of the interstitial atoms and consequently the time extent of the TED. This was used as a fitting parameter for the simulations.

On fig. 2a we show ΔV_T for fast and slow ramp furnace rates as a function of channel length. We notice that simulation predicts fairly well the trend of experimental results although there is a discrepancy around the peak of the curve. This discrepancy is more pronounced when the second set (B) of devices is considered.

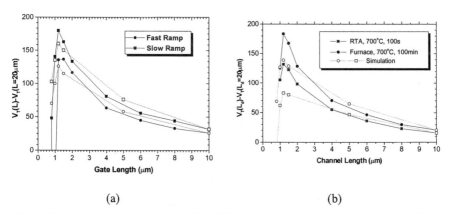

(a) (b)

Figure 2. Comparison between experimental (solid lines) and simulation (dot lines) results for (a) Fast and Slow furnace annealed devices and (b) Furnace and RTA annealed devices at 700°C for 100min and 100sec respectively.

Figure 2b shows the comparison with the simulations of devices annealed at 700°C for 100min or 100sec. For this case the simulations were performed using the same set of parameters with the previous devices. The discrepancy between simulation and experiments exists because the first order model that was used for the simulations does not describe the initial phase of silicon interstitial clustering.

As mentioned above there exist two phases of TED during clustering formation. An initial phase of ultrafast TED, for a short time period (f.e. 100s for 700 °C), where the supersaturation of interstitials is high due to the ripening of very small interstitial clusters, precursors to the nucleation of {113} defects, followed by a second phase of lower TED that persists for much longer time, until the transient diffusion ends. From our experiments (especially from RTA annealed devices at 700°C for 100sec) the influence of this phase on the boron profile cannot be ignored. This effect is reduced when the devices are annealed to temperatures higher that 700°C (f.e fast and slow furnace ramp) since the transition time from the ultrafast TED phase to the phase governed by the dissolution of the {113} defects is reduced.

4 Conclusions

We have probed the influence of sub-amorhizing damage in the S/D on the RSCE. Simulations using existing models for the dissolution kinetics of {113} defects are able to successfully predict the trend of the RSCE. However, our results demonstrate the need for more advanced models able to describe not only {113} kinetics but also of their precursors small clusters. Attempts towards this direction are in progress.

5 Ackowledgements

We acknowledge EU for financial support through the project RAPID and Stanford Un. for ALAMODE donation.

References

1. Rafferty C. S., Vuong H. –H, Eshraghi S. A., Giles M. D., Pinto M. R., and Hillenius, S J., Explanation of Reverse short Channel effect by defect Gradients, Proceeding of IEDM 1993, pp. 311-314
2. Cowern N. E. B., Mannino G., Stolk P. A., Roozeboom F., Huizing H. G. A., and Berkum J. G. M., Energetics of Self-Interstitial Clusters in Si, Physical Rev. Lett. **82** (1999) pp. 4460-4463
3. Rafferty C. S., Gilmer G. H., Jaraiz M., Eaglesham D. and Gossmann H. –J, Simulation of cluster evaporation and transient enhanced diffusion in silicon, Appl. Phys. Lett. **68** (1996) pp. 2395-2397.

NOISE MODELING OF INTERDIGITATED GATE CMOS DEVICES

E. F. TSAKAS AND A. N. BIRBAS

Department of Electrical and Computer Engineering, University of Patras, Patras 26500,
Greece
E-mail: birbas@ee.upatras.gr

The shrinkage of the MOSFET device dimensions along with the relatively wide gate electrode devices needed to accommodate RF applications lead to reconsideration of the noise properties of submicron MOSFET's. In this paper we present the noise properties associated with interconnect resistors of an interdigitated structure and the resulting noise source (strong function of the number of fingers) is evaluated against the other noise sources present in the device such as channel thermal noise, induced gate noise and resistive gate voltage noise. Short channel effects have been taken into account for the evaluation of these noise sources and two-port analysis performed in order to calculate minimum noise figure and optimum input resistance for noise matching.

1 Noise Sources In Submicron Interdigitated MOSFETS

The dramatic improvement of MOSFET properties and performance points to a new era in RF integrated circuit design especially at the low GHz range where the dominant technology is currently silicon bipolar. While MOSFET devices are dramatically scaled down, new materials are used extensively in the fabrication steps and new fabrication techniques are unavoidably employed in order to accommodate the ultra-small device dimensions. This progress leads us to reconsider some basic properties of the new miniaturized devices.

The noise contribution of the first stage of an amplification system influences dominantly the overall noise performance. CMOS inadequacies in device-level performance often have to be circumvented by innovations at the circuit architectural level. The circuit topology of the first stage of a CMOS implemented low noise amplifier (LNA) is usually a common source topology with inductive degeneration [1]. The most significant noise sources associated with the input MOSFET of an LNA are the channel thermal noise ($\overline{i_d^2}$), the induced gate noise ($\overline{i_g^2}$), the gate resistance voltage noise, the substrate resistance noise and the noise associated with the series drain/source resistances.

When the interdigitated structure is being used, there are no specific rules to decide the exact number of fingers to employ. This parameter is usually set to an appropriate value in order to reduce occupied area by suitably setting the shape of the device. Recently in the literature, the number of fingers was used to match the optimum noise impedance to the source impedance [2]. The n parallel-finger transistors are treated as two-port networks and the known formula of the NF was recalculated. The NF$_{min}$ was considered unchanged when we move from one finger to a multi-finger scheme.

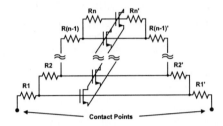

Figure 1. Equivalent circuit of the interdigitated structure. An uncorrelated voltage noise source is associated with each interconnect resistance.

In order to calculate an accurate NF$_{min}$ for an interdigitated MOSFET, we take into account another noise source originating from the gate interconnect material of interdigitated structures [3,4]. We consider an interdigitated transistor with n fingers and 2n interconnect resistances (Fig. 1). The n parallel transistors are equal and the total transconductance of the device is the sum of the transconductances of all the fingers. The voltage noise that each interconnect resistance is contributing is given by:

$$\overline{V_j^2} = 4k_B T R_{int} B \qquad (1)$$

where R_{int} is the value of the interconnect resistance and B is the effective bandwidth. The channel current fluctuation caused by this resistance, as defined for one finger, is:

$$\overline{\delta I_{int_{ij}}} = g_{m_i} \overline{\delta V_{g_i(x)}} \qquad (2)$$

where g_{mi} is the transconductance of each finger and $\overline{\delta V_{g_i(x)}}$ is the average voltage fluctuation along the particular finger caused by the ith interconnect resistance. This average can be defined as the mean value of the fluctuations at the two ends of the finger. Then the average channel current fluctuation for one finger is:

$$\overline{\delta I_{int_{ij}}^2} = g_{m_i}^2 \left(\frac{\delta V_{g_i(w)} + \delta V_{g_i(0)}}{2\overline{\delta V_j}} \right)^2 \overline{\delta V_j^2}$$

$$= g_{m_i}^2 \left(\frac{\delta V_{g_i(w)} + \delta V_{g_i(0)}}{2\overline{\delta V_j}} \right)^2 4k_B T R_{int} B = 4k_B T R_{int} B M_{ij}^2 g_{m_i}^2 \qquad (3)$$

$$M_{ij} = \frac{\delta V_{g_i(w)} + \delta V_{g_i(0)}}{2\delta V_j} \quad (4)$$

and finally:

$$\overline{I_{int}^2} = 4KTBR_{int}\left(\frac{g_m}{n}\right)^2 2\sum_{j=1}^{n}\left(\sum_{i=1}^{n} M_{ij}\right)^2 \quad (5)$$

where $g_m = n \cdot g_{m_i}$ is the total device transconductance and the factor 2 accounts for the two rows of interconnect resistors (both sides of the fingers).

2 NOISY TWO-PORT DESCRIPTION

With the addition of the interconnect resistance noise we have suitably described all the noise sources appearing in a submicron interdigitated MOSFET. Nevertheless, it is very useful to calculate the RF noise properties of this device, such as noise figure (NF) and optimum noise resistance (R_{opt}), in the context of all these noise sources. Fig. 2a shows the noise equivalent circuit of the interdigitated structure. The referred to the input correlated noise sources are:

$$\upsilon_A = \upsilon_g - \frac{i_d + i_{int}}{g_m} + (R_g + r_i)_g - j\frac{f}{f_t}(R_g + r_i)(i_d + i_{int}) \quad (6)$$

$$i_A = i_g - j\frac{f}{f_t}(i_d + i_{int}) \quad (7)$$

where r_i is the channel charging resistance [2], i_d is the channel thermal noise, i_{int} is the interconnect resistance noise, i_g is the induced gate noise, υ_g is the gate voltage noise due to the resistive gate material perpendicular to the direction of channel current and f_t is the unity current gain frequency. Following a standard network analysis we can divide the voltage noise source to one uncorrelated and another correlated part aiming at the calculation of the minimum noise figure and the optimum input resistance for noise matching. The resulting optimum noise resistance and minimum noise figure are given by:

$$NF_{min} = 1 + 2\left(\frac{f}{f_t}\right)^2 \gamma g_{d0} C_2 (r_i + R_g) + \sqrt{\left(\frac{f}{f_t}\right)^2 C_3} \quad (8)$$

$$R_{opt} = \sqrt{\left(\frac{C_2}{C_1}\right)^2 (R_g + r_i)^2 + \left(\frac{f_t}{f}\right)^2 C_4} \quad (9)$$

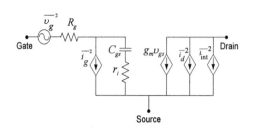
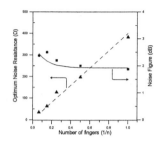

The parameters C_1 - C_4 are suitably described parameters dependent strongly on the number of fingers n and are very complicated functions. For this reason we can have only numerical evaluation of NF_{min} and R_{opt}. For n = 1 to unity parameter C_1 is less than one and if the induced gate noise is neglected is equal to unity. Parameters C_2 and C_3 are also very close to unity for n = 1 but their values are increasing rapidly to a few hundreds for n = 50. The results are shown in Fig. 2b where NF_{min} and R_{opt} are plotted versus 1/n. The interconnect resistance noise were calculated for two contact points and the presented experimental measurement points were reprinted from [2]. It is apparent from this curve that as the number of fingers increases, the noise contribution of interconnect resistances increases dramatically. For n equal to unity, or R_{int} equal to zero, equation (8) reverts to the classical expression of the noise figure plateau, which is independent of n. For the process parameters used in our calculations this noise floor lies somewhere below 2 dB.

3 DISCUSSION AND CONCLUSIONS

The main goal of this paper was to provide a complete noise analysis of interdigitated structures at the device level, compatible with the most recent advancements of MOS process technologies, by introducing an easy to use formulation of the interconnect resistance noise. Two-port noise properties were also treated providing new guidelines in design of RF frond-ends.

References

1. D. K. Shaeffer, and T. H. Lee, "A 1.5-V, 1.5-GHz CMOS Low Noise Amplifier," *IEEE J. Solid-State Circuits*, vol. **32**, no. 5, pp. 745-759, 1997.
2. T. Manku, "Microwave CMOS—Device Physics and Design," *IEEE J. Solid-State Circuits*, vol. **34**, no. 3, pp. 277-285, 1999.
3. R. P. Jindal, "Noise Associated with Distributed Resistance of MOSFET Gate Structures in Interdigitated Circuits," *IEEE Trans. Electron Devices*, vol. ED-**31**, no. 10, pp. 1505-1509, 1984.
4. K. K. Thornber, "Resistive-Gate-Induced Thermal Noise in IGFET's," *IEEE J. Solid-State Circuits*, vol. **SC-16**, no. 4, pp. 414-415, 1981.

SUBTHRESHOLD CHARACTERISTICS OF 0.15μm SOI-MOSFETs AFTER HOT-CARRIER STRESS

P. DIMITRAKIS, G. J. PAPAIOANNOU
University of Athens, Solid State Physics Section, Panepistimiopolis,15784 Zografos, GREECE
E-mail: pdimit@cc.uoa.gr

J. JOMAAH, F. BALESTRA
LPCS-ENSERG, INPG, 24 Av. Des Martyrs, BP257, 38016 Grenoble cedex 1, France
E-mail: balestra@enserg.fr

0.15μm gate length MOSFETs on Low Dose (LD) SIMOX substrates were stressed by hot-carriers under "off-state" operation conditions. Subthreshold device characteristics were studied. It was found that the lower the effective doping of the Si film the higher the device lifetime. The subthreshold swing and $\Delta C_{it}/C_{it}$, are used to monitor the ageing of the devices. Under hot-electron stressing, these exhibited logarithmic time dependence due to the quick saturation of the defects. Back-channel characteristics were affected by a parallel conduction in the subthreshold region, originated probably by the "edge transistor". After hot-hole stressing this conduction is eliminated.

1 Introduction

Sub-threshold behaviour of MOSFETs is of special interest at the present time because the operating voltages are lowering towards 1.5V and gate lengths are scaled down to sub-quarter microns. SOI N-MOSFETs play a key role for low power/high speed applications since they have a two generations advantage over bulk Si devices [1].

The aim of the presented work is to thoroughly investigate the device degradation mechanisms when it operates in the sub-threshold region. In real operating conditions, this is similar to a microprocessor or a memory chip operating in the standby mode.

2 Experimental Procedure

Low dose SIMOX MOSFETs with W/L=20/0.15, 4.5nm gate oxide, 80nm buried oxide and 100nm silicon film were tested. The devices were partially depleted (PD) with edges. During stress, the source electrode was grounded, the drain bias was V_{ds}=+3V and the front gate bias V_{gf} near the threshold voltage of the front channel V_{thf} and back gate voltage V_{gb} (A) 0V (depletion), (B) +20V (inversion) and (C) − 20V (accumulation). The total stress time was 50000s. MOSFETs with two different channel doping concentrations were tested (1) $8 \times 10^{17} cm^{-3}$ and (2) $3 \times 10^{17} cm^{-3}$. I-V measurements were performed under the following conditions: front channel characteristics were obtained with back interface in accumulation, while for back channel they are obtained with $V_{gf}=-3V$, i.e. accumulated front interface. So, device names with subscript "1" denote MOSFETs with high effective doping and names with subscript "2" low doped ones.

Subthreshold degradation was monitored by the subthreshold swing shift with time. Additionally, the introduction rate of interface states was calculated by the relative shift of their capacitance, $\Delta C_{it}/C_{it0}$. This was achieved by calculating the slope of the output conductance $g_d=dI_d/dV_d$ in weak inversion [2].

3 Results and Discussion

The shift of threshold voltage ΔV_{th}, transconductance $\Delta g_m/g_{m0}$ and saturation drain current $\Delta I_{dss}/I_{dss0}$, exhibited a logarithmic time dependence [3]. The calculated lifetimes are in agreement with those previously published for devices with similar gate lengths and stress conditions [4-5]. It was found that the lower the effective channel doping the higher the lifetime of the device.

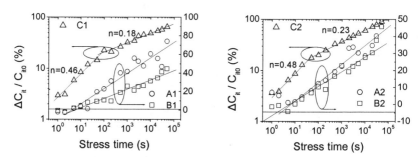

Figure 1. Relative shift of interface state capacitance after different stress times. Devices with a higher doping concentration (1) are more vulnerable than low doped devices (2).

The interface state capacitance is proportional to their density, Dit $(cm^{-2}eV^{-1})$. Results for the C_{it} evolution, hence the D_{it}, are presented in figure 1. In highly doped devices, the interface degradation is more severe than for low doped channel MOSFETs. If fact, D_{it1} is doubled after 50ks and $D_{it1}=2D_{it2}$. The introduction of D_{it} for both devices after simultaneous stress of both channels by hot-electron stress is a "logarithmic" process, i.e. $\Delta C_{it}/C_{it0}=Q+P \cdot \log(t)$. When the back channel is stressed by hot-holes, the front interface aging process is a two-step and power-law, i.e. $\log(\Delta C_{it}/C_{it0})=Q+n \cdot \log(t)$. This is clear from data presented in figure 1. There a two-slope process was found for the D_{it} evolution of C1 and C2 devices. The first process has an exponent value n=0.46 (0.48 for C2) and might be related with the generation of interface states, while the second process has n=0.18 (0.23 for C2) and is related with carrier trapping at the interface. Although, the tested devices were partially depleted there is a direct influence of the front channel degradation on the back channel bias conditions.

Subthreshold swing [6] was also calculated by the following equation

$$S=(d\log_{10}I_d/dV_{gf})^{-1}=2.3\phi_t(C_{ox}+C_d+C_{it})/C_{ox}$$

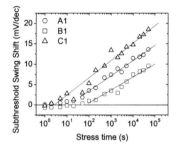
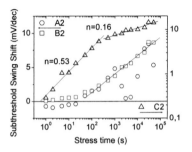

Figure 2. Shift of subthreshold swing after different stress times. Again like C_{it}, devices with higher doping concentrations are more vulnerable than low doped devices.

In figure 2, we present the data for the S shift vs the stress time. Since S is proportional to C_{it}, one is expecting for S a behavior similar to that of $\Delta C_{it}/C_{it0}$ with respect of stress time. In general this is true, with an exception for C1 devices. For these devices, there exists a logarithmic degradation process instead of a two-step power-law process. However, for C2 devices the aging mechanisms corresponding to the n-exponents are the same with those observed for $\Delta C_{it}/C_{it0}$.

Back and front channels were stressed simultaneously, under their individual biases. All devices suffer from "edge" transistor action at the back channel [7]. This parasitic conduction is evidenced by the presence of a "hump" in the subthreshold region of I_d-V_g curves. In figure 3a, we present I-V curves after hot-electron stress. The "hump" exists after stress in both normal and reverse mode. On the contrary, this has disappeared after hot-hole stress (fig. 3b). The positive charge trapped at the interface between the Si film and the buried oxide, in combination with the lowering of the lateral electric field due to the damage near the drain caused the elimination of parasitic "edge" transistor conduction.

4 Conclusions

Enhancement mode, partially depleted, low dose SIMOX N-MOSFETs exhibited a very quick saturation of the defects created by hot-carriers. The subthreshold operation of these devices is degraded very rapidly. A two-stage process was investigated to govern the aging of the front interface. The back channel defects induced by hot-holes change the electric field near source/drain and Si film junctions causing the elimination of the parasitic "edge" transistor.

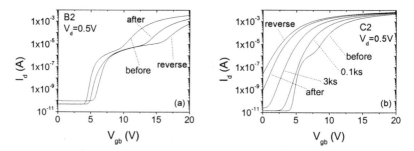

Figure 3. Back channel I-V curves after stress with (a) hot-electrons and (b) hot-holes. The "edge" transistor action is eliminated after 3ks of stress with hot-holes.

5 Acknowledgements

This work has been financially supported by the GSRT and French Embassy in the framework of bilateral common exchange program "PLATO" and the University of Athens (Proj. No.70/4/4079).

References

1. A. O. Adan, T. Naka, T. Ueda, Y. Seguchi, A. Kagisawa, in *SOI Techn. and Devices*, ed. by S. Cristoloveanu, R. Wilson, P. Hemment, K. Izumi (The Electrochemical Society Proceedings Series, Pennington, NJ, 1997), PV 97-23, pp.340-351
2. P. Dimitrakis, J. Jomaah, F. Balestra and G. J. Papaioannou, in *SOI Techn. and Devices*, ed. by S. Cristoloveanu (The Electrochemical Society Proceedings Series, Pennington, NJ, 2001), accepted for publication
3. P. Dimitrakis, J. Jomaah, F. Balestra, G. J. Papaioannou, in *Proceedings of Advanced Semiconductor Devices and Microsystems*, ed. by J. Osvald, S. Hascik, J. Kuzmik, J. Breza, (The IEEE, Piscataway, NJ, 2000) pp.59-62.
4. S. H. Renn, J. L. Pelloie, F. Balestra, in *IEDM Technical Digest '96*, (The IEEE, Piscataway, NJ, 1996), pp.877-880.
5. S. H. Renn, C. Raynaud, J. L. Pelloie, F. Balestra, in *Proceedings of the IRPS'98*, (The IEEE, Piscataway, NJ, 1998), pp.203-208.
6. S. Cristoloveanu, S. S. Li, *Electrical Characterization of Silicon-On-Insulator Material and Devices*, (Kluwer Academic Publishers, Boston, 1995), p.249
7. *ibid*, p.235-240

A LOW VOLTAGE BIAS TECHNIQUE TO INCREASE SENSITIVITY OF MOSFETS DOSIMETERS

G. FIKOS, S. SISKOS
Electronics Lab., Aristotle Univ. of Thessaloniki, 54006 Thessaloniki, Greece
E-mail: siskos@physics.auth.gr   gfiko@skiathos.physics.auth.gr

A. CHATZIGIANNAKI
AHEPA General Hospital of Thessaloniki, , 54006 Thessaloniki, Greece
E-mail: ana@skiathos.physics.auth.gr

G. SARRABAYROUSE
LAAS du CNRS, 7 Av. du Colonel Roche, 31077 Toulouse, France
E-mail: sarra@laas.fr

Metal-Oxide-Semiconductor (MOS) dosimeters in a stacked configuration have been already used to increase their sensitivity to radiation. In this work, a new bias technique is presented, allowing stacking a higher number of MOS dosimeters resulting in a dosimetric system with higher sensitivity in a large range of doses. The proposed low voltage technique consists in the biasing of each MOS dosimeter of the chain at the limits of the saturation region. So the output voltage of the stacked MOS dosimeters may take a lower value comparing with the classical stacked configuration allowing a greater number of stacked dosimeters. This configuration is aimed to be used for low dose personal dosimetry.

1 Introduction

Metal-Oxide-Semiconductor (MOS) dosimeters are MOS transistors with a specially processed gate insulator in order to make them radiation soft [1]. MOS dosimeters have advantages such as low cost, small size and weight, robustness, accuracy, large measurable dose range, sensitivity to low energy radiation (<10keV), real-time or delayed direct read out, information retention, possibility of monolithic integration with other sensors and/or circuitry capable realizing measurement, signal conditioning and data processing, possibility of use without power supply.

The sensitive part of the MOS transistor is the gate insulator. Irradiation creates electron-holes pairs throughout the whole volume. Depending upon the electric field in the insulator some generated pairs recombine and the remaining are separated. Electrons are rapidly swept out of the insulator and holes move slowly towards the cathode (Gate or Silicon substrate) and some of them are trapped in a narrow region near the cathode on previously existing hole traps generated during insulator processing. Also interface states are created during irradiation at the insulator-silicon interface.

The net result is the creation of a permanent positive charge, which shifts the threshold voltage V_T of the transistor by an amount ΔV_T. This shift depends upon the absorbed dose D, the gate bias during irradiation, the type and the energy of the radiation and the insulator thickness Dox. ΔV_T is the measured dosimetric parameter and is obtained during or after irradiation [2].

Sensitivity of the dosimeter can be increased: a) through technology interventions by increasing the oxide thickness (increase of the electron - hole pairs generation within the oxide) b) during irradiation by biasing the gate (increase of the hole trapping efficiency) and c) with measurement techniques after irradiation by using stacked dosimeters and/or taking advantage from the body effect.

It has been mentioned that the response to radiation may be improved by using several MOS transistors in a series (stacked) configuration. The stacked configuration uses MOS transistors in diode connection with [3] or without [4] backbias. This configuration results to a very high output voltage value depending on the number of the stacked transistors. The output voltage is increased more due to the body effect preventing the stacking of a great number of MOSFETs.

The proposed measurement technique is based on the reduction of the output voltage value by biasing each transistor of the chain at the limits of the saturation region allowing to increase the number of the stacked transistors and therefore increase the sensitivity.

2   Low voltage bias technique

The basic idea to increase sensitivity of the dosimetric system is to stack as many MOS transistors as possible. Fig. 1a shows the classical stacked configuration proposed in [3]. Fig. 1b shows the chain of the stacked dosimeters with the bias circuit. For this purpose a voltage shifter is designed using current mode devices.

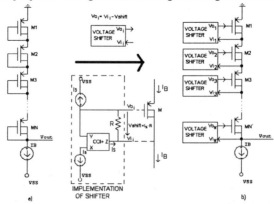

Figure 1. Stacked MOS dosimeters a) classical configuration b) proposed configuration

In the stacked configuration all MOS dosimeters must operate in the saturation region ($V_{DS} \geq V_{GS} - V_T$). Then the output voltage for N stacked transistors is given by:

$$V_{out} = N \bullet V_{DS}.$$

Since V_{out} is proportional to V_{DS}, it is clear that for a given upper limit of V_{out} (due to the power supplies), the number of stacked transistors can be increased by reducing V_{DS}.

In the case of the classical configuration (Fig. 1a) considering that I_B is the drain current of the MOSFETs and $V_{GS}=V_{DS}$ (diode connected), the output voltage is given by

$$V_{outClassic} = N(\sqrt{\frac{I_B}{K_P}} + |V_{TP}|)$$

In the proposed configuration (Fig. 1b) the reduction of the V_{DS} is achieved by appropriate biasing of each MOS transistor. The low voltage biasing technique uses a voltage shifter between gate and drain in order to bias the MOSFET at the limits of the saturation region where V_{DS} is slightly greater than $V_{GS}-V_T$. Therefore V_{DS} of each transistor is less than in the diode connection by a constant voltage shift. In this manner the output voltage of the MOS transistor is reduced significantly allowing stacking a greater number of dosimeters and increasing the sensitivity of the whole system. Such a dosimetric system is aimed to measure doses in the range of decades of mrads.

The voltage shifting circuit must respect the following condition in order to ensure that each transistor is saturated : $V_{shift} \leq |V_T|$. Then, the output voltage of the chain of dosimeters is given by

$$V_{outProposed} = N(\sqrt{\frac{I_B}{K_P}} + |V_{TP}| - V_{shift})$$

The difference of V_{out} between the two configurations is

$$V_{outClassic} - V_{outPproposed} = N \bullet V_{shift}$$

The quantity that measures dose, is the difference of V_{out} before and after irradiation. This quantity is the same for both configurations and equals :

$$\Delta V_{outClassic} = \Delta V_{outPproposed} = N \bullet \Delta V_T$$

Therefore the advantage of the proposed configuration is the reduction of V_{out} by $N \bullet V_{shift}$, while the rest properties of the classical configuration remain intact in the proposed one.

The dosimeters are fabricated with PMOS transistors with thick gate oxide (1.6um) and $V_T=-6V$, without backbias. The shifter circuits are mounted using discrete components - the current conveyor was previously fabricated using MIETEC 2um CMOS technology [4]. The second generation current conveyor, CCII+ is a known, current mode, three port, analog building block with the following characteristics: $V_x=V_y$, $I_x=I_z$ and $I_y=0$.

The current Is is set by a current source and produces the V_{shift} across resistor R between gate and drain of the PMOS transistor.

3 Results and discussion

Measurements were computer aided, using a 16 bit data acquisition module. Irradiation is performed in a Co-60 beam, using a 5mm layer of plexiglass to provide electronic equilibrium. The PMOS dosimeters were unbiased during irradiation.
The whole system may increase sensitivity more than 5 times comparing to the conventional system, it is more accurate since using a greater number of transistors slight differences in the threshold voltage values are statistically annealed and it may measure doses in the range from 0.1 mGy to 10 Gy. In order to keep output voltage value small the measurement system may be reset, measuring only the new threshold voltage shift of the MOS dosimeters. With this bias technique we may use low doped substrates in order to reduce the temperature dependence of the MOS dosimeters. It is also possible to exploit the shifter circuit designed with an appropriate temperature coefficient with back-gate biasing in order to achieve a zero temperature coefficient bias for the overall circuit.
Integrated stacked MOS dosimeters will be fabricated in the near future to evaluate the real performance of the whole system.

References

1. Poch W. J. and Holmes-Siedle A. G., RCA Engineer **16** (3), (1970) pp. 56
2. Sarrabayrouse G., Siskos S., Radiation dose measurement using MOSFETs, IEEE Instr. Measur. Magaz., **1**(3), (1998), p. 26-34.
3. Vychytil F., Cechak T., Gerndt J., Petr J., Increasing of the sensitivity of MOS dosimeters in series configuration, Jaderna Energie **24** (11), (1978) pp. 419-421.
4. O'Connell B., Kelleher A., Lane W. and Adams L., Stacked Radfets for increasing radiation sensitivity, IEEE Trans. Nucl. Sci **43** (3), (1996), pp. 985-990.
5. Laopoulos Th., Siskos S., Givelin Ph., Bafleur M, CMOS Current Conveyor, Electronic Letters, **28** (24), (1992), pp. 2261-2262.

HIGH PRECISION CMOS EUCLIDEAN DISTANCE COMPUTING CIRCUIT

G. FIKOS, S. SISKOS
Electronics Laboratory, Department of Physics,
Aristotle University of Thessaloniki, Thessaloniki 54006, Greece.

In this paper a novel voltage-mode MOS circuit is presented, which calculates the Euclidean norm of an input vector of unipolar voltages. It is based on a very simple structure, utilizing a one-transistor cell. Therefore, it is easily expandable to multiple inputs and very fast, exhibiting high linearity and precision. These characteristics were confirmed by simulations and experimental results with commercial transistor arrays (CD4007).

1 Introduction

The Euclidean norm of a n-dimensional vector $X=(x_1,x_2,...,x_n)$, defined as

$$\| X \| = \sqrt{\sum_{i=1}^{n} x_i^2} \quad (1),$$

is necessary for calculating the unitary vector in the direction of a given vector X ($X/\|X\|$) which is used for calculating the angular similarity function of two vectors [1], for *kernel orthonormalization* [2], and to implement neural algorithms such as *competitive learning* [3]. In the case of $x_i = y_{i,ref} - y_i$ in eq. (1), the Euclidean norm is used by *vector quantization* and *nearest neighbor* classification algorithms as a direct measure, for finding the reference vector $Y_{ref}=(y_{1ref},y_{2ref},...,y_{nref})$ that is closest to an input vector $Y=(y_1,y_2,...,y_n)$ [4].

In this work, a simple novel MOS circuit is presented, that has a unipolar voltage vector V_{in} as input, and outputs a voltage V_O that is proportional to the Euclidean norm of the input, $V_O \propto \|V_{in}\|$.

2 The proposed circuit

The proposed circuit (fig. 1a) exploits the quadratic expression of the drain current of a n-MOS in saturation region,

$$I_D = k(V_G - V_S - V_{TN})^2 \quad (2),$$

(V_G, V_S, V_{TN}, are the gate, source and threshold voltages respectively and k is the transconductance parameter of the n-MOS) to implement simple voltage squaring elements, with V_G as input and I_D as output, provided that:

$$V_S = -V_{TN} \quad (3)$$

Solving eq. (2) for V_G, a square-rooting operation with I_D as input, arises:

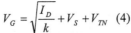

$$V_G = \sqrt{\frac{I_D}{k}} + V_S + V_{TN} \quad (4)$$

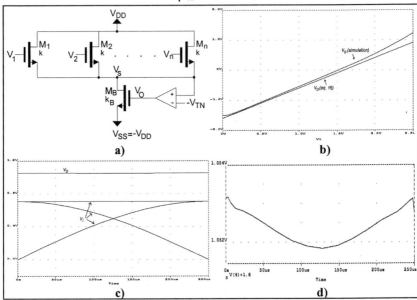

Figure1: a) The proposed circuit, **b)** V_O versus V_I for $V_2=V_3=0$, simulated and calculated results (eq. 14), **c)** V_O for the inputs of eq. 16, **d)** the small variations of V_O not seen in fig. 1c.

The above analysis is valid for the proposed circuit (fig. 1a). Through the negative feedback established using the opamp, eq. (3) is valid. Assuming that transistors M_i, i=1,2,..,n and M_B are in saturation and applying K.C.L. at node S:

$$I_B = \sum_{i=1}^{N} I_{Di} \quad \overset{(2) \wedge (3)}{\Rightarrow}$$

$$V_O = V_{TN} + V_{SS} + \sqrt{\frac{k}{k_B}} \sqrt{\sum_{i=1}^{N} V_i^2} \quad (5)$$

Eq. (5) is the circuit's output and it can be seen that it consists of an offset $V_{OFFSET}=V_{TN}+V_{SS}$ and of a quantity proportional to $\|V_{in}\|$, $\sqrt{\frac{k}{k_B}}\sqrt{\sum_{i=1}^{N} V_i^2}$.

All transistors must be saturated, meaning that for each one the relations
$$V_{GSi} \geq V_{TN} \quad (6), \quad V_{GDi} \leq V_{TN} \quad (7), \quad i=1..n$$
should be valid. Applying eq. (3) on eqs. (6), (7), the limitations derived are:
$$V_i \geq 0, \quad i = 1,2,..,n \quad (8)$$

$$V_O \leq V_S + V_{TN} \overset{(3)}{=} 0 \overset{(5)}{(9)} \Rightarrow$$

$$\sum_{i=1}^{n} V_i^2 \leq \frac{k_B}{k}(V_{DD} - V_{TN})^2 \quad (10).$$

Body Effect: In the case of n-well technology, V_{TN} of M_i, i=1,..,n, is given by

$$V_{TN} = V_{TN0} + \gamma\left(\sqrt{2\Phi_B + V_{DD} - V_{TN}} - \sqrt{2\Phi_B}\right) \quad (11)$$

V_T extraction is then much more sophisticated and external application of V_{TN} requires prior exact knowledge of Φ_B, V_{TN0}, γ. So, the practical solution for n-well technology would be to design the complementary circuit using p-MOSFETs.

Channel Length Modulation: The drain current of a n-MOS in saturation, taking into account channel length modulation, is given by

$$I_D = k(V_G - V_S - V_{TN})^2(1 + \lambda V_{DS}) \quad (12)$$

and following the same steps as in deriving eq. (5), the output is then given by:

$$V_O = V_{TN} + V_{SS} + \sqrt{\frac{k}{k_B}}\sqrt{\frac{1 + \lambda(V_{DD} + V_{TN})}{1 + \lambda(V_{DD} - V_{TN})}}\sqrt{\sum_{i=1}^{N} V_i^2} \quad (13)$$

So, channel length modulation slightly modifies coefficients k, k_B, with $\{1 + \lambda(V_{DD} + V_{TN})\}$, $\{1 + \lambda(V_{DD} - V_{TN})\}$, respectively.

3 Simulation and experimental results

The circuit was simulated with Spice, using 2μ MIETEC CMOS model parameters (V_{TN0}=0.9V, KP=57μA/V^2). All transistors had equal dimensions (W/L)=5/20, V_{DD}=|V_{SS}|=2.5V. Setting V_2=V_3=0 and sweeping V_1 from 0 to 2.5V, the output voltage V_O (fig. 1b) was captured. Substituting on eq. (5) we calculate:

$$V_O = V_{TN} + V_{SS} + V_1 = -1.6 + V_1 \text{ (V)} \quad (14)$$

and substituting eq. (14) on eq. (9), for correct operation:

$$0 \leq V_1 \leq 1.6 \text{(V)} \quad (15)$$

Eq. (14), as shown in fig. 1b, is in fairly good agreement with the simulated results.

The depart of V_O (fig.1b) from eq. (14) in the region of zero, is owed to the subthreshold current resulting in the simple quadratic expression for the MOSFET's drain current being accurate only if $V_{GSi} > V_{TN} + 3kT/q$. Consequently, at room temperatures, V_{in} should exceed 75mV for the quadratic expression to be valid [4]. Better performance at this region can be expected by setting (W/L)$_B$=Σ(W/L)$_i$.

The slightly different slope between the two curves is due to channel length modulation. A least square approximation straight line fit to V_O, resulted in $Slope_{(simul)}$=1.0406V^{-1}, while the slope for λ=0.05V^{-1} (technology used) from eq. (13) is $Slope_{(eq.13)}$=1.0408V^{-1}, verifying our starting assumption. The linearity error is found to be less than 0.17% and the simulated upper limit of V_1 for correct operation (V_O=0V) is 1.58V, being in good agreement with eq. (15) (1.25% error).

The circuit's transient response to the input vector of constant norm of eq. (16), was captured, showing all inputs and the output voltage (fig. 1c) and the output alone (fig. 1d) where one can see the small deviations of the output voltage from its mean value. In such a case, the output should remain constant.

$$V_1 = \frac{1}{\sqrt{2}}\sin 2\pi ft, \; V_2 = \frac{1}{\sqrt{2}}\cos 2\pi ft, \; V_3 = \frac{1}{\sqrt{2}} \Rightarrow \sqrt{\sum_i V_i^2} = 1 \; (f = 1\text{KHz}) \quad (16)$$

The value of the output expected by eq. (13) (subtracting the offset $V_{TN}+V_{SS}$) is $V_{O(theor)}=1.041$V while from fig. 1d, $V_{O(mean)}=1.052$V and the maximum deviation from this mean value is 0.665mV or 0.06%.

Simulating the frequency response of the circuit with three different opamps, a slow one (LM324), a faster one (TL082/TI), and an ideal one (BW=0-∞), the bandwidth of the circuit was found to be $BW_{3db(LM324)}=607$kHz, $BW_{3db(TL082)}=2788$kHz, while the $BW_{3db(ideal)}=22.804$MHz, proving that the bottleneck of the proposed circuit is in the opamp since the rest circuitry has the property of very fast response, provided the opamp can support it.

Testing the transient response to the input vector of constant norm of eq. (16) on a 3-input implementation with cheap commercial products (CD4007 for identical n-MOS, LF444 opamp) produces a deviation of the output from its mean value less than 0.5%. The difference from the corresponding simulation result (0.06%) is owed to transistor mismatch and not perfectly constant norm of the inputs.

4 Conclusion

A novel, one-trasistor cell, Euclidean distance calculator was proposed. Its accuracy and speed were verified through simulation and experimental results. The circuit is expected to be useful in hardware implementation of neural algorithms.

References

1. Cao, J., Ahmadi, M., and Shridhar, M.: "Handwritten numeral and machine printed multiple font character recognition using neural network classifier", *J. of Circuits, Systems, and Computers*, Vol. **6**, No. 6 (1996), pp. 569-580..
2. Kaminski, W., and Strummilo, P.: "Kernel orthonormalization in radial basis function neural networks", *IEEE Transactions on Neural Networks*, September 1997, Vol. **8**, No. 5, pp. 1177-1183.
3. Simpson, P. K.: "Foundations of Neural Networks", in *Artificial Neural Networks: Paradigms, Applications and Hardware Implementations*, by Sanchez-Sinencio, E., and Lau, C., (eds), IEEE Press, 1992.
4. Collins, S., Brown, D.R. and Marshall, G.F.: "An Analogue Vector Matching Architecture", *Analog Integrated Circuits and Signal Processing*, Vol. **8**, 1995, pp.247-257.

Microsystems

PRESSURE SENSORS BASED ON 3C-SiC ON Si-ON-INSULATOR FOR HIGH TEMPERATURE APPLICATIONS.

S. ZAPPE, M. EICKHOFF[1] AND J.STOEMENOS[2]
Technical University of Berlin, Secr. TIB 3.1, Gustav-Meyer-Allee 25, 13355 Berlin, Germany
[1]*DaimlerChrysler Research and Technology, Munich, Germany*
[2]*Aristotle University of Thessaloniki, Physics Department, 54006 Thessaloniki, Greece*

The fabrication and characterization of a membrane type piezoresistive pressure sensor with piezoresistors consisting of 3C-SiC with maximum temperature ~ 400°C, is presented. The sensitivity at RT is S=0.5mV/V bar. The membrane of the device was defined by etching of a UNIBOND SOI wafer. The technological problems related with the realization of SiC/SOI pressure sensors are discussed.

1 Introduction

Cubic Silicon Carbide (3C-SiC) has extraordinary electronic and chemical properties. The 3C-SiC is a semiconductor suitable for high-temperature, high-frequency and high-power electronic applications because of its wide band gap, the high-saturated electron velocity and the high-breakdown electric field. The epitaxial growth of 3C-SiC on Si-substrate by chemical vapour deposition (CVD) permits the development of sensors working at high temperatures and in harsh environments. Thus, a large number of parameters can be accurately determined, in order to optimize combustion processes in industry. For such extreme range applications the SiC/Si system is suitable as a sensing element.
The epitaxially growth of 3C-SiC on Si has also the advantage of bulk micro-machining processing and large wafer size capability. Despite of these advantages the SiC/Si system has a major problem. At temperatures greater than 200°C the SiC/Si heterojunction becomes leaky, short-circuiting the SiC sensing element through the Si substrate [1]. In contrast epitaxially grown 3C-SiC on Si On Insulator (SOI) eliminates the leakage problem because the existing buried oxide layer isolates the overgrown SiC from the Si-substrate [2]. Moreover the 3C-SiC films can be deposited on large area Si wafers because the isolating SiO_2 layer acts as a buffer, absorbing the strain developed due to the differences of thermal expansion coefficients of the 3C-SiC overgrown and the Si-substrate [3]. Therefore the 3C-SiC/SOI system fulfils the requirements for applications up to 500°C. On the other hand, the state of the art silicon and SOI technologies are far better than the SiCOIN technology. Therefore it appears of decisive importance, for using high temperature sensors, to bring up the state of development of SiCOIN to those of silicon and SOI

technology. In the present paper we present the technological problems related with the realization of SiC/SOI pressure sensors.

2 Fabrication of the sensor device

A cross section of a complete sensor device is given in Fig.1. A UNIBOND SOI wafer with a buried oxide (BOX) thickness of 400nm and Si-overlayer (SOL) 200nm was used for the realization of the membrane structures. For a nominal pressure of 10 bar, a membrane thickness of 60 µm was chosen. A gap of about 4.5µm between the centerboss and the silicon substrate was formed by isotropic deep reactive ion etching, as shown in Fig.1. The chip size is 3mm x 3mm, chip thickness is 300µm and the diameter of the cavity is 1mm. Applying a pressure of 10 bar the deflection of the centerboss in z-direction is $\delta z = -3.8$µm. Maximum tensile and compressive stresses in the 3C-SiC top layer are about $\sigma = \pm 280$ MPa as shown in Fig.2a. The relative change of the resistance $\Delta R/R_o$ at 10 bar is shown in Fig.2b. The sensitivity at RT is S=0.5mV/V bar and the maximum working temperature is 400C°.

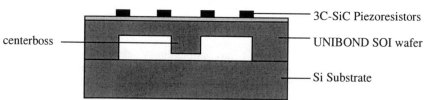

Fig. 1: Cross-section of the sensor cell. A UNIBOND SOI wafer is structured by ICP etching. The etched wafer is bonded to a Si-substrate and the 3C-SiC layer was deposited afterward onto the SOI wafer.

The UNIBOND membrane wafer and the Si substrate were joined together by applying a Si fusion bonding process at 1200°C for 30 min. For the formation of the 3C-SiC piezoresistors two different methods were applied:
a) Etching of the 3C-SiC overlayer and the SOL by Reactive Ion Etching
b) Selective deposition of the 3C-SiC
In both cases the piezoresistors were passivated by thermal oxidation. Contact holes were opened and contacts were formed using TiWN capped with a Au layer. The width and the length of the piezorezistors are 15μm and 150μm, respectively. The sensor chip is shown in Fig.3a.

3 Technological Problems related with the realization of the 3C-SiC/SOI sensors

3.1 - Stability of the SOI structure during SiC deposition.

The stability of SOI structure at high temperatures is very important because the

SiC/SOI system is very sensitive in the formation of cavities in the Si-overlayer at the Si\SiC interface during 3C-SiC deposition resulting to the decomposition of the BOX. This is attributed to the following reasons:
a) Due to the very low oxygen partial pressure.
Under low oxygen partial pressure SiO_2 decomposition at the Si/SiO_2 interface occurs according to the reaction [4]

$$Si + SiO_2 \rightarrow 2SiO_{(g)} \qquad (1)$$

The decomposition is enhanced at higher temperatures and lower partial oxygen pressures.
b) Due to the presence of hydrogen
Hydrogen is used as carrier gas during SiC deposition; this easily diffuses through the 3C-SiC overgrown and the SOL, finally reacting with the BOX according to the reaction (2) [5].

$$H_2 + SiO_2 \rightarrow H_2O_{(g)} + SiO_{(g)} \qquad (2)$$

c) The ball-up of the Si in the SOL at the edges of the cavities,
This is attributed to the poor wetting of Si on SiO_2 at high temperatures. Thus a wetting angle of 87° is experimentally observed at the Si melting temperature between Si and SiO_2 [6], which renders wetting possible but extremely unstable since at 90° the Si-overlayer is beading up.
In order to avoid decomposition of the BOX due to instability of the SOI, the 3C-SiC deposition temperature was reduced to 1200°C. At this temperature methylsilane (MeSi) H_3C-SiH_3 was used as precursor gas [7] and the deposition was carried out in a cold wall, low pressure CVD reactor with lateral gas flow resulting in the growth of good quality 3C-SiC films.

3.2- Formation of cavities in Si at the SiC/Si interface

A substantial step of the 3C-SiC epitaxial growth on Si by CVD is the carbonization process at the early stage of growth. During this process only hydrocarbon is released, which reacts with the Si at the surface, forming a thin SiC layer, which acts as seed for the subsequent growth of the 3C-SiC film. Significant mass transport from the Si substrate to the SiC surface occurs during the carbonization process, suggesting Si out-diffusion from the bulk Si for the formation of the SiC buffer layer. Pits having the shape of reversed tetragonal pyramids are formed in this case. Loss of Si from the SOL is expected due to Si out diffusion, as in the case of SiC deposition in bulk Si [8]. This loss of Si occurs also during the SiC deposition until a 3C-SiC film about 200nm is formed. Above this thickness no significant changes of the cavities are observed, revealing that the cavities are formed at the early stage of growth after the carbonization process. The density and the size of the cavities, which are formed in the Si-overlayer (SOL) during SiC deposition on SOI wafers, are higher than in the bulk silicon, because in bulk Si the substrate is the reservoir for the Si supply. In contrast in SOI only the thin Si-overlayer (SOL) is the Si supplier. In SOI during the early stage of SiC growth, the cavities in the SOL have the typical shape of the reversed tetragonal pyramids. When they touch the Si/SiO$_2$ interface, start to extend laterally and become trapezoidal as shown in Fig.3b. The formation of cavities is enhanced for thinner SOL because, for the same loss of Si, the area covered by cavities will be larger.

A two-step growth process was applied in order to suppress the cavities. Deposition started with a very fast flow rate in order to form a 500nm thick SiC layer. Due to the fast deposition no significant loss of Si occurs. The deposition continues with the standard flow rate up to the total thickness of about 3μm. The two-flow rate method gives good quality SiC films suppressing also the cavities in the SOL.

3.3 Residual thermal strain

Residual thermal strain due to the difference of the thermal expansion coefficients of Si and SiO$_2$ also exists [9]. When the thickness of the SOL is comparable with this of the BOX the strain is distributed in both the SOL and the BOX. As the SOL becomes thinner the BOX relaxes and all the strain is transferred to the SOL, which degrades [9]. The strain increases during SiC deposition as the thickness of the overgrown SiC increases resulting to the bending of the substrate or ruptures the 3C-SiC film at the interface. For the realization of the microsensors most of the deposited 3C-SiC is etched off, demonstrating that deposition of 3C-SiC on the complete wafer surface layer leads to an unnecessary shrink in applicability. Therefore the selective deposition on prestructured Si-substrates has two main advantages in comparison to the standard deposition processes [10]: Reduces the area covered by the 3C-SiC, decreasing the mean strain imposed to the substrate, also simplifies the device fabrication by dispensing the etching of the 3C-SiC. The quality of the 3C-SiC overgrown is good comparable with the SiC grown under the same condition on standard SOI wafers. Lateral growth at the edge of the 3C-SiC film is observed, which is comparable with the thickness of the SiC film. Selective deposition gives good quality films even at the edges where lateral growth occurs. Also reduces the area covered by the 3C-SiC decreasing the mean strain imposed to the substrate.

4 Pressure Measurements

The fist pressure measurements were carried out using a climate control system with a maximum operating temperature of 160°C. The temperature dependent output characteristics of a sensor are shown in Fig.4a. At 10bar the centerboss touches the ground. The sensitivity is decreased in the overload range from the nominal sensitivity S=0.5mV/Vbar to the reduced value of S=0.03mV/Vbar. Output characteristics at various temperatures are shown in Fig.4b. The decrease of the sensitivity with temperature in the rage 25 to 150°C is 7%. The drift of the offset voltage within the same temperature range is 20mV. High temperature measurements up to 400°C with a different measurement setup are in progress.

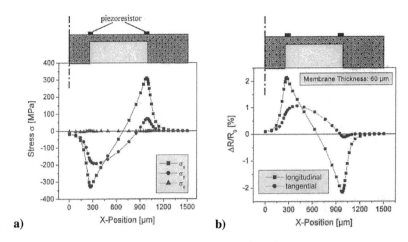

Fig.2: a) Radial stress σ_x and tangential stress σ_y on top of the membrane in the 3C-SiC film. b) Relative change of resistance at 10 bar of a longitudinal and a transversal stressed piezoresistors versus resistor .position

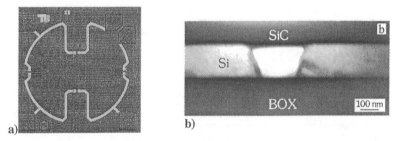

Fig.3. a) Photograph of the sensor chip. b) The cavities in the SOL become trapezoidal as soon as they touch the Si/SiO$_2$ interface

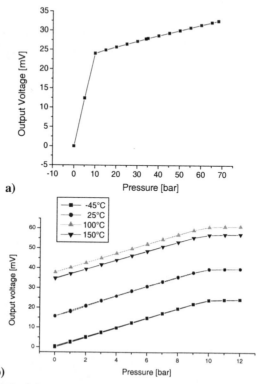

Fig. 4: a) Output characteristic at room temperature. At a pressure of 10 bar, the centerboss touches the ground. b) Output charcteristics at various temperatures. High temperature measurements (up to 400 °C) are currently in progress. (Supply voltage during measurements: 5 V)

5 Acknowledgment

This work was supported by the EU, BRITE-EURAM project CT96-0261

References

[1] W. Reichert, E. Obermaier and J. Stoemenos, Diam. Related Mat. **6**, (1997),1448
[2] G. Krötz, H. Möller and M. Eickhoff, 2nd European Conf. On Silicon Carbide and related Materials, Sep. 2-4,1998 Montpellier, France, paper K-2
[3] F. Namavar, P. Colter, A. Cremins-Costa, C-H. Wu, E. Gagnon, D. Perry and P. Pirouz. Mat. Res. Soc. Symp. Proc. Vol. **423** (1996) p409

[4] Smith F.W. and Ghidim G. J., Electroch. Soc. **129** (1982) 1300
[5] Ferro G., Planes N., Papaioannou V., Chaussende D., Monteil Y., Stoemenos J., Camassel J., Mat.Sci. Eng. **B61- 62** (1999) 586
[6] E. Yablonovich and T. Gmitter, J. Electrochem. Soc. **131**, 2625 (1984)
[7] H. Möller, M Eickhoff, M.Rapp, H.W. Grueninger and G. Krötz, Appl. Phys. A68, (1999) 461
[8] G. Ferro, Y. Monteil, H. Vincent, F. Cauwet, J. Bouix, P. Durupt, J. Olivier and R. Bisaro, Thin Solid Films **278**, (1996), p. 22
[9] J. Camassel, N. Planes, L. Falkovski, H. Möller, M Eickhoff and G. Krötz, Electronic Let. **35**, (1999)1284
[10] M. Eickhoff, H. Möller, M.Rapp, G. Krötz, Thin Solid Films **345**, (1999) p197

MICROMACHINED L.T.GAAS/ALGAAS MEMBRANES AS SUPPORT FOR 38 GHZ AND 77 GHZ FILTERS

G. DELIGEORGIS*, M. LAGADAS*, G. KONSTANTINIDIS*, N. KORNILIOS, A. MÜLLER**, S. IORDANESCU**, I. PETRINI**, D. VASILACHE**, P. BLONDY***

*FORTH IESL Heraklion, PO BOX 1527 Heraklion, Greece,
**IMT Bucharest, PO Box 38-160, Bucharest Romania,
***IRCOM Limoges, 123, Av. A. Thomas, Limoges Cedex, France
Contact author : G. Deligeorgis: deligeo@physics.uoc.gr

This paper presents the manufacturing of micromachined filters for 38 and 77 GHz having as support a 2.2 µm thin GaAs/AlGaAs membrane. The membranes were manufactured using selective dry etching techniques with AlGaAs as the etch-stop layer. On wafer measurements of the filter structures were performed. Losses less than 1.5 dB at 38 GHz and less than 2 dB at 77 GHz have been obtained.

1 INTRODUCTION

An alternative solution to the limitation of millimeter wave circuit performances due to the substrate (high dielectric and radiative losses, dispersion effects), consists in using micromachining techniques. Antennas, filters and transmission lines can be integrated on thin dielectric membranes [1-3]. The GaAs semiconductor membrane as support for microwave circuits represents an interesting solution due to the possibility for integration of passive elements with active elements manufactured on the same substrate [4]. Low Temperature Molecular Beam Epitaxy (LT.MBE) growth of III-V materials was used to fabricate the GaAs/AlGaAs heterostructure. The AlGaAs layer was used as the etch-stop [5] In this work, thin GaAs membranes were fabricated utilizing conventional contact lithography to define the filter pattern while backside alignment and reactive Ion Etching was used to fabricate the thin membrane.

2 FILTER DESIGN

Two band pass filters were designed with central frequencies at 38 and 77GHz respectively. The first one consisted of a four cascaded/opposited, double folded co-planar waveguide (CPW) open-end series stubs. In order to decrease the overall length of the filter, quarter wavelength double folded stubs were used. The second filter was based on a four cascaded, standard CPW, open-end series stubs. In the latter case there was no need to resort to folded stubs since the length of the filter is smaller compared to the first one. Electromagnetic wave propagation modeling was used in order to simulate the response of the filters, taking into account necessary corrections for the open-end CPW's. Other than the above, the simulation assumed ideal CPW's.

Furthermore, in order to improve the characteristics of the filters and to partially correct the effect of the abrupt change in the dielectric constant (from bulk material to membrane) quarter-wavelength matching stubs were used at both ends of the filters.

3 FABRICATION

The structure depicted in Fig.1 was grown in a VG80 MBE system on semi-insulating GaAs {100} oriented wafers. L.T. GaAs was chosen since it possesses high resistivity. This ensures minimal losses and dispersion effects of the RF signal and prevents enlarged cross talk between neighboring metallizations. Both are important in creating a filter exhibiting high efficiency and selectivity.

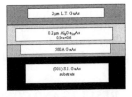

Fig1. Structure Grown with

As metallization, 700nm Au over 50nm Ti were e-beam deposited and patterned using lift off technique, thus creating the filters on the front side of the wafer. The wafer was subsequently thinned using CMP lapping technique down-to 150μm thickness. The membrane's dimensions were defined using backside contact photolithography. Samples were loaded facedown in a Vacutec 1350 RIE chamber for subsequent etching.

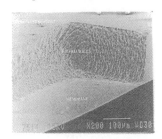
Fig2. SEM View of the etched membrane

Fig3. Detailed SEM view of the GaAs/AlGaAs membrane

The employment of a base pressure at 10^{-5}mbarr, gas flow of 10sccm CCl_2F_2 and process pressure of 75mTorr with an RF power of 75Watts, resulted in high selectivity between GaAs substrate and AlGaAs etch stop layer.(Al fraction more than 0.5) The etch rate ratio between GaAs and AlGaAs is more than 500, thus eliminating all non-uniformity created by the deep etching. As a result a smooth surface is created on the backside of the released membrane and the thickness is controlled to nm scale accuracy. Some roughness is created on the side walls of the etched pattern but this has no impact on the functional characteristics of the filter.

The etching process applied, produces etch rates in the range of 3μm/min and virtually vertical side wall profiles (Fig.2), making the fabrication of the membrane a relatively fast and straight forward step. The key towards the implementation of this technique for creating thin GaAs membranes is the pre-grown AlGaAs etch stop layer with an Al fraction of more than 0.5.

In order to detect the point, at which the etching process was completed, an optical end-point detection system was used. As soon as the AlGaAs layer was exposed, roughness on the bottom of the etched pattern was eliminated and high reflectance was observed. Given the extremely low etch rate of AlGaAs (in the order of 1nm/min), over-etching poses no threat to the membrane, since the etch stop layer can withstand more than 40min of etching. During this time, any residual GaAs is removed from the surface of the membranes. SEM inspection of the exposed AlGaAs surface validates the above (Fig.3).

4 RESULTS AND CONCLUSION

Microwave measurements were performed at IRCOM-Limoges, France using an on-wafer measuring set-up. As shown in (Fig.4) a minimum insertion loss of 1.46dB as well as a maximum return loss of 34.2dB at the central frequency (36.4GHz) were obtained for the first filter The bandwidth exhibited (3dB limit) was 15.8GHz. A fairly good agreement between simulation and measurements was achieved and the 4% shift in central frequency can be attributed to uncertainty in the value of effective permitivity of the membrane.

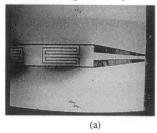 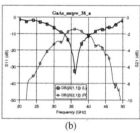

(a) (b)

Fig4. a) Top view Photo of a 38GHz filter and b) corresponding RF characteristics

For the 77GHz filter, a minimum insertion loss of 1.87dB at 72.4GHz and a maximum return loss of 17.4dB at 74.8GHz where obtained. The slight mismatch in the frequencies mentioned above is attributed to a misplacement of the edges of the membrane compared to the matching stubs. (Fig.5)

 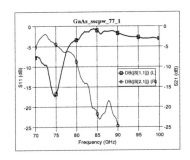

(a)   (b)

Fig5. a) Side view photo of a 77GHz filter and b) S parameter characteristics

To conclude, micromachined filters were fabricated on thin GaAs/AlGaAs membranes. The membranes were obtained using selective RIE etching, MBE structure and conventional photolithographic techniques. Admirably low losses were obtained.

The main advantage of the process presented, consists in the possibility for integrating active modules on GaAs with passive elements suspended on membranes, for microwave applications in the millimeter and sub-millimeter wavelength range.

5 ACKNOWLEDGEMENTS

This work has been supported by the Commission of the European Communities under the INCO-COPERNICUS PROJECT 977131 "MEMSWAVE"

References

1. S.V Robertson, L.B.Katehi, G.M.Rebeiz, IEEE Trans. Microwave Theory Tech., vol. **44**, no. 4, pp. 598-605, April 1996.
2. A. Müller, D. Dascalu, D. Neculoiu, S. Iordanescu, I. Petrini, V. Avramescu, D. Vasilache, R. Marcelli, Proc. 9th Micromechanics Europe Workshop MME'98, Ulvik in Hardanger, Norway, pp. 151-154, June 1998.
3. A. Dehe, H. Klingbeil, C. Weil, H.L. Hartnagel, IEEE Microwave Guided Wave Lett., vol. **8**, no. 5, p. 185, May 1998.
4. A.Müller, I.Petrini, V.Avramescu, S.Iordanescu, R.Marcelli, V.Fogllietti, M.Dragoman, Proc. SPIE, Micromachining and Microfabrication Process Technology III Conf., Austin, Texas, Sept., vol. **3223**, p. 208, 1997
5. K.Hjort, J. Micromech. Microeng., no. **6**, 1996, p. 370.

ALTERNATIVE SIGNAL EXTRACTION TECHNIQUE FOR MINIATURE FLUXGATES

P. D. DIMITROPOULOS, J. N. AVARITSIOTIS

Microelectronic Sensor Laboratory, Department of Electrical & Computer Engineering, National Technical University of Athens – 9 Heroon Polytechneiou Str., 15773 Athens, Greece – Email: abari@cs.ntua.gr

Despite their potential properties, fluxgate sensors present serious difficulties when miniaturizing is attempted, preventing them from being used in applications, like magnetic anomaly scanning where utilization of sensor arrays is required. In this work an alternative signal extraction technique is being analytically examined for its ability to be utilized in miniature fluxgates. The application of this technique makes sensor response less sensitive in core cross-section and in inferior characteristics of integrated planar coils. A fluxgate sensor with sensitivity as higher as four decades compared with conventional micro-systems is presented.

1 Introduction

Fluxgate sensors are considered to be the most suitable magnetic field sensors for applications requiring vector measurements of fields in the range between 10nT up to 1mT with overall accuracy as high as 100pTFluxgates cover the gap between the complicated SQUID magnetometers and the inexpensive integrated *Hall* effect and *Magneto-resistance* effect devices, which in general, exhibit lower accuracy. Alternative technologies, competitive to fluxgates, based on the *Anisotropic Magneto-resistance, Giant Magneto-resistance* and *Magneto-impedance* effects have been recently applied for the construction of precise magnetometers. Fluxgate systems prevail over these competitive technologies not only due to their higher sensitivity, but mainly due lower noise level and their robustness against *cross-field* effect [1].

Although currently available technology enables magnetic field measurements of satisfactory accuracy levels, applications requiring magnetometer arrays able to scan simultaneously the magnetic field on a surface are prevented due to the lack of ultra-sensitive integrated devices. One highly important application of this kind is the imaging of the magnetic dipole distribution on an area below the scanned surface. That enables, for example, the capturing of magnetic signature of ferromagnetic structures and alternative imaging techniques of current distributions in the human body which may provide important biomedical diagnosis tools.

Integrated fluxgate devices could boost such applications. Unfortunately, despite many, extensive manufacturing attempts [2], it has been impossible, up to now, to attain miniature fluxgate sensors with system sensitivities higher than 1000 Volt/T. As main degradation reasons one could mention, a) the extremely narrow cross-section of miniature sensor core, that is usually deposited by electroplating or sputtering of Permalloy compounds onto silicon chip surface, b) the inferior integrated coil features and c) the "noisy", non-repeatable magnetization process observed in such cores. A novel fluxgate micro-system is presented here that is based not on the conventional secondary harmonic signal extraction technique, but on a technique that makes its response significantly less sensitive on the above mentioned degrading parameters. The system provides generic linearity, very low noise level and sensitivity of approximately 40000 Volt/T, having a very simple signal conditioning circuitry.

2 Modeling the Fluxgate

Fluxgate sensors consist of two identical ferromagnetic cores that are driven deeply to saturation by a magnetic field waveform, with period T_o denoted with $H_e(t)$, which is applied to the cores with a phase difference of 180°. Excitation field is induced by a primary coil wound around the two cores inversely. Under the influence of the excitation waveform, the core is subjected to an oscillation according a hysteresis loop, whose shape represents the coil-core setup and depends on T_o, core material, core shape and orientation. Due to core saturation, two identical receiving coils wound around the cores provide two periodical pulse-series $V_1(t)$ and $V_2(t)$, that are added to form the sensor output signal $V(t)$, which is low-pass filtered up to its secondary harmonic that is up to frequency $2/T_o$. Let H_c denote the coercive force of the hysteresis loop and $2H_r$ denote the field intensity region within it the core jumps from negative to positive saturation and vice-versa. Ambient field, H, is added to excitation field, H_e, forcing the hysteresis loop to move horizontally against positive or negative values depending on the sign of applied field, H. This move forces the peaks of pulse-series $V_1(t)$ and $V_2(t)$ to move in time within period T_o, making the signals $V_1(t)$ and $V_2(t)$ asymmetric and thus giving rise to a sinusoidal signal, $V(t)$, with period $2/T_o$, whose amplitude carries the ambient field information. Ideally, when H=0 peaks of pulse-series $V_1(t)$ and $V_2(t)$, having inverse signs cancel each other and $V(t)$ equals to zero. It can be proven that time instances $T_{H,1}$, $T_{L,1}$ where positive and negative peaks of $V_1(t)$ and $T_{H,2}$, $T_{L,2}$ where positive and negative peaks of $V_2(t)$ occur, are as given in equations (1) and (3). Corresponding peak widths are given in equation (2). The output signal of the conventional fluxgate excited, as usually done, by a pulse or a triangular waveform, $H_e(t)$, is calculated in equation (4), where T_R is the pulse raise time, that becomes $T_o/2$ in case of triangular excitation, A is the core cross-section, B the core saturation magnetization and N the amount of secondary coil windings.

$$T_{H,1} = \tfrac{1}{2} \cdot (T_{a,1} + T_{b,1}) \text{ and } T_{L,1} = \tfrac{1}{2} \cdot (T_{c,1} + T_{d,1}) \quad (1)$$

$$T_{H,2} = \tfrac{1}{2} \cdot (T_{a,2} + T_{b,2}) \text{ and } T_{L,2} = \tfrac{1}{2} (T_{c,2} + T_{d,2})$$

$$W_{H,1} = T_{b,1} - T_{a,1} \text{ and } W_{L,1} = T_{d,1} - T_{c,1} \quad (2)$$

$$W_{H,2} = T_{b,2} - T_{a,2} \text{ and } W_{L,2} = T_{d,2} - T_{c,2}$$

$$T_{a,1} = H_e^{-1}(-H + H_c - H_r) \text{ and } T_{a,2} = (-H_e)^{-1}(-H + H_c - H_r)$$
$$T_{b,1} = H_e^{-1}(-H + H_c + H_r) \text{ and } T_{b,2} = (-H_e)^{-1}(-H + H_c + H_r) \quad (3)$$
$$T_{c,1} = H_e^{-1}(-H - H_c + H_r) \text{ and } T_{c,2} = (-H_e)^{-1}(-H - H_c + H_r)$$
$$T_{d,1} = H_e^{-1}(-H - H_c - H_r) \text{ and } T_{d,2} = (-H_e)^{-1}(-H - H_c - H_r)$$
$$V(H) = \frac{16 \cdot T_o}{A \cdot N \cdot B} \cdot \mathrm{sinc}(2 \cdot \frac{T_R}{T_o} \cdot \frac{H_r}{H_o}) \cdot \sin(\pi \cdot 2 \cdot \frac{T_R}{T_o} \cdot \frac{H_r}{H_o}) \quad (4)$$

A general feature of the conventional fluxgate is that whatever the hysteresis loop and excitation waveform shapes are, the absolute maximum sensor sensitivity, before amplification, cannot exceed the value of $16T_o/ANB$ that is excellent in macro-systems, but can become rather unsatisfactory when A takes values as low as $10^{-9} m^2$ in actual micro-systems, given that excitation frequency cannot increase by more than few hundreds of kHz due to the damping of core magnetization rotation.

3   Miniature fluxgate based on an alternative signal extraction technique

A prototype fluxgate sensor has been developed, consisting of only one pair of excitation and receiving coils. In this case pulses provided by the secondary coil are used by a sampling circuitry as sampling moments of the excitation current, $I_e(t)$ flowing through the primary coil. It is $I_e = k_{cs} H_e$, where k_{cs} is a proportionality factor. The current is sampled over a resistor R_{cc}. The voltage, $I_e R_{cc}$, samples are added and proven to carry ambient field intensity i0nformation. The sensor output is calculated theoretically in equation (5). A simplified functional schematic of the sensor os presented in fig. (1).

$$V = R_c \cdot k_{cs} \cdot \left\{ H_e(T_H + \frac{W_H}{2}) + H_e(T_L + \frac{W_L}{2}) \right\} = R_{cc} \cdot k_{cs} \cdot \left\{ H_e(T_b) + H_e(T_d) \right\} = -2 \cdot R_{cc} \cdot k_{cs} \cdot H \quad (5)$$

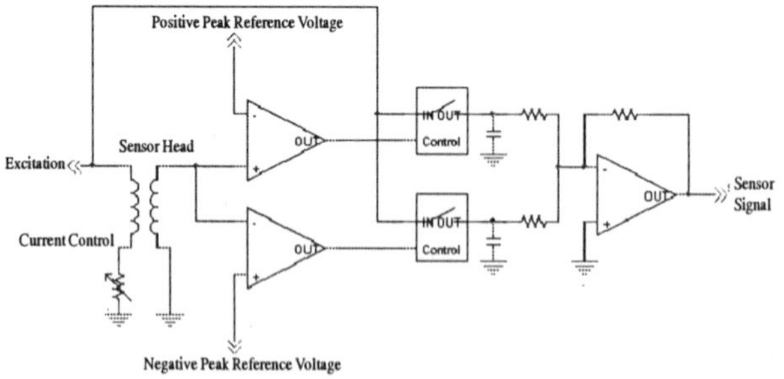

Figure 1: Simplified functional schematic of the sensor.

Each coil is formed by combination of one pair of adjacent rectangular planar coils, manufactured by standard *Printed Circuit Board* techniques. Pairs are positioned vertically, forming in that way a two-layer sensor head in the middle of which the core is positioned. Two additional layers are being used for interconnection between planar coil pins. The planar coils of each pair are connected in series. The planar coils of the excitation and receiving pair have correspondingly 25 and 10 windings each. The winding pitch is 350 microns. As ferromagnetic core a standard $Fe_{78}Si_7B_{15}$ amorphous wire is used that is kindly provided by the *National Institute of Research and Development in Technical Physics, Iasi Romania*. The core is of 125 microns in diameter and 25 mm in length. Scaling down is directly possible, without degradation. Planar coil pairs can be formed by sputtering, having a pitch of approximately 50 microns, according standard CMOS techniques. Core may be formed by standard electroplating of *Permalloy 80* compound.

Sensor sensitivity is measured to be approximately 40kVolt/T before amplification. Maximum field sampling rate is set by the output signal filter to be 10Hz. Measurement accuracy has been statistically derived to be 25nT and linear working Span is verified to exceed up to approximately 200 uT. Calibration data is presented in Fig. (2).

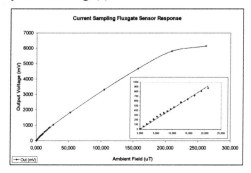

Figure 2: Calibration data for the current sampling Fluxgate sensor.

4 Discussion

An analytical model of the conventional fluxgate sensor proves its generic disability of forming ultra sensitive micro-systems. An alternative fluxgate signal extraction technique is examined theoretically and practically. The sensor signal provided by this technique seems to be rather robust against integrated coil winding amount, against core Barkhausen noise and hysteresis loop shape. Sensor sensitivity is set simply by a resistor. Long term stability of sensor sensitivity and zero offset are assured due to the generic independence of the output signal on the parameters of the ferromagnetic core, that are rather sensitive in temperature changes. The absolute response linearity and the very simple signal conditioning circuitry required are essential for micro-fabricated devices.

References

1. P. Ripka, Review of fluxgate sensors, Sensors and Actuators **A33** (1992) 129-141.
2. T. M. Liakopoulos, Chong H. Ahn, A micro-fluxgate magnetic sensor using micromachined planar solenoid coils, Sensors and Actuators **A77** (1999) 66-72.

INTEGRATED GAS FLOW SENSOR FABRICATED BY POROUS SILICON TECHNOLOGY

G. KALTSAS AND A. G. NASSIOPOULOU

Institute of Microelectronics, NCSR "Demokritos",
P.O. Box 60228, 15310 Aghia Paraskevi Attikis, Athens, GREECE
E-mail: G.Kaltsas@imel.demokritos.gr

An integrated thermal CMOS compatible gas flow sensor was designed, fabricated and tested. The thermal isolation is achieved through a thick porous silicon layer. Different sensor geometries and material combinations were tested with and without flow. The responsivity dependence of the different structural layers and the geometrical parameters also studied. Two flow velocity ranges with different sensor behavior were studied with sensitivity per heating power equal to 6.0 mV/(m/sec)W and 12.39 mV/(m/sec)$^{1/2}$W respectively for low and high flows. The sensor shows very rapid response and the time constant is below 1.5 msec.

1 Introduction

Flow sensors gain an increasing demand from both industry and the market. This work describes design, fabrication and characterization of an integrated thermal flow sensor. The sensor fabrication is based on porous silicon technology. The obtained parameters are discussed in detail.

2 Design and Fabrication

Figure 1. Image of the flow sensor

The sensor principal of operation is based on the differential temperature measurement on both sides of a heated resistor, under flow. The heater is a polysilicon resistance and the temperature sensing elements are realized with Al/p-polysilicon thermopiles. The thermal isolation between the active elements and the silicon substrate is critical for all thermal sensors. In the present sensor the isolation is realized with a thick porous silicon layer, which shows important advantages compared to other standard techniques (see reference [1]). A top view of the sensor is illustrated in fig. 1, where the different parts of the sensor are shown: heater, thermocouples, wire bonding contacts and porous silicon isolation area. A more detailed description of the flow sensor design and fabrication was published elsewhere [1].

3 Results and Discussion

3.1 Sensor characterization under zero flow conditions (static)

All the main parts of the sensor were studied extensively. The heating resistance was found to increase linearly with temperature and with input power. The corresponding slope was found to be 2.24 $\Omega/^{\circ}C$ and 3.16 Ω/mW, respectively. The above values lead to a heater temperature increase rate with power of 1.41 $^{\circ}C/mW$.

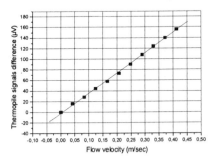

Figure 2. Sensor response as a function of flow velocity.

The effectiveness of porous silicon thermal isolation was examined by using a second polysilicon resistor on bulk silicon with the same characteristics as the heater on porous silicon. The rate of temperature increase with power was in that case equal to 0.16 Ω/mW, that is 20 times lower than that of the resistor on porous silicon. By thermal simulation of the sensor using the Ansys® simulation package, the thermal conductivity of porous silicon was derived. The obtained value was 1.2 $W/m/^{\circ}K$, which is 120 times lower than that of bulk silicon. This value is very close to others reported in literature [2]. It approaches the thermal conductivity of SiO_2 and it assures an efficient thermal isolation of the device.

Different geometries and layer structures were tested in order to find the optimum sensor configuration. The thermopiles were found to respond linearly with the input power of the heater. The slope of this line corresponds to the responsivity of the sensor. For a sensor consisted of 23 thermocouples per thermopile, with electrical isolation assured by 0.5 µm TEOS and with 20 µm distance between the heater and the hot thermopile contact (D_{th}), the responsivity was 1.8 V/W. From the above analysis the Seebeck coefficient of the Al/p-polysilicon thermocouples can be calculated. The extracted value was 48 µV/C.

The time response of the sensor was studied extensively for different electrical isolation layers and various distances D_{th}. The time constant was found to be 1.5 ms, while the rise (fall) time and the rise (fall) delay were determined as 1 ms and 220 µs respectively. These values were typical for devices with 0.5 µm TEOS electrical isolation and with a distance from thermocouples to heater of 20 µm.

3.2 Sensor characterization under flow conditions (dynamic)

The experimental set-up and the measurement settings were described in details elsewhere [1]. In all cases the measurements were performed under pure nitrogen flow. A constant current of 5 mA was supplied to the heater, which corresponds to 67 mW input power. The flow was applied in a direction perpendicular to the heater and parallel to the sensor. The sensor response was monitored through a specially designed data acquisition system. All measurements were carried out under laminar flow conditions. Different sensor configurations have been tested for low and high flow velocities (0 – 4.0 m/sec).

3.2.1   Results at low flow velocities (0 – 0.4 m/sec)

As it was predicted from theory the fluid flow causes the upstream thermopile response to decrease and the downstream thermopile response to increase. The difference of these two responses was taken as sensor signal in order to compensate any fluctuation in the heater temperature. A typical sensor response as a function of flow velocity is illustrated in figure 2. The specific device consists of 12 thermocouples per thermopile, 0.5 µm TEOS oxide electrical isolation and D_{th} = 40 µm. It is clear that the sensor responses linearly with the flow velocity with a slope of 0.4 mV/(m/sec). This value represents the sensitivity (S) of the sensor. The corresponding sensitivity per heating power can be expressed as:

$$S_n = \frac{S}{P_{in}} = 6.0 \quad mV/((m/sec)W)$$

Table 1. Sensor sensitivities

Electrical isolation		0.5 µm TEOS			0.2 µm Si$_3$N$_4$ / 0.2 µm TEOS		
Distance D_{th} (µm)		20	40	60	20	40	60
	Velocity range						
Sensitivity /	(0 – 0.4 m/sec)	6.7	33	8.7	4	10	7.8
Thermocouple ($\times 10^3$)	(0 – 4.0 m/sec)	47	175	55	28	90	46.2

Table 1 summarizes the dynamic characterization results for different device parameters. Clearly the TEOS electrical isolation of 0.5 µm thickness, leads to improved sensitivities for all values of the distance D_{th}. The best performance in both cases of electrical isolations is achieved for D_{th} = 40 µm.

3.2.2   Results at high flow velocities (0 – 4.0 m/sec)

The sensor response to higher flow velocities is presented in figure 3a. The response in this case is not linear with flow. The theory predicts [3] that the signal should be analogue to the square root of the flow velocity. Figure 3b illustrates the sensor response as a function of the square root of the flow velocity. Obviously the relationship is now linear for flows larger than 1 m/sec. From the slope of the linear region the sensitivity for this flow range and the corresponding sensitivity per heating power can be calculated. The extracted values are:

$$S = 0.83 \quad mV/\sqrt{m/sec} \qquad S_n = 12.39 \quad mV/(\sqrt{m/sec}\ W)$$

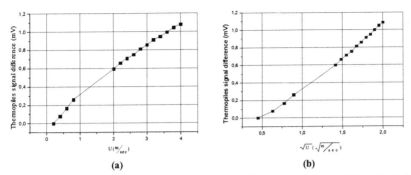

Figure 3. Sensor response, as a function of the flow velocity (a) and the square root of flow velocity (b).

The above results were obtained from a device consisted of 15 thermocouples per thermopile, electrical isolation 0.5 μm TEOS and D_{th} = 60 μm. Different device parameters were also studied in this case and the characterization results are summarized in Table 1. The general behavior is the same as in the previous flow range. The best performances are obtained by TEOS electrical isolation, 0.5 μm thick, while the optimum distance between the hot thermopile contact and the heater is D_{th} = 40 μm in both cases of electrical isolation.

4 Conclusion

An integrated CMOS compatible gas flow sensor based on porous silicon thermal isolation was fabricated and tested. Extensive characterization of the sensor was performed under static and dynamic conditions. Different types of isolation layers and geometries were tested in order to find the optimum sensor configuration. Two flow velocity ranges with different sensor behavior were studied, corresponding to low and high flow velocities. The sensor showed superior characteristics compared to other existing sensors and especially it was among the fastest existing devices. It is so very attractive for many contemporary applications.

References

1. Kaltsas G., Nassiopoulou A. G., Novel C-MOS compatible monolithic silicon gas flow sensor with porous silicon thermal isolation. *Sensors and Actuators A*, **76** (1999) pp. 133-138.
2. Lang W., Thermal conductivity of porous silicon. In L. Canham (ed.) *Properties of Porous Silicon*. EMIS Datareviews Series, No 18, on INSPEC publication, UK (1997) pp. 138-141.
3. Incropera F. P., De Witt D. P., Fundamentals of heat and mass transfer, Wiley, Canada, (1990).

LINEAR ARRAYS OF POLY SI-GE UNCOOLED MICROBOLOMETERS WITH CMOS READOUT AS LONG WAVELENGTH INFRARED SENSORS.

SPYROS KAVADIAS, PIET DE MOOR, MARTIN GASTAL AND CHRIS VAN HOOF

IMEC, Kapeldreef 75, 3001 Leuven, Belgium

Email : kavadias@imec.be

An Infra-Red sensor based on a linear array of poly Si-Ge bolometers is presented. The bolometers are surface micromachined devices employing a suspended structure to achieve thermal insulation from the substrate. Linear arrays of 64 pixels have been connected to a CMOS readout chip by means of a Multi Chip Module. The readout chip provides a pulsed bias for the bolometers, amplification of the signal and multiplexes the output in an analog line. At room temperature an NETD of 300mK has been achieved, while the maximum readout speed is 1kHz.

1 Introduction

Thermal Infra Red (IR) sensors can operate at room temperature with relatively good noise performance. This makes them attractive devices in comparison with narrow bandgap photon detectors especially in applications where the required overhead of detector cooling cannot be tolerated. Applications where a linear imaging array is required are space based pushbroom earth sensing, spectral environmental monitoring and industrial process control.

In recent years there has been a wide interest in the development of surface micromachined bolometers [1]. This technology allows the fabrication of structures with good thermal isolation. In order to realize a temperature dependent resistor with high sensitivity, thin film VO_x is mostly used [2]. An alternative material is the polycrystalline SiGe that allows easier fabrication of the bolometers and is compatible with CMOS technology. This later, makes poly SiGe an attractive material for the implementation of monolithic systems. Additionally, the support structure and the resistive element are made with the same material.

Together with the bolometer array, a sensor system must incorporate units for the biasing of the resistor elements and to some extent, processing of the signal delivered by each element. This processing must at least provide a multiplexing to a common output line. Various readout schemes have been proposed in the literature. These include the integration of the current flowing through the bolometer [1,3], the modulation of the phase of an RC-oscillator due to the resistance change [4] and chopper stabilization techniques for the reduction of the 1/f noise [5,6].

In this work a sensor based on poly SiGe bolometers fabricated using surface micromachining technology will be presented. The sensor is composed by two chips interconnected by means of a multi chip module, therefore a hybrid approach is realized. The first chip contains a linear array of 64 bolometers, each having dimensions of 50μm × 50μm. The second chip contains the readout circuitry. It is fabricated using a CMOS technology with a minimum feature size of 0.7μm.

The next paragraphs describe the sensor and the readout chips followed by the measurement results.

2 The sensor chip

A part of the linear array of microbolometers is shown in figure 1. A detailed description can be found in [7] together with fabrication details. A thin poly SiGe layer is grown on a sacrificial TEOS layer. The poly SiGe layer is dry etched to form the pixel and the supporting beams. At the end of the process the sacrificial TEOS layer is etched away in vapor HF environment resulting in a suspended device. An absorber with high absorption efficiency in the 8-12μm wavelength region can be obtained using a quarter wavelength absorber on top of the SiGe or a thin film absorber. To produce devices with response time shorter than 30ms, as required in most imaging applications, the heat capacity of this absorber limits the thermal conductivity of the weak link and hence, the sensitivity of the bolometer. Therefore, a thin film absorber with an efficiency up to 70% and a negligible heat capacity has been used [7].

By using deposition conditions that minimize the stress, fabrication of long (50μm) and thin (0.6μm) legs that support the very thin released bolometer (0.2μm) 2μm above the substrate is feasible. With this structure a low thermal conductivity below 10^{-7} W/K is obtained at 10^{-3} mbars pressure. In addition, an optimal doping resulted in the Temperature Coefficient of Resistance (TCR) of 1% together with acceptable noise levels.

Figure 1. SEM photograph of a part of the linear array of 64 microbolometers. Each has dimensions of 50μm × 50μm.

3 The readout chip.

The readout chip can be used with sensors having up to 144 pixels. The addressing of the pixels is performed by applying 8 bits to the address decoder.

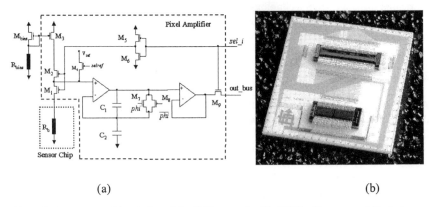

(a) (b)

Figure 2. a) Schematic of the pixel amplifier .b) Photograph of the MCM with a sensor of 64 pixels.

The schematic of the pixel amplifier is shown in figure 2a. It is designed on a pitch of 50µm in order to match the pixel pitch, for future integration with the bolometers in a monolithic sensor. The readout operation proceeds as follows. Before the readout of a line a calibration phase proceeds where no pixel is selected. Signal *phi* is pulsed high and *selref* is pulsed low. Therefore, the voltage V_{ref} is sampled across the capacitor C_2. Then, the readout phase takes place. When a certain pixel is selected, the bolometer is biased with a current determined from the current source consisting of M_3, M_{bias} and R_{bias}. Now, *phi* is pulsed low setting the gain of the amplifier to $(C_1+C_2)/C_1=10$. Therefore, the pixel output is equal to the difference of the voltage across the bolometer and V_{ref}, multiplied by the gain. The power consumption per pixel is 1mW.

4 Measurements and discussion

Table 1 summarizes the most important characteristics of the sensor. The dominant noise source is the 1/f generated at the bolometer therefore, the minimum resolvable resistance variation and consequently, the Noise Equivalent Temperature Difference (NETD) are mainly determined by this noise source.

The advantage of this readout scheme is the high dynamic range. Depending on the actual bolometer resistance the bias current can be adjusted in order to set the operating point within the input range of the pixel amplifiers (0.5-4 V). Since a continuous measurement of the pixel signal is provided, self-heating effects can be fully studied. Additionally, this allows the implementation of signal processing methods for noise reduction such as chopper stabilization in order to reduce the 1/f noise emerging from the bolometer.

Table 1. Characteristics of the sensor.

Pixel Dimensions	50µm × 50 µm
Pixel Array	1 × 64
Wavelengths	8-12µm
Maximum Line Scan Frequency	1kHz
Bolometer thermal time constant	≈30ms
Bolometer Electric Time constant	3µs
TCR	1 %
Minimum Resolvable Resistance Variation	0.9 Ohms
NETD	$F^2 \times 0.3$ K (F: f-number of optics)

References

1. P. W. Kruse and D. D. Skatrud, Uncooled Infrared Imaging Arrays and Systems, Semiconductors and Semimetals, **47** (1997), Academic Press.
2. B. E. Cole, R. E. Higashi and R. A. Wood, Monolithic Two-Dimensional Arrays of Micromachined Microstructures for Infrared Applications, *Proceedings of the IEEE*, **86** (1998), pp.1679-1686.
3. A. Tanaka, S. Matsumoto, N. Tsukamoto, S. Itoh et. al., Infrared Focal Plane Array Incorporating Silicon IC Process Compatible Bolometer, *IEEE Trans. on Electron Devices*, **43** (1996), pp.1844-1850.
4. U. Ringh, C. Jansson, C. Svensson and K. Liddiard, CMOS RC-Oscillator Technique for Digital Readout from an IR Bolometer Array, *The 8th International Conference on Solid-State Sensors and Actuators and eurosensors IX*, Stockholm, Sweden, June 25-29, 1995, pp.138-141.
5. U. Ringh, C. Jansson and K. Liddiard, Readout concept employing a novel on-chip 16 bit ADC for smart IR focal plane arrays, *Proc.SPIE*, **2745** (1996), pp.99-110.
6. C. Menolfi and Q. Huang, A Low-Noise CMOS Instrumentation Amplifier for Thermoelectric Infrared Detectors, *IEEE Journal of Solid State Circuits*, **32** (1997), pp.968-976.
7. P. De Moor, J. John, S. Sedky and C. Van Hoof, Linear arrays of fast uncooled poly SiGe microbolometers for IR detection, *Proc. SPIE*, **4028** (2000), pp.27-34.

SILICON CAPACITIVE PRESSURE SENSORS AND PRESSURE SWITCHES FABRICATED USING SILICON FUSION BONDING

S. KOLIOPOULOU, D. GOUSTOURIDIS, S. CHATZANDROULIS, D. TSOUKALAS

Institute of Microelectronics, NCSR 'Demokritos', 15310 Aghia Paraskevi, Greece
E-mail: stavros@imel.demokritos.gr

A simple single crystal silicon process for fabricating capacitive type pressure sensors and pressure switches is described. The process relies on silicon fusion bonding for the sealing of the pressure sensor cavity and device construction. The pressure sensor cavity is etched in a thick oxide, thus allowing freedom in cavity design. Pressure sensors and pressure switches have successfully been fabricated using this process. Results are presented of a capacitive type element operating in the medical pressure regime (0-300mmHg) with a sensitivity to pressure of 1.5 fF/mmHg or 305 ppm/mmHg, and of a pressure switch operating at 5 bar for use in truck tyres.

1 Introduction

Pressure sensing devices are of interest for a wide variety of applications in the fields of biomedical, automotive and industrial engineering. Micromachined sensors present inherent advantages over conventional devices because of their low cost, based on both batch fabrication techniques and their higher performance [1]. In this work a pressure sensing element that may also function as a pressure switch with a slight process variation, has been fabricated using the silicon fusion bonding (SFB) technique [2]. The device consists of a cavity etched in a thick wet oxide, a fixed electrode and a flexible electrode (fig. 1). When pressure is applied the flexible electrode deflects towards the fixed electrode and the device capacitance changes. In capacitive type pressure sensors it is this change that is of interest. Rather in pressure switches the value of pressure at which the flexible electrode will touch the fixed electrode is the important parameter.

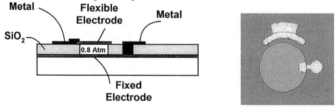

Figure 1. (a) Schematic view of a pressure sensor. (b) Photograph of a circular sensor

2 Fabrication

The fabrication process, depicted in figure 2, begins with two silicon wafers which are to be fusion bonded. Wafer A is a 4-inch n-type (100) silicon wafer, from now on called the substrate wafer, and wafer B, a 4-inch silicon wafer with a stress compensated boron doped $Si_{1-x-y}Ge_xB_y$ epitaxial layer (2.4um thick) [3]. Wafer A will constitute the main body of the final device, while wafer B is all etched except the highly boron doped epitaxial layer. The fabrication process begins with wafer A undergoing a phosphorus doping, in order to form what will eventually become the fixed electrode. This step is followed by a wet oxidation to form a 1um thick oxide. The oxide is patterned to form the pressure sensor cavity and the necessary openings for the fixed electrode contacts. Because the cavity of the sensor is formed in the thick oxide it can take any arbitrary shape, but only sensors with circular and rectangular form were made. After oxide etch the two wafers are bonded and annealed at 1000°C for 1 hour. The bonded structure is then thinned mechanically, from the epitaxial $Si_{1-x-y}Ge_xB_y$ wafer side, to leave a film 80 um thick on the substrate wafer. The rest of this wafer is then etched in an EDP solution so as to leave only the heavily doped epitaxial layer of 2.4 um thickness. The obtained roughness of the remaining silicon membrane, after the chemical etching, is below 500Å.

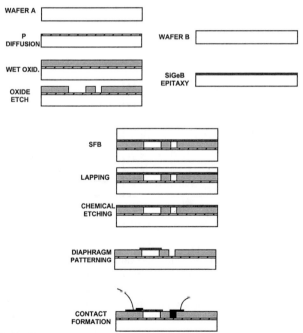

Figure 2. Fabrication Process.

Next the epitaxial layer is patterned over the oxide cavities, in SF_6 plasma, so as to leave a suspended diaphragm over a sealed cavity on the substrate wafer. Patterning also reveals the openings for the contacts to the fixed electrode. Aluminum deposition and patterning of the contacts to the diaphragm and the fixed electrode, is the final step in the fabrication process.

When a pressure sensor is to be fabricated a thin insulating film (eg. SiO_2) is grown prior to bonding to prevent shortening of the two electrodes when they come in contact. On the other hand a pressure switch may be fabricated with the same process. In this case no SiO_2 insulating film is grown. In addition attention should be taken to remove all remaining SiO_2 from the cavity bottom and the flexible electrode prior to bonding the two wafers.

3   Results

Testing of the pressure sensing devices was performed in a custom pressure control chamber using a 4192 HP bridge. In fig. 3(a) the capacitance to pressure (C-P) response of a circular 395 µm in diameter pressure sensor is shown, while in fig. 3(b) that of a rectangular sensor with a side of 375 µm. The devices display a temperature dependence due to the expansion of the trapped gas, as the sealing is performed in ambient. A temperature coefficient of offset (TCO) of a 900ppm/°C for the circular sensor and 513 ppm/°C for the rectangular of about the same size is measured. This temperature dependence may be eliminated by sealing the device in vacuum.

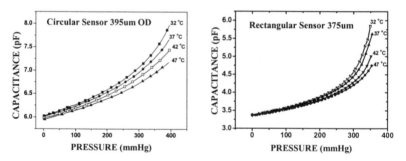

Figure 3. C-P response of a circular (a) and a rectangular (b)

In the case of pressure switches the value of pressure at which the flexible electrode will touch the fixed one is of interest. At this point a current flows between the two electrodes if a bias voltage is applied across the switch. In fig. 4(a) the current is plotted against pressure for three different switches with diameters of 110, 120 and 150 µm, while in fig. 4(b) the resistance of a 120 µm OD vs pressure is plotted. The devices operate in the 3 to 8 bar pressure range and thus are suitable for use in automotive applications. In one such application a pressure switch is used to ensure that the pressure in truck tyres remains above 5 bar at which point it switches and the accompanying electronic interface alerts the driver [4].

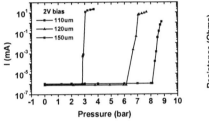 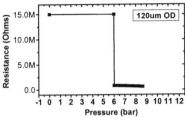

Figure 4. (a) The current flowing through the switch when a 2V bias is applied for three device sizes is shown. (b) The resistance of the switch vs pressure is depicted.

4 Conclusions

A process for the fabrication of single crystal silicon pressure sensors and pressure switches was presented. The process relies on the silicon fusion bonding of two silicon wafers to seal the sensor cavity and construct the device. Pressure sensors operating in the medical pressure regime (0-300 mm Hg) with sensitivity to pressure of 1.5 fF/mm Hg or 305 ppm/mm Hg were fabricated. These devices may be connected to a specially designed Capacitance to Frequency converter electronic circuit to implement a complete blood pressure measuring system [5]. Pressure switches (for use in truck tyres) were also fabricated and tested.

5 Acknowledgements

The authors would like to acknowledge the financial support of EU through the ESPRIT project MICROMEDES (no 8902), CASE project INCO-COPERNICUS 960136 and CRAFT project-BE-S2-5511.

References

1. J.Bryzek, K.Petersen, W.McCulley, IEEE SPECTRUM, May 1994, p.20
2. D. Tsoukalas, C. Tsamis and J. Stoemenos, Appl.Phys.Lett.**63** (1993) p. 3167
3. H-J.Herzog, L.Csepregi and H.Seidel, J.Electrochem. Soc., vol **131**, no 12, p. 2969,Dec 1984
4. CRAFT project-BE-S2-5511 Technical report
5. S.Chatzandroulis, D.Tsoukalas, ICECS' 99, Sep. 5-8, Pafos, Cyprus

LOW-POWER SILICON MICROHEATERS ON A THIN DIELECTRIC MEMBRANE WITH THICK-FILM SENSING LAYER FOR GAS SENSOR APPLICATIONS

V. GUARNIERI, S. BRIDA, B. MARGESIN, F. GIACOMOZZI, M. ZEN,
ITC-IRST, Via Sommarive 18,38050 Povo (TN), ITALY e-mail: guarni@itc.it

A.A. VASILIEV, A.V. PISLIAKOV
Institute of Molecular Physics, RRC Kurchatov Institute, 123182, Moscow, Russia

G. SONCINI, G. PIGNATEL
Dep. Of Materials Engineering, University of Trento, 38050, Mesiano TN, Italy

D. VINCENZI, M.A. BUTTURI, M. STEFANCICH, M.C. CAROTTA, G. MARTINELLI
Physics Dep. and INFM, University of Ferrara, via Paradiso 12, 44100 Ferrara, Italy

We report on the design, fabrication, and characterisation of a microheater module for chemoresistive, metal-oxide semiconductor gas sensors. The microheater consists of a dielectric stacked membrane with a polysilicon resistor heater element as well as a polysilicon temperature-sensing element. The geometry of both, the membrane and the heater have been optimised by means of finite element computer simulation in order to maximise heating efficiency. These devices complete of the sensing layer require only 30 mW to achieve a temperature of about 400 °C, while conventional thick film sensors fabricated on alumina substrates require typically more than 500 mW to reach the same working temperature. The proposed micromachining technology allows low-power microheaters to be fabricated at low cost and compatible with mass production technology; furthermore silicon micromachining is potentially suitable for the integration of the sensing and the heating element as well as the required electronics into the same battery-operated portable microsystem. The calibration curve of the sensor prototypes for various gases will be presented together with some future development about the microheater structure.

1 Introduction

The thick film technology combined with alumina substrates as been successfully exploited for hybrid circuits since many years. Recently this technology has been applied to gas sensing since it allows the deposition of high specific-surface films with reliable low cost processes [1,2]. The devices fabricated with this technology have been reworked during last years; at the moment require up to 500 mW of input power, which is a high consumption for a battery-operated system. The exploitation of silicon micromachining technology has recently demonstrated that free-standing dielectric membranes can be successfully used to build up micro-hotplates with low power consumption thanks to the high thermal insulation they provide [3,4,5,6].

Nevertheless, no example of thick-film micromachined gas sensors has been presented up to now. This is a consequence of the hard matching between screen-printing technique and micromachining. In this paper we report on the fabrication of a microheating module designed to enable a working temperature of 400°C with only 30mW of power consumption upon which we deposited by numerically controlled screen printing a SnO_2 based sensing film [7]. The G/G0 vs. concentration curves of the devices are also reported for CO, CH_4 and NO_2. The future developments will allow the enhancement of the fabrication process for the metal contact and the deposition of different sensing layers such as sol-gel In_2O_3 and TiO_2.

2 Fabrication of the micro-hotplate.

Even if micromachined hotplates have been already realised since some years ago, their exploitation for thick-film gas sensors has seemed to be unlikely because of the stresses that take place during the film deposition and firing. To overcome this problem we designed a micro-membrane with multilayer structure characterised by a great resistance to mechanical and thermal stresses. The starting material was a 4-inch, 500-μm-thick, p-type, (100) oriented, double polished silicon wafer. Onto the substrate we grew by low-pressure chemical vapour deposition (LPCVD) a 500-nm-thick silicon dioxide layer. Then a further 300-nm-thick layer of silicon dioxide has been grown by thermal wet oxidation, followed by a 200-nm-thick stoichiometric Si_3N_4 film obtained by LPCVD from ammonia and dichlorosilane. The heating element and the temperature sensors have been realised by polysilicon and embedded into the membrane itself. Thus we grew a 450-nm-thick in-situ Boron doped polysilicon by LPCVD from silane. The polysilicon was subsequently patterned to shape the resistors for the heater and the temperature sensors. A further insulating multilayer (50-nm undoped oxide, 400-nm boron-phosphorous-doped silicate glass (BPSG), and 50-nm undoped oxide) was deposited. After a reflow at 925°C we have grown a Ti/TiN diffusion barrier by sputtering deposition upon which we evaporated Cr and Au for the ohmic contacts. Silicon was then anisotropically removed from the backside by tetramethyl ammonium hydroxide (TMAH) etching, leaving a 1-mm×1-mm thin diaphragm. A SEM micrograph of a sample is presented in Fig. 1. The Ti/TiN/Cr/Au metal for the pads and the sensing layer contacts showed some problems when we heated the samples even at moderate temperatures (about 250 °C); in fact the chromium layer, after an annealing at temperatures above 650 °C, was completely migrated towards the surface and thus the adhesion layer resulted weakened and in that conditions were not possible to bond the pads. Leaving the Ti/TiN metal as an adhesion layer on the contact pad we adopted platinum metals for the contacts of the sensing layer in order to avoid all the diffusion problems. The platinum layer has been patterned by a lift-off process, we spinned a 2.1-mm-thick layer of ma-N 1420 resist from Micro Resist Technology

(Germany); the resolution shape of the Pt metal after the lift off process was very good showing also a sufficient adhesion to the SiO_2 substrate.

3   Thick-film deposition

The thickness of the different dielectric layers are calculated in order to obtain a low tensile stress of the membrane to achieve a completely flat surface at working temperature. In Tab.1, the measured intrinsic stress of the different used materials is shown.

Table 1: Intrinsic stress measurements (- compressive; + tensile)

Material	Thermal Oxide	TEOS	TEOS + Oxidation	LTO	Nitride
Stress [MPa]	-274 ±11	36 ±17	-275 ± 1.7	153±7	837 ± 15

This is an important task in order to allow high-resolution printing and good adhesion of the paste to the substrate. Even the membranes are designed to resist to thermal and mechanical stresses that may take place during the screen-printing process, it is mandatory to apply negligible force on them during the film deposition. For this reason we used a stencil-type screen which, thanks to its extreme stiffness, distributes the squeege pressure over the whole wafer leading to an average pressure of 6.4 g/cm^2. After the film deposition the chuck is maintained under the wafer for 15 s and then it is removed at a rate of 15 mm/s. This expedient is useful to complete the adhesion of the paste to the substrate. The smallest sensing layer we deposited has been 250-μm×350-μm in size and about 40-μm thick.

We implemented an own ink obtained mixing Pd-catalysed powder grown by sol-gel with low-melting glass frit, α-terpineol and ethyl-cellulose. After the deposition the films are dried at 200 °C for 10 minutes and then fired at 650 °C for 1 hour by direct heating through the micro-hotplate itself.

The overall fabrication yield is greater than 80% thus the entire process can be easily scaled to a batch production.

4   Electrical characterisation

After the firing, the devices were wire-bonded into a plastic package and lodged in a sealed chamber (300 cm^3 in volume); the test gases were mixed with the carrier and flowed at a rate of 500 cm^3/min. The typical current flowing into the sensing film is 700÷800 nA at 5 V bias, thus the contribution to the heating of the substrate is largely negligible and the response to gas was recorded by calculating the ratio between the current flowing into the film when test gas is fed and the current flowing when the sensor is exposed to technical air. From the response of the sensor towards several concentrations of CO, CH_4 and NO_2 we extrapolated the calibration curves (Fig 2, 3, and 4 respectively). We performed several long-term stability tests of the microheating structure, and we found that the heater undergoes to a shift in resistance that is less than 3% after 1 month at working conditions. The shift tends to be lower and lower after the first week of operation and it is almost stable after 1 month. The temperature sensing elements, being operated at very low current, have a lower shift (within 1 %), thus the working temperature can be controlled very precisely even after long periods.

We didn't perform a long-term stability test of the sensing layer because it was already done for sensors deposited onto alumina substrates (which exploits the same sensing layer). The selectivity of the sensing film we exploited in this work (SnO_2 - Pd doped) is in general quite low (as can be seen from the response curve presented), but for environmental monitoring they can fulfil all the requirements needed to monitor pollutants as it is shown in literature.

The functional comparison between these microsensors and ceramic-based ones has been carried out in a previous work [8,9,10].

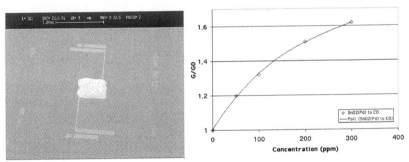

Figure 1. SEM micrograph of a 500-μm×500-μm sensing film. **Figure 2.** Calibration curve of the response toward CO.

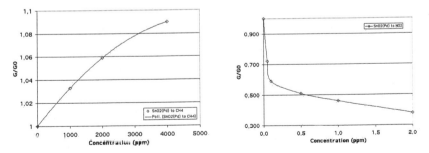

Figure 3. Calibration curve of the response toward CH_4. **Figure 4.** Calibration curve of the response toward NO_2.

5 Conclusions

A screen-printed thick-film gas sensor with micromachined substrate has been designed, fabricated and characterised. It combines the advantages of a micromachined hotplate (low power consumption and small thermal inertia) with the exploitation of a simple and reliable technique such as screen-printing. The overall power consumption is as low as 30 mW at a film temperature of about 400°C. The calibration curves for steady state mode are presented for CO, CH_4 and NO_2.

References

1. G. Martinelli and M.C. Carotta, Thick film gas sensors. *Sensors and Actuators* B **23** (1995) 157-161.
2. G. Martinelli and M.C. Carotta, A study of the conductance and capacitance of pure and Pd-doped SnO2 thick films. *Sensors and Actuators* B **18-19** (1994) 720-723.
3. D. Nicolas, E. Souteyrand, and J.R. Martin, Gas Sensor characterization through both contact potential difference and photopotential measurements. *Sensors and Actuators* B **44** (1997) 507-511.
4. D.D. Lee, W.Y. Chung, M.S. Choi, and J.M. Baek, Low-Power micro gas sensor. *Sensors and Actuators* B **33** (1996) 147-150.
5. A. Pike and J.W. Gardner, Thermal modelling and characterisation of micropower chemoresistive silicon sensors. *Sensors and Actuators* B **45** (1997) 19-26.
6. C. Rossi, E. Scheid, and D. Esteve, Realization and performance of thin SiO2/SiNx membrane for microheater applications. *Sensors and Actuators* A **63** (1997) 183-189
7. M.C. Carotta, C. Dallara, G. Martinelli, and L. Passari, CH_4 thick-film gas sensors: characterization method and theoretical explanation. *Sensors and Actuators* B **3** (1991) 191-196.
8. Carotta M.C., Martinelli G., Crema L., Gallana M., Merli M., Ghiotti G., Traversa E., Array of thick film sensors for environmental monitoring application. *Sensors and Actuators* B **68** (2000) 1-8
9. Martinelli G., Carotta M.C., Malagú C., Physics and technology of thick film sensors and their applications for environmental gas monitoring. *Electron Technology* 33 (1/2) (2000) 40-44
10. Vincenzi D., Butturi M.A., Guidi V., Carotta M.C., Martinelli G., Guarnieri V., Brida S., Margesin B., Giacomozzi F., Zen M., Giusti D., Soncini G., Vasiliev A.A., Pisliakov A.V., Gas-sensing device implemented on a micromachined membrane: A combination of thick-film and very large scale integrated technologies. *Journal of Vacuum Science and Technology* B **18** (2000) 2441-244

MICROSYSTEMS FOR ACOUSTICAL SIGNAL DETECTION APPLICATIONS

D. K. FRAGOULIS AND J.N. AVARITSIOTIS

National Technical University of Athens, Department of Electrical and Computer Engineering, 9 Heroon Polytechneiou St., 157 73 Zographou, Athens, Greece
E-mail: dfrag@central.ntua

The aim of the presented work is the design of an application specific microsystem for the detection of acoustical signals the frequency components of which vary according to specific periodic patterns. In this category belong signals usually produced by the siren of an emergency vehicle. By analyzing prerecorded data from a typical ambulance siren, it is illustrated how a filter bank of proper length can be used to obtain signal identification. Subsequently, an implementation of the filter-bank is proposed using a microsystem that consists of a number of beam-like elements. The whole analysis reveals how a simple, low-cost microsystem can replace an electronic analog or digital system for the implementation of a filter bank.

1 Introduction

Until today various ways have been proposed for the realization of a system that controls vehicle traffic lights. One idea is to place a special transmitter on each emergency vehicle in order to allow priority passage through intersections [1,2]. Another is to use traffic light controllers equipped with detectors capable of detecting flashing lights (usually special strobe lights) mounted on each emergency vehicle [3]. However, it seems that a better approach to the problem can be the detection of sounds produced by emergency vehicle sirens.

2 Description of the System

The proposed siren detection system comprises a microsystem implementing a filter bank and a simple logic circuit for counting the time between successive activation of neighboring filters. Essentially, the basic operation of the proposed system is to track the frequency characteristic curve of the siren, which represents the actual spectral content of the siren signal in variation with time. A typical characteristic curve of a common siren sound is illustrated in Fig. 1.

The recognition of the frequency curve of the siren can be achieved using a bank of band pass filters whose center frequencies are distributed in the spectral range from 400Hz to 1000Hz. As the instantaneous frequency of the siren signal changes, it passes successively from the pass bands of each filter of the bank, thus activating one filter at the time. A filter is considered as activated when the input sound level in its pass band exceeds a specific threshold. This threshold is related to the sensitivity of the detector in order to reject sounds below a corresponding intensity level. By counting the duration between the activation of two neighboring filters of the bank, it is easy to track the frequency characteristic curve of the siren, as shown in Fig. 1.

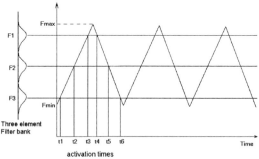

Figure 1. A three element Filter bank used for tracking a siren frequency characteristic curve.

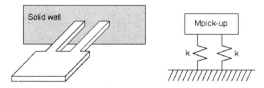

Figure 2. Actual structure of one element of the proposed microsystem and the equivalent system.

3   Structure of the Microsystem

The microsystem, ideally, consists of a number of beam-like elements that implement the band pass filters of the bank. For the current application, just 3 elements are sufficient to obtain a high detection performance. A basic design specification for each element is to comprise a small plate for collecting a sufficient amount of energy from the sound source [4,5]. Moreover, the whole element has to operate as a band pass filter i.e. to have a frequency characteristic curve with a dominant peak. A structure that accomplishes these restrictions is presented in Fig. 2. The two cantilever beams holding the plate, act as two parallel springs that have attached at their free end a pick-up. At each beam an equivalent spring constant can be assigned according to the following equations:

$$I = (b \cdot h^3)/12 \qquad (1)$$

$$k_{equiv} = (3 \cdot E \cdot I)/L^3 \qquad (2)$$

where L is the length, b is the width, h the thickness, E the young's modulus of elasticity and I the moment of inertia of each beam [6,7]. By choosing proper values of the above parameters, the elasticity of the beams can easily be adjusted. The resonant frequency of the element can be calculated by:

$$\omega = \sqrt{k_{equiv}/m_{pick-up}} \qquad (3)$$

where $m_{pick-up}$ is the mass of the plate. In Fig. 3, the resonance curves corresponding to four brass elements of different dimensions, which have been obtained in the laboratory using an electromagnetic transducer, is illustrated. The oscillation of the elements may also be observed with an electrostatic sensor. It is noticed, that the surface area of the plate affects the damping coefficient of the element, determining thus its behavior as a band pass filter.

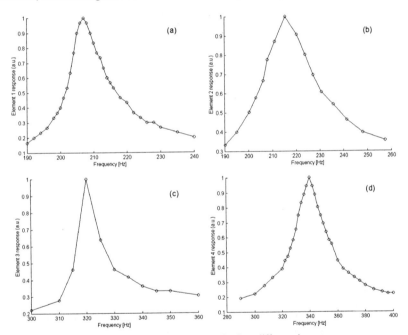

Figure 3. Resonance curves for three different elements.

4 Discussion

As analyzed above, the frequency of resonance of the filters can be determined by altering the dimensions of the supporting beams. However there are some limitations related to the weight and to the elasticity of the structure. Another point of examination concerns the damping, which clearly affects the width of the resonance curve. The value of damping can be adjusted by altering the surface area of the pick-up load, thus changing the resistance of the atmosphere at the movement of the plate. The results illustrated in Fig. 3a and 3b, have been obtained by testing two elements of different dimensions that appear almost the same frequency of resonance. A scale down procedure can be applied to the proposed structure in order to implement a microsystem [8]. The decrease in sensitivity caused due to the reduction of the plate's surface area can be compensated by increasing beam's elasticity as described in Eq. (1) and (2).

5 Conclusion

According to the above, it can be stated that the main advantage of the proposed approach is the ability to construct a reliable siren detector by reducing significantly the realization cost of such a system. This is accomplished by replacing a microprocessor or an analog system with the proposed microsystem. By implementing a filter-bank, the microsystem is able to perform the necessary signal processing for the specific task.

References

1. Mitchell W. L. Traffic light control for emergency vehicles (US4443783, 1984).
2. Obeck C. J. Traffic signal control for emergency vehicles (US5014052, 1991).
3. Markson R. J. and Govaert J. A. Optical warning system (US5057820, 1991).
4. Cremer L. and Heckl M. Structure-born sound (Springer Verlang, Berlin, 1988).
5. Morse P.M. Vibration and Sound (McGraw Hill, NY, 1948).
6. D.G. Fertis and G.E. Chidiac, Dynamically Equivalent Systems for Beams and Frames, Proc. of the 21st Midwestern Mechanics Conference, Michigan Technological University, Houghton, Michigan Vol. 15 (1989), pp. 399-401
7. D.G. Fertis, Vibration of Beams and Frames by using Dynamically Equivalent Systems, Proc. of the structural Dynamics and Vibration Symposium, Energy-Sources Technology Conference and Exhibition, ASME, New Orleans, LA, January 1994
8. R.T. Howe, Resonant Microsensors, Technical Digest, 4th Int. Conf. on Solid-State Sensors and Actuators (Transducers '87), Tokyo, Japan, pp. 843-848 June 1987

CAPILLARY FORMAT BIOANALYTICAL MICROSYSTEMS

K. MISIAKOS
[1] *Microelectronics Institute, NCSR "Demokritos", 15310 Athens, Greece*

C. MASTICHIADIS, S.E. KAKABAKOS
[2] *Immunoassay Lab., IR-RP, NCSR "Demokritos", 15310 Athens, Greece*

Here we describe a bioanalytical microsystem that is based on a capillary format immunosensor. The microsystem developed could perform either in heterogeneous (after washing of the immunoreactants) or in homogeneous detection mode. The advantages of the proposed microsystem are exemplified through the development of dual band heterogeneous immunosensor for the determination of two pesticides (pirimicarb and hexaconazole) Results obtained using the homogeneous version of the proposed bioanalytical microsystem are also provided. Short analysis time, simultaneous determination of different analytes as well as low consumption of sample and antibody are some of the advantages of the proposed immunosensing microsystem.

1 INTRODUCTION

Analytical techniques based on antibody-antigen reaction have found widespread application for the measurement of analytes in biological fluids, food, agricultural, and environmental samples. During last years, the need for rapid and low cost analysis and the trend for miniaturization of bioanalytical devices have promoted research in the field of immunosensing microsystems [1,2,4,5].

Here we describe a bioanalytical microsystem that is based on a capillary format immunosensor. According to the principle of the capillary immunosensor [6] several distinct bands are formed in the internal surface of a plastic capillary tube (1 mm internal diameter) by immobilizing appropriate antigens or antibodies. Then, the sample containing the analytes of interest is introduced into the capillary together with a mixture of fluorescein-labeled antigens or analyte-specific antibodies. After a short incubation time (5 to 15 min) the capillary is washed and the fluorescence bound onto the respective bands is measured. A laser light beam (485 nm) strikes the capillary perpendicularly and excites the fluors. Part of the photons emitted by the fluors is waveguided through the capillary walls and is quantified by a light sensor placed at the end of the capillary.

The feasibility of this device to determine simultaneously more than one analytes in the same sample is demonstrated through the development of a dual analyte capillary optical fluoroimmunosensor for the determination of pirimicarb (PMC) and hexaconazole (HEX). PMC is a carbamate insecticide that acts as a cholinesterase inhibitor, while HEX is a broad spectrum triazole fungicide that interferes with fungal sterole synthesis. The results obtained with this device are well correlated with those provided by single analyte fluoroimmunoassays developed in white opaque microtitration wells.

2 Methods

2.1 Materials

Poly-melthyl-pentene capillary tubes and white opaque microtitration wells were from Nunc A/S, (DK). Anti-PMC antiserum, anti-HEX antiserum, PMC and HEX γ-globulin conjugates were kindly provided by Zeneca Agrochemicals (UK). Fluorescein labeled anti-rabbit IgG were from Chemicon (USA). Bovine serum albumin (BSA), bovine γ-globulins, and casein were from Sigma, (USA).

2.2 Instrumentation

The nitrogen-dye laser system and the band pass filters at 480 nm (excitation) and 538 nm (emission) were from Oriel Instruments (USA). The instrumentation included a digital oscilloscope (LeCroy) for data acquisition and processing, a photomultiplier (Hamamatsu H5773-04) and a low noise transimpedance amplifier (Amptek 250). The excitation and detection set up is presented in Figure 1.

The measurement of fluorescence in microtitration plates was carried out at 538 nm after exciting the labeled molecules at 480 nm using the Ascent Fluoroscan plate reader (Labsystems OY, Finland).

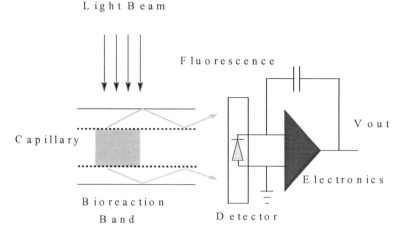

Figure 1. Experimental set-up showing the optical excitation, the capillary-fiber and the detection devices. The light striking the detector is the light that has been trapped and waveguided in the capillary walls. The excitation and detection filters are not shown.

2.3 Coating of the Capillaries

Distinct bands of biomolecules (3-mm long) are created into the capillaries by introducing 5 µL volumes of appropriate antibody or antigen-conjugate solutions in 0.05 M carbonate buffer, pH 9.2, with a microsyringe. Then, the capillaries are washed and blocked either with 1% casein or 1% BSA.

2.4 PMC – HEX panel assay

Mix two volumes of water sample or standard solution with one volume of anti-Hex/anti-PMC mixture prepared in 0.15 M PBS buffer, pH 7.0, containing 0.05% Tween 20. Fill the capillaries with the solution and incubate for 5 min at RT. Wash the capillaries with 4X2 mL of 0.1 M Tris-HCL buffer, pH 8.25, containing 0.05% Tween 20. Fill the capillaries with a 5 mg/L solution of anti-rabbit IgG-fluorescein conjugate in 0.15 M Tris-HCL buffer, pH 8.25, containing 1 g/L bovine serum albumin and 0.5 g/L bovine γ-globulins. Incubate the capillaries for 30 min at RT. Wash the capillaries with the above washing solution and measure the fluorescence bound to the solid by scanning the capillary.

2.5 Single analyte FIA for HEX and PMC

Single analyte FIA for the substances of interest were developed in white opaque microtitration wells following protocols similar to those published previously (3). The total assay time for HEX and PMC was 2.5 h.

3 Results and Discussion

Here we present a bioanalytical microsystem based on plastic capillary tubes that is appropriate for multi-analyte determinations for biological and environmental samples. The capillary serves as test tube for the immunochemical reactions and as optical biosensor through fiber and critical angle optics. This feature allows both heterogeneous (Figure 2A) and homogeneous-real time (Figure 3) detection modes. The capillary solid-support provides significantly higher surface to sample volume ratio, compared with other solid-supports (e.g. four times higher compared with the microtitration plates) used in immunoassays, thus providing the ability for rapid and high sensitivity analysis (Figure 2B). The cost for the determination is reduced due to the small amounts of biological reagents (antigens, antibodies, samples) required and to the multi-analyte feature of the proposed bioanalytical microsystem.

Real-time determination (Figure 3) could be achieved by adding appropriate amount of black ink into the immunoreaction solution. We found that a 20-fold decrease of the bulk fluorescence signal was achieved using the black ink whereas the specific signal was decreased by 40-50%. Thus, the ink serves as internal filter trapping mainly the bulk fluorescence photons. Following this approach the analysis time is decreased since the washing step after immunoreaction is omitted.

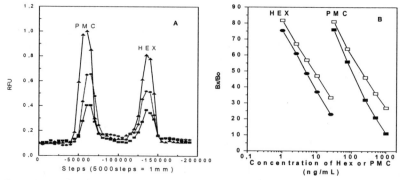

Figure 2. (A) Signal obtained by scanning three capillaries assayed with samples containing different PMC and HEX concentrations. (▲) zero standard; (●) 2.5 ng/mL Hex and 100 ng/mL PMC; (■) 10 ng/mL Hex and 250 ng/mL PMC. (B) Typical calibration curves for HEX and PMC assays obtained using the dual-band immunosensor (solid symbols) or microtitration plates (open symbols).

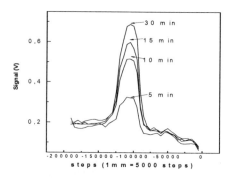

Figure 3: Time evolution of the signal obtained by scanning a one band rabbit IgG capillary immunosensor during reaction with fluorescein- labelled anti-rabbit IgG.

4 Acknowledgements

This work was supported by the EU, project number BRPR CT97 0393 (BOEMIS).

References

1. Ekins, R. and Chu, F.W., *Clin. Chem.* **37** (1991) pp. 1955-67.
2. Kakabakos, S.E., Christopoulos, T.K., Diamandis, E.P., *Clin. Chem.* **38** (1992) pp. 338-342.
3. Kakabakos, S.E., Georgiou, S., Petrou, P.S., Christofidis, I., *J. Immunol. Methods* **222** (1999) pp. 183-187.
4. Kricka, L.J., *Clin. Chem.* **44** (1998) pp.2008-14.
5. Marco, M.P. and Barceló, D., *Meas. Sci. Technol.* **7** (1996) pp. 1547-62.
6. Misiakos, K. and Kakabakos, S.E., *Biosensors & Bioelectronics* **13** (1998) pp. 825-830.

ULTRATHIN NANOPOROUS AND MICROPOROUS SIO$_2$ COATINGS FOR GAS/VAPOR SEPARATION AND SENSOR APPLICATIONS

K. BELTSIOS, N. KANELLOPOULOS
Inst. of Physical Chemistry, NCSR Demokritos, Agh. Paraskevi, 15310
Greece
E-mail:kgbelt@e-mail.demokritos.gr

E. SOTERAKOU, G. TSANGARIS
Dept. of Chemical Engineering, NTUA, Zografou Athens, 15780
Greece

Ultrathin nanoporous and microporous coatings are of particular interest as top layers of asymmetric porous structures for gas/vapor separations, microreactor and gas / vapor sensors. In this work we find that fabrication of such coatings is possible through the Langmuir-Blodgett deposition of sesquioxane polymers on nanoporous silica tubes and mesoporous and macroporous alumina discs and subsequent plasma. Pore size belongs to the 6-10 Å range and thickness to the 200 Å range.

1 Introduction

Ultrathin nanoporous (pore diameter 20-40 Å) and microporous (pore diameter 4-20 Å) coatings are of particular interest as top layers of asymmetric porous structures for gas/vapor separations, microreactors and also as top layers of gas or vapor sensors offering improved the signal-to-noise ratio. The deposition of *ultrathin* nano/microporous layers on *porous* substrates constitutes a double fabrication challenge. An earlier effort to develop such layers was based on the press-on application of a crosslinked silicone on a porous Vycor silica (pore diameter: 40 Å) substrate and subsequent oxidative plasma treatment [1]. This method led to pore sizes of 10 to 34 Å but it can only be applied on a silica substrate. Recently, it was demonstrated that top porous layers can become thinner, pore size smaller and the porous substrate need not be limited to silica, through a multiple Langmuir-Blodgett ordered deposition of a fatty acid salt and subsequent plasma oxidization of the deposited material [2]. In this work it is shown that the latter approach can be modified and pore size can be reduced further by the employment of a sesquioxane LB substance (Figure 1a). Following plasma oxidation this approach leads to a near defect-free microporous silica ultrathin layer. It must be noted that silicones also form LB films but they can not be repeatedly transferred on a substrate, as the deposited material behaves as a disordered fluid for depositions in excess of two or three monolayers.

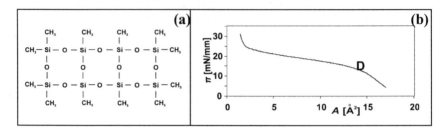

Figure 1 (a) Example of a methylsesquioxane oligomer, (b) LB isotherm (surface pressure (π) vs. surface area (A) per monomer) for sesquioxane at 20°C. Point D corresponds to deposition coordinates.

2 Experimental

Methylsilsesquioxane (obtained from UCT) with an estimated molecular weight in the range of 1,500 is dissolved in chloroform and spread in a LB apparatus (KSV-5000) with Millipore water at 20°C as a subphase. A LB isotherm is determined (Figure 1b) and deposition coordinates are selected [2,3]. Deposition substrates are nanoporous Vycor silica tubes and mesoporous (d = 200 Å) and macroporous (d=1000 or 2000 Å) Anopore alumina discs. 10 to 25 sesquioxane monolayers are deposited on all substrates. The obtained composite structures are subjected to radio frequency (RF) plasma oxidation at an oxygen pressure of 10 mTorr, a power of 300 W and a DC-bias of (-) 80-100 Volt. The reactor-sample arrangement employed allows for chemical treatment without excessive sputtering. Integral permeability measurements (gases: He, N_2, CO_2, CH_4 and CO_2) are differential permeability measurements (CO_2 in the pressure range of 0.5 to 60 bar) performed as discussed in detail elsewhere [1].

3 Results and discussion

Deposition on all substrates is completed successfully. In the case of Anopore substrates full monolayers are deposited after a few partial (ca. 80%) depositions. In the case of a Vycor substrate the monolayer transfer remains partial (ca. 70% -75%) for all transfer steps, but without a sign of decay. Previous experience with LB transfer on Vycor substrates suggests that partial deposition ca. 70% without decay in the degree of transfer does not undermine the integrity of the final oxide layer [2].

Integral permeability values for a composite membrane with a Vycor tube substrate (wall thickness = 11 mm, tube outer diameter = 70mm), thirteen deposited sesquioxane layers and a temperature of 100.0°C are as follows: 5.27×10^{-4} cm^2/s (He), 2.10×10^{-4} cm^2/s (N$_2$), 2.05×10^{-4} cm^2/s (CO$_2$), 2.82×10^{-4} cm^2/s (CH$_4$) and 2.74×10^{-4} cm^2/s (C$_4$H$_{10}$). It follows that permeabilities compared to those for Knusden flow (on the basis of the He value) are enhanced by a factor (N$_2$: 1.05, CH$_4$: 1.07, CO$_2$: 1.18, C$_4$H$_{10}$: 1.98) which increases with the critical temperature of the gas employed, in agreement with the tendency observed and discussed elsewhere [3], for the same gases and metal oxide LB/plasma coatings.

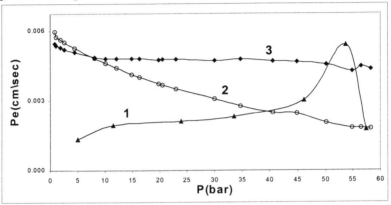

Figure 2 Permeability vs. pressure for differential permeability of CO$_2$ at T=35°C for 3 samples with a Vycor substrate and a different number (n) of LB layers. Sample 1: n=0, sample 2: n=13, sample 3: n=25. Permeability of 1 is divided by 5.

Differential permeability values for a Vycor tube substrate with 0 monolayers (non-coated, reference Vycor sample), 13 and 25 monolayers are presented in Figure 2. As it is clear from this figure, the developed ultrathin coating alters completely the response of the membrane. Non-modified Vycor shows a major peak at a pressure of 53-56 bar, indicative of the presence of d = 40 Å pores, while permeability drops steadily as pressures become smaller. On the other hand, the two modified membranes show traces only of the d = 40 Å peak, while permeability increases steadily at lower pressures to peak at a pressure in the range of 1 bar. Peaks at pressures less than ca. 15 bar are a sign of microporous (d < 20 Å) surface and, for the same material, the lower the peak pressure the smaller the pore size. Following comparison with the differential and integral permeability data for LB/plasma processed membranes with metal oxide coatings [2,3] our new coatings are judged to be microporous with a d = 6-10 Å and contain the smallest concentration of defects achieved so far through the LB/plasma route. On the basis of the number of layers and size of molecules deposited we conclude that a near defect-free SiO$_2$ coating is achieved at a thickness not exceeding 200 Å. Our new microporous coatings are superior to those made earlier through the press-on crosslinked silicone route [1], in terms of thickness, range of potential substrates, defect-concentration and pore size. In addition the new coatings present structural features similar to those achieved by LB/plasma deposition of metal oxides [2,3], with the exception of defect concentration which is smaller, as judged by the magnitude of the defect peak at a pressure in the range of 55 bar.

Ultrathin microporous SiO_2 coatings are highly desirable for nanotechnology applications involving the separation of a gas or vapor component from the air or a gaseous stream [3]. Potential applications include fabrication of separation membranes (aim of coating: removal of a gaseous or vapor component), components of microreactors for gas reactions (aim of coating: preferential removal of one gaseous or vapor product) and coatings of gas/vapor sensors (aim of coating: increase of the signal-to-noise ratio).

In conclusion we have demonstrated that sesquioxanes can be used for the fabrication of near defect free, ultrathin microporous SiO_2 coatings through the LB/plasma route. Pore size and coating thickness and perfection are very attractive for nanotechnology applications.

References

1. Beltsios K., Charalambopoulou G., Romanos G. and Kanellopoulos N., A Vycor membrane with reduced-size surface pores. I. Preparation and characterization, *J. of Porous Materials* **6** (1999) pp. 25-31.
2. Soterakou E., Beltsios K. and Kanellopoulos N., Asymmetric ceramic membranes from Langmuir-Blodgett deposition precursors. Deposition of fatty acid salts on porous ceramic substrates, *J. Eur. Cer. Soc.* **20** (2000) pp. 1105-1113.
3. Beltsios K., Soterakou E. and Kanellopoulos N., Composite Ceramic Membranes from Langmuir-Blodgett and self-assembly precursors. In *Recent advances in gas separation by microporous ceramic membranes*, ed. by N.K. Kanellopoulos, (Elsevier, Amsterdam 2000), pp.417-434

STRUCTURE CONTROL OF MICROPOROUS CARBON COATINGS FOR GAS/VAPOR SEPARATION AND SENSOR APPLICATIONS

K. BELTSIOS, G. PILATOS, F. KATSAROS AND N. KANELLOPOULOS
*Institute of Physical Chemistry, NCSR Demokritos, Aghia Paraskevi, 15310
Greece
e-mail: kgbelt@mail.demokritos.gr*

A. ANDREOPOULOS
*Department of Chemical Engineering, NTUA, Zografou Athens, 15780
Greece*

Resole resin precursors can be converted to continuous microporous carbon layers (pore diameter: 13 Å) of variable microporosity. A novel combination of X-ray powder spectra and pore size measurements suggests that the narrow pore size distribution is the result of formation of carbon nanodomains of two types having similar order and different size, with activation removing preferentially the smaller type. Finally, reference is made to ways of generating asymmetric carbon structures with microporous top layers and to potential applications of such structures.

1 Introduction

Controlled carbonization of resole and novolac resin precursors leads to carbons that can be activated to yield precisely tailored microporous structures with a pore diameter of 13 Å and a microporosity ε of 0.31 to 0.53 [1]. Similarly processed novolac and resole samples of the type studied in the present and earlier work [1] are known to yield identical structures, at least at the <100 Å scale. In terms of nanostructure evolution and control, it is of great interest to understand the origin of pores with a size surprisingly insensitive to the porosity level. According to an earlier interpretation, the carbon structure contains two types of nanodomains differing in the degree of order and activation removes the less ordered of them. A novel combination of porosimetry data and X-ray diffraction applied here supports a different structural picture that is apparently unique, so far, for nanostructured materials: Carbon before activation consists of two types of nanodomains with a similar degree of order and a dissimilar size and activation removes preferentially the smaller of the domains. Resole precursors can be solution processed and asymmetric structures based on microporous carbon can be fabricated in a number of ways and find nanotechnology applications involving the separation of a gas or a vapor.

2 Experimental

24 novolac and 3 resole samples are carbonized (14 samples), or carbonized and activated (13 samples) as discussed previously [1], but with the application of a broader range of preparation conditions. The maximum heat treatment temperature is either 800 or 900°C. Microporosity is determined by a combination of nitrogen porosimetry, mercury porosimetry and He density measurements for all samples. Subsequently, powder diffraction spectra are determined for all samples and instrumental background is subtracted on the basis of a graphite-powder spectrum. Surface morphology of resole samples is studied by atomic force microscopy (AFM).

3 Results and discussion

Porosimetry and He density data are combined for the evaluation of the (local) microporosity (ε), which is calculated as the void volume fraction of the structure, when all pores with a diameter larger than 20 Å are discounted (for the samples of interest ε includes only pores of a 12-13 Å diameter [1]). Non-activated and activated samples lead to ε values covering the 0.31 to 0.53 range. Sample thickness does not affect results. The surface of an activated resole sample appears in Figure 1a. The surface porosity is ca. 25% of the microporosity of the bulk material, suggesting that the openings seen are pore-mouths, unless surface pores have a distinct structure.

The corrected areas under powder diffraction spectra are split into the five portions shown in Figure 1b and the area ratio $A_1/2B_2$ is calculated. By an extension of earlier ideas [2], area A_1 is proportional to the more organized fraction of carbon and area $2B_2$ contains contributions of the less organized part that may also include scattering from the periphery of pores. We now proceed to associate $A_1/2B_2$ with ε for various structural models, assuming that the non-oxidizable (C_1) and oxidizable (C_2) parts of carbon occupy each 50% of a reference material without pores, as the stability of pore size breaks down beyond $\varepsilon = 0.53$ (~0.5). Model I [1] assumes that C_1 and C_2 are characterized by different order. Hence, A_1 is proportional to C_1 and $2B_2$ is proportional to C_2. It follows that:

$$2B_2/A_1 = 1 - 2\varepsilon \quad (1)$$

Model II (a variation of I) assumes that A is proportional to C_1 and $2B_2$ is a sum of a contribution proportional to C_2 and another proportional to ε. It follows that:

$$2B_2/A_1 = 2(\lambda-1)\varepsilon + 1 \quad (2)$$

Constant $(\lambda -1)$ is either positive or negative.

New model III assumes that C_1 and C_2 have the same order and both contribute to A_1, while $2B_2$ is proportional to ε. It follows that:

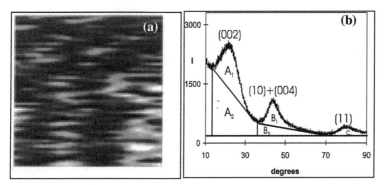

Figure 1. (a) AFM micrograph of the surface of an activated resole layer. Linear dimension of image 20 Å. (b) The splitting of a microporous carbon X-ray spectrum into five regimes (I: intensity in counts per second).

$$2B_2/A_1 = \lambda' \, \varepsilon / (1- \varepsilon) \quad (3)$$

Constant λ' is positive. A plot of data according to equation (1), shows that the latter is fully inapplicable ($2B_2/A_1$ increases instead of decreasing with increasing ε). Equation (2) may also be rejected, as an extrapolation of the linear plot of $2B_2/A_1$ vs. ε for $\varepsilon = 0$, yields $2B_2/A_1 < -2$, instead of a positive value of order 1. Data follow equation (3), though with substantial scatter. Upon splitting data into a group for activated samples and a group for non-activated samples, the data of the former group show an improved fit to (3), while a similar splitting of data for equations (1) and (2) does not remove pertinent objections cited earlier. We conclude that the assumption of an oxidizable and a non-oxidizable portion of carbon having the same order is in a much better agreement with the data, than the alternative assumption that C_1 and C_2 differ with respect to the order (Figure 2a, A vs. B) Hence, C_1 and C_2 may differ only with respect to size.

We now proceed to evaluate the relative sizes of C_1 and C_2 domains. The full width at half-maximum (FWHM) of a diffraction peak is inversely proportional to the domain size, with peaks (002) and (11) corresponding to the L_c and L_a domain sizes. Also, for $\varepsilon = 0.50$ domains C_2 are absent and the FWHM of the peaks corresponds to C_1 only, while C_1 and C_2 contribute equally to FWHM for the extrapolated $\varepsilon = 0$ value of a 1/FWHM vs. $\varepsilon /(1-\varepsilon)$ plot for activated samples. Finally we take that approximately:

$$1/FWHM_{C1} + 1/FWHM_{C2} = 2/FWHM_{C3} \quad (4),$$

where C_3 is a mixture of equal volumes of C_1 and C_2. Upon application of the above on 1/FWHM vs. $\varepsilon /(1-\varepsilon)$ plots (Figure 2b) we find that the two types of domains have a ratio of sizes of 1.9 for L_c (: dimension perpendicular to graphitic planes) and 2.9 for L_a. (: dimension parallel to graphitic planes). It follows that each large domain has a volume of approximately 16 times that of a small domain.

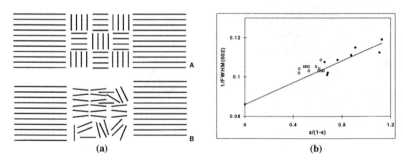

Figure 2. (a) Structural picture of two domains. A corresponds to model III and B corresponds to models I and II. (b) 1/FWHM vs. $\varepsilon/1-\varepsilon$ for (002) peak. Straight line is drawn on the basis of activated carbon data (circles). $FWHM_{C1}$ and $FWHM_{C3}$ are calculated from $\varepsilon=0.50$ and $\varepsilon=0.0$ respectively. Non-activated samples (squares) show a nearly constant FWHM value, possibly because of some structural correlation between adjacent large and small domains; this correlation is lost upon initiation of activation.

Microporous carbon layers from a solution processed resole precursor can become part of an asymmetric structure by deposition on a macroporous novolac-derived substrate [1], by using intense asymmetric activation, or by employing a mesoporous substrate generated through carbonization of a compact resorcinol-formaldehyde precursor. Potential applications of microporous carbon coatings include gas/vapor separation membranes (aim of coating: separation of one gas or vapor component), microreactor components (aim of coating: preferential removal of one gaseous or vapor reaction product) and top layers of gas-vapor sensors (aim of coating: increase of the signal-to-noise ratio).

In conclusion, we have found that the carbonization of novolac or resole materials leads to the generation of two types of nanodomains of similar order and a volume differing by a factor of ca. 16. Activation removes smaller domains and yields a microporous layer with precisely tailored pores having size insensitive to microporosity, for a microporosity level of 0.3 to 0.5. Such layers are of interest for nanotechnology applications involving separation of a gas or a vapor.

References

1. Steriotis T., Beltsios K., Mitropoulos A., Kanellopoulos N., Tennison S., Wiedenman A., Keiderling U., On the Structure of an Asymmetric Carbon Membrane from a Novolac Resin Precursor, *J. Applied Polymer Science*, **64** (1997), pp. 2323-2346.
2. Warren B.E. and Bodenstein P., The diffraction pattern of fine particle carbon blacks, *Acta Cryst.*, **18** (1965), pp.282-286.

2-D SIMULATION OF ON-CHIP BAW RESONATORS

E. D. TSAMIS and J. N. AVARITSIOTIS

Department of Electrical and Computer Engineering National Technical University of Athens, 9 Heroon Polytechneiou Street 15773 Zografou-Athens, GREECE
E-mail: abari@cs.ntua.gr

Bulk Acoustic Wave devices (BAW) are state of the art technologies developed to address the current trends towards miniaturization and chip level integration of communications systems. The substrate plays a key role in device behaviour due to the need to provide support for the resonator thin piezoelectric film (thickness a few microns). The substrate then becomes an integral part of the resonator affecting its performance with extra losses, scattering and possibly cross talk between two adjacent resonators. The results of 2-D simulation of the substrate effect in two BAW resonator geometries are presented aiming at the derivation of "design rules". Our method provides more insight on the device behaviour given that existing methods are 1-D approaches going back to Mason's work. The developed model is based on finite difference analysis and some realistic approximations for the resonators.

1 Introduction

Over the past decade, the electromagnetic community in order to simulate EM wave propagation, resonators, and scattering has carried out extensive development of the finite difference time domain (FDTD) method. This method also has been applied to the simulation of acoustic wave propagation and scattering [1]. The second order, centered time and space (leapfrog) finite difference scheme is the most commonly used.

In this paper a second order FDTD scheme for simulating the time evolution of stress-velocity fields in loss free BAW resonators [2] is developed for two dimensions. Our aim is to simulate the substrate effect on device behavior and to extract useful "design rules" for on-chip HF bandpass filters.

2 Mathematical formulation

The acoustic field equations in loss free media can be written following Auld [3] as:

$\bar{v} = \frac{\partial \bar{u}}{\partial t}$ (1) $\nabla \cdot \bar{T} = \frac{\partial (\rho \bar{v})}{\partial t} - \bar{F}$ (2) $\nabla_s \bar{v} = \frac{\partial \bar{S}}{\partial t}$ (3) $\bar{T} = [c] \cdot \bar{S}$ (4)

In piezoelectric media the constitutive relation (4) is replaced with these equations:

$\bar{T} = [c^L] \cdot \bar{S} - [e] \cdot \bar{E}$ (5) $\bar{D} = [e]^t \cdot \bar{S} + [\varepsilon^S] \cdot \bar{E}$ (6)

The electric field must satisfy the equations given bellow:

$\nabla \bar{D} = 0$ (7) $\bar{E} = -\nabla \phi$ (8)

Here the "quasi-static" approximation is assumed for the electric field.

Assuming only x,z dependence, from the above equations we arrive at the following difference equations for the time evolution of the field [1]:

$$v_x^{n+2}(2i,2j+1) = v_x^n(2i,2j+1) + \frac{\Delta t}{\rho}\left(F_x^{n+1}(2i,2j+1) + \frac{T_{xx}^{n+1}(2i+1,2j+1) - T_{xx}^{n+1}(2i-1,2j+1)}{\Delta x}\right.$$

$$\left. + \frac{T_{xz}^{n+1}(2i,2j+2) - T_{xz}^{n+1}(2i,2j)}{\Delta z}\right)$$

$$v_z^{n+2}(2i+1,2j) = v_z^n(2i+1,2j) + \frac{\Delta t}{\rho}\left(F_z^{n+1}(2i+1,2j) + \frac{T_{zz}^{n+1}(2i+1,2j+1) - T_{zz}^{n+1}(2i+1,2j-1)}{\Delta z}\right.$$

$$\left. + \frac{T_{xz}^{n+1}(2i+2,2j) - T_{xz}^{n+1}(2i,2j)}{\Delta x}\right)$$

$$T_{xx}^{n+1}(2i+1,2j+1) = T_{xx}^{n-1}(2i+1,2j+1) + \Delta t \cdot c_{11} \frac{v_x^n(2i+2,2j+1) - v_x^n(2i,2j+1)}{\Delta x}$$

$$+ \Delta t \cdot c_{13} \frac{v_z^n(2i+1,2j+2) - v_z^n(2i+1,2j)}{\Delta z} - \Delta t \cdot e_{z1} \frac{dE_x}{dt}\bigg|_{n,(2i+1,2j+1)}$$

$$T_{zz}^{n+1}(2i+1,2j+1) = T_{zz}^{n-1}(2i+1,2j+1) + \Delta t \cdot c33 \frac{v_z^n(2i+1,2j+2) - v_z^n(2i+1,2j)}{\Delta z}$$

$$+ \Delta t \cdot c_{13} \frac{v_x^n(2i+2,2j+1) - v_x^n(2i,2j+1)}{\Delta x} - \Delta t \cdot e_{z3} \frac{dE_z}{dt}\bigg|_{n,(2i+1,2j+1)}$$

$$T_{xz}^{n+1}(2i,2j) = T_{xz}^{n-1}(2i,2j) + \Delta t \cdot c_{44} \frac{v_x^n(2i,2j+1) - v_x^n(2i,2j-1)}{\Delta z}$$

$$\Delta t \cdot c_{44} \frac{v_z^n(2i+1,2j) - v_z^n(2i-1,2j)}{\Delta x} - \Delta t \cdot e_{x5} \frac{dE_x}{dt}\bigg|_{n,(2i,2j)}$$

in case of piezoelectric media with hexagonal crystal structure and similar for the rest media constituting the resonator.

3 Simulation cases

Simulation based on the above equations for the BAW resonator geometry depicted in Figure 1 was conducted.

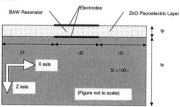

Figure 1. Layout of BAW Resonator on top of a Silicon substrate.

The piezoelectric film is ZnO and the exact dimensions for the simulated cases are tabulated in Table 1. The discretization steps in time and space were calculated according to [1] using material constants from [3].

Table 1. Geometry data for the simulated cases

CASE	#1	#2	UNITS
t_p		3	
t_s		10	
d_1	20	20	
d_2	50	100	µm
d_3	20	20	
Comments	Single BAW resonator	Wider resonator from case #1	

The following assumptions were made for the piezoelectric layer because of the large width to thickness ratio:
1. Under the resonators the electric field is constant throughout the material and equal approximately to the externally applied field.
2. In the rest areas the constant electric displacement approximation is valid resulting in "stiffening" of the material.
3. The electric current flowing in resonator upper and bottom electrodes is calculated numerically based on relation:

$$I = \pm \frac{\partial}{\partial t} \int_{Electrode} \overline{D} \cdot \overline{n} \cdot ds \qquad (9)$$

The required excitation for the simulation was a gaussian shaped applied electric field, z directed with appropriate frequency content. The ratio of the FFTs of applied field waveform and calculated electrode current waveform gives us an estimation of a BAW resonators *input electrical impedance*. Also the FFTs of stress components at selected material points was used to identify resonance frequencies.

Due to limited computer resources and for clarity the substrate was only ten microns thick. ZnO film's thickness was chosen so that main resonance occurs about 1GHz. One dimensional modeling was also conducted so that we can compare the two methods qualitatively. This was based on relations from Sitting [4].

4 Results

In figure 2(a) we can see the results of the one-dimensional modeling of the resonator layout case #1. In figure 2(b) we see the same result calculated using our method. Clearly there are many extra eigen frequensies although the basic ones remain common. This happens because our method models the complete acoustic field inside the simulated structure including edge effects, material's anisotropy and lateral wave propagation completely ignored in 1-D modeling. Many authors suggest from practical experience to increas resonator's lateral dimensions in order to reduce such spurious resonancies. This is seen in case #2 (Fig. 2c) as predicted by our 2-D analysis.

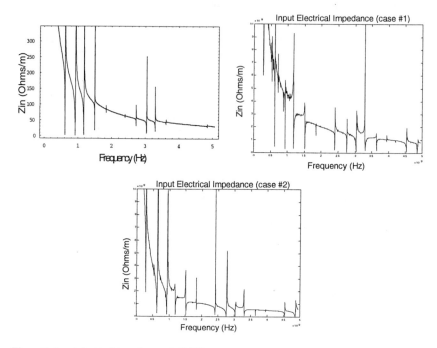

Figure 2. Input electrical impedance of a BAW resonator. (a)One dimensional modeling, (b)our method case #1 and (c) for case #2

5 Acknowledgements

This work was supported by ICCS, Project "ARCHIMEDES".

References

1. Virieux J., P-SV wave propagation in heterogeneous media: Velocity-stress finite difference method, Geophysics **51** (1986) pp.889-901.
2. Lakin K. M., Thin film resonators and filters, IEEE Ultrasonics Symposioum (1999), pp. 895-906.
3. Auld B. A., Chapter 8 Piezoelectricity, Acoustic Fields and Waves in Solids **I** (1973) pp. 265-346.
4. Sitting K. E., Design and technology of piezoelectric transducers for frequencies above 100 MHz, Physical Acoustics, Academic Press (1972) pp.221-272

THEORETICAL CALCULATION OF THE X-RAY PHOTON DETECTOR RESPONSE FABRICATED ON SI GALLIUM ARSENIDE

V.G. THEONAS*, P. DIMITRAKIS AND G. PAPAIOANNOU

Solid State Physics Section, Physics Department, National and Kapodestrian University of Athens, Panepistimiopolis Zografos 15784 Athens, Greece

Contact author's, e-mail: vtheonas@cc.uoa.gr

Radiation detectors are implemented in medical, environmental, industrial, geological and other applications. Especially X-ray detectors gave the ability to produce, high resolution digital radiography systems. In the present work we conducted theoretical calculations on the transient response of SI GaAs X-ray detectors. The detector under consideration is a typical reverse biased Schottky diode on high purity undopped GaAs or compensated material. The electrical characteristics of the "dark" equilibrium state derived by the Poisson and continuity equations, indicate that in the compensated detector a wider active region is established.

1 Calculation of cce and energy resolution

We considered that an X-ray pulse incidents the detector on either the Schottky or the ohmic contact region. The energy of the X-ray photons is assumed to be in the range of 15 to 200 KeV, suitable for digital radiography. The cce achieved by the detector is calculated by the Ramo's theorem [1], taking under consideration the absorption possibility of a photon at each point within its volume (Fig.1a).

$$cce = \frac{1}{L_d^2} \cdot \int_0^{L_d} \left[\int_x^{L_d} \exp\left(-\int_x^{x'} \frac{dx'}{\lambda_e(x')}\right) dx + \int_0^{\bar{x}} \exp\left(-\int_x^{\bar{x}} \frac{dx'}{\lambda_h(x')}\right) dx \right] \cdot \exp(-a \cdot \bar{x}) d\bar{x} \quad (1)$$

where L_d is the detector's width, $\lambda_{e,h}$ the mean free paths of the charge carriers respectively and **a** the absorption coefficient of the material for the specific photon energy E_Φ.

The evaluation of energy resolution is based on the calculation of the mean value of the induced charge to the detector's contacts **Q*** and it's typical deviation σ_{Q^*} (Fig.1b).

$$\text{Resolution} = 100 \cdot \frac{FWHM}{\langle Q^* \rangle} = 100 \cdot \frac{\sigma_{Q^*}}{\langle Q^* \rangle} = 100 \cdot \frac{\sqrt{\langle (Q^*)^2 \rangle - \langle Q^* \rangle^2}}{\langle Q^* \rangle} \quad (2)$$

$$Q^* = q \cdot a \cdot \frac{E_\Phi}{\varepsilon} \cdot \exp(-a \cdot \bar{x}) \left[\int_{\bar{x}}^{L_d} \exp\left(-\frac{1}{\tau_e} \int_x^{x'} \frac{dx'}{u_e(x')}\right) dx + \int_0^{\bar{x}} \exp\left(-\frac{1}{\tau_h} \int_x^{\bar{x}} \frac{dx'}{u_h(x')}\right) dx \right] \quad (3)$$

where ε is the GaAs ionization energy, $u_{e,h}$ the charge carriers velocities and $\tau_{e,h}$ the charge carriers lifetimes respectively.

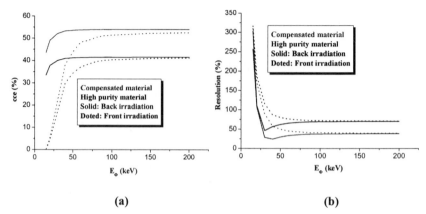

Figure 1: cce and energy resolution achieved by the high purity and compensated detector, when they are irradiated by X-ray photons from the Schottky and the ohmic contact.

From the above figure it becomes obvious that there is a difference in the results for cce and energy resolution when the detector is irradiated from the front (Schottky) and the back (ohmic) contact, especially when the incident photon energy is in the range of 15 to 60 KeV. The difference is significant in this energy range, due to the dissimilar distribution of the photo-carriers within the detector and the width of the specific active region established. The best results are accomplished when the back contact is irradiated by X-ray photons with energies in the range of 30 to 60 KeV (according to the width of the detector's active region), hence the choice of the irradiated contact is meaningless for photon energies greater than 100 KeV. One other important notice is that the best energy resolution is not attained at the energy that the cce is maximum.

2 Calculation of the detector's transient response

We consider a short pulse of X-ray photons that incidents the detector. The dynamic behavior of the detector is obtained by the use of finite element analysis, taking under consideration the generation – recombination processes, the drift of the carriers to the collecting contacts and their collection, i.e. the current continuity equation:

$$\frac{dn(x)}{dt} = G - R + \frac{1}{e} \cdot \frac{dJ(x)}{dx} \qquad (4)$$

where **G** is the charge generation rate, **R** the charge recombination rate, **J** the electron current and **n** the electron concentration of the conduction band. The current pulse response of a high purity LEC GaAs detector is represented in Fig.2.

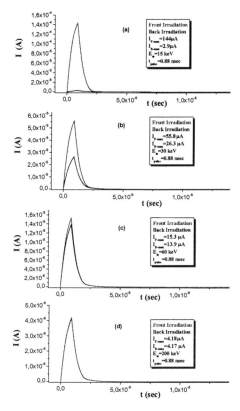

Figure 2: Pulse formation for a detector fabricated by high purity LEC GaAs, when irradiated from the front (red curve) or the back (blue curve) contact, by X-ray photons of energy (a) 15 KeV, (b) 30KeV, (c) 60KeV, (d) 200 KeV. The X-ray pulse duration is 0.88 nsec.

Figure 2 indicates that the current pulse formed during the front irradiation has greater height and FWHM than the one formed during the irradiation of the back contact. This difference is reduced as the energy of the incident photons is increased and becomes negligible when $E_\Phi \geq 60$ KeV. This is due to the to the dissimilar distribution of the photo-carriers generated within the detector when the photon energy is low. Due to the relatively large magnitude of the GaAs absorption coefficient for low energy photons, the majority of carriers are generated near the irradiated contact. Thus, during the front irradiation most of the carriers have their origin inside the detector's active region and they consequently drift to the collecting contacts having high velocities. On the other hand during the back irradiation, the distance that must be traveled by the carriers until their collection is relatively large and a significant number of them are not collected because of recombination phenomena. So the current pulse formed during the front irradiation has a larger height and width. During the back irradiation most of the carriers have their origin outside the active region and they consequently drift to the collecting contacts having low velocities. The distance that (high mobility) electrons have to travel is very small and their drift is slightly affected by the recombination. So the back irradiated detector responds with a current pulse having low height and width. As the photon energy is increased, the charge carrier contribution becomes more homogeneous and above the energy of 60 KeV the choice of the irradiated contact implies no practical difference to the detector's response. In the present simulation the "sharper" pulse is formed when the detector is irradiated from the ohmic contact by X-ray photons with energy of 30 KeV, which is in agreement with the results deduced for energy resolution. Generally the best energy resolution is accomplished by irradiating from the back by photons with energies in the range of 30 to 60 KeV (according to the width of the detector's active region).

As the active region of the compensated detector is wider than the one of the high purity material, it is obvious that it will achieve more satisfactory transient response.

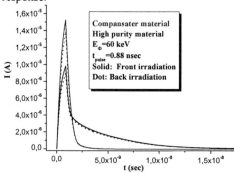

Figure 3: Comparison between the transient response of a detector fabricated by compensated GaAs (red) and a high purity one (blue). The detectors are irradiated by an X-ray pulse with photon energy 60 KeV and duration 0.88 nsec. The response of the compensated GaAs detector is obviously sharper, indicating that better energy resolution is achieved.

Finally, we derived a set of results related to the physical – fabricational parameters of the detector, in order to achieve the best detection results. The high purity detector must have the minimum possible concentration of deep donors (located as deep as possible in the bandgap), otherwise we must use highly compensated material. We must provide contacts that allow the detector to be biased by high voltage and produce high reverse bias current. Generally the reverse bias current must be as high as possible provided that it will allow the discrimination of the current pulse formed by the incident photons. The detector's width must be low enough to achieve the formation of an active region that exceeds the maximum possible fraction of its width but not very low in order to attain the desirable photon absorption.

3 Acknowledgement:

This work is partially supported by the GSRT in the frame of the common joint research between Greece and Slovakia.

References

1. T.E.Schlesinger–R.B.James: "Semiconductors for Room Temperature Nuclear Detector Applications", Semiconductors and Semimetals **V.43**, Academic Press (1995)
2. G.F.Knoll: "Radiation Detection and Measurement", Third Edition John Wiley & Sons (2000)
3. J.W.Chen et al.: Nuclear Instruments and Methods in Physics Research, **A 365**, 273–284 (1995)
4. V.L.Dalal et al.: Applied Physics Letters, **16**, 489–495 (1970)

EFFECTIVENESS OF LOCAL THERMAL ISOLATION BY POROUS SILICON IN A SILICON THERMAL SENSOR

D. N. PAGONIS, C. TSAMIS , G. KALTSAS AND A.G. NASSIOPOULOU

Institute of Microelectronics, NCSR "Demokritos",
15310 Aghia Paraskeui
Athens, Greece

The purpose of this work is to investigate the effectiveness of porous silicon as an isolation material for a silicon integrated gas flow sensor based on a heated resistance and two series of thermocouples. Thermal analysis of the devices is performed for different geometrical designs and different thermocouple materials using commercial software. Simulation results for the optimization of sensor design and comparison with experimental data will be presented.

1 Introduction

The fabrication of low power thermal devices requires the optimization of the sensor's design in order to reduce thermal losses, which constitute one of the main factors limiting device performance. Porous silicon as a material for thermal isolation on bulk crystalline silicon exhibits many advantages over bulk micromachining techniques[1,2,3]. It does not require double side alignment. It is a low cost process and compatible with CMOS technology, which is very important for the fabrication of integrated sensors.
In this paper we present simulation results on thermal analysis of a silicon micromachined thermal sensor with porous silicon for local thermal isolation using the commercial software package MEMCAD 4.8

2 Some information about the sensor

The design, principle of operation and characterization of the sensor are found in references[4,5] Fig.1 shows a top view of the sensor. The heater and hot parts of the thermopiles lie on a thermally isolated area (shown in black color). Depending on the effectiveness of the thermal isolation used, by passing a given current (say 5mA) through the heater, a difference in temperature between the confined (*hot region*) and the rest area (*cold region*) is created. Thus, a potential difference is developed at each thermocouple situated at both sides of the heater.

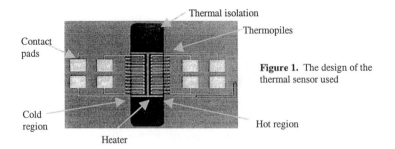

Figure 1. The design of the thermal sensor used

When there is a gas flow over the sensor, a temperature difference is developed between the upstream and downstream thermopile. The difference of the voltage developed at each thermopile gives the output of the sensor.

3 Simulation package - Exact conditions explored

All the simulations were performed using the commercial software *MEMCAD v.4.8*, MICROPROSE simulation package. The models used for each simulation were created with the aid of *IDEAS Master Series 6* software. Both packages were operated on a *SUN ULTRA 10* workstation. The process flow for each simulation carried is summarized in figure (2).

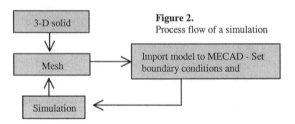

Figure 2. Process flow of a simulation

In the first set of simulations, a 40 μm layer of porous silicon is used as the thermal insulating layer, while in the second case, silicon is etched and the sensor structure lies on a 2 μm free-standing bridge made of Polysilicon, Oxide or Nitride (fig. 3). In each case, the effect of the presence of the thermocouples was explored together with the effect of thermocouple material (p^+/ Au, p^+ / Al and p^+/ n^+). The same boundary conditions were applied in all cases. A fixed room temperature (300 Kelvin) was set at the bottom of the structure while a heat flux of 8.57×10^6 W/m^2 was set at the top of the heater. The specific flux produces the same heating effect as an equivalent of 5 mA current passing through the heater. The main goal of the simulations was to determine the maximum temperature reached on the heater for each of the cases mentioned.

Figure 3. A model used in the second set of simulations. A 2 μm free standing bridge after the hypothetical removal of porous-Si is assumed

4 Simulation results

Fig. 4 shows the temperature distribution on the device surface for the case of the

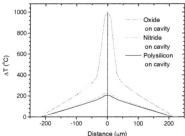

Fig. 4 Temperature distribution for different type of free-standing bridges.

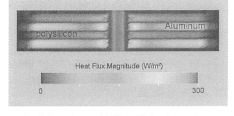

Fig. 5 Heat flux on Al / Poly-Si thermocouple situated on a polysilicon bridge.

free-standing bridge after removal of porous silicon. Notice the significant increase of the maximum temperature for different bridge materials. Fig. 5 shows graphically the heat flux on two different thermopiles made of Aluminum and Polysilicon. The effect of thermocouple material is seen in table 1.

	Temperature increase on heater for different thermopile material (Kelvin)		
Thermal Isolation	p^+ / Au	p^+ / Al	p^+ / n^+
Porous (40 μm)	108.9	109	109.8
2 μm polysilicon membrane	201.7	210.3	252.1

Table 1. Maximum temperature reached on the heater for both types of isolation depending on the thermopile material.

5 Experimental structures

Devices with one and two thermopiles of Al/p$^+$ polysilicon thermocouples were successfully constructed in the laboratory. The thermal isolation employed was 40 µm porous silicon and their response was monitored. As it is shown in fig. 4, the thermal losses due to the second thermopile were negligible as predicted from the corresponding simulations.

6 Conclusion

From the simulation results we conclude the following:
1. In thermal isolation by air in the case of a cavity, thermal losses are minimized so the presence of thermocouples is not negligible as it concerns thermal losses.
2. When porous silicon is used for isolation the influence of thermocouples to the maximum local temperature is practically not detectable.

7 Acknowledgments

This work was partially supported by the *General Secretariat of Research and Technology*

References

1. W. Lang, Thermal conductivity of Porous Silicon, in: Properties of Porous Silicon, Ed. L. Canham, EMIS Datareview Series, No. 18, INSPEC, London 1997 pp. 138-141
2. Ph. Roussel, Y. Lysenko, B. Remaki, G. Delhomme, A. Dittmar and D. Barbier, Sensors and Actuators A 74, 100 (1999)
3. A.G. Nassiopoulou and G. Kaltsas, Porous Silicon as an Effective Material for Thermal Isolation on Bulk Crystalline Silicon, phys. stat. sol. (a) 182.307, 2000 pp. 307-311
4. G. Kaltsas, A.G. Nassiopoulou, Sensors and Actuators A, 76 (1999), p.133
5. Patent N° OBI 1003010, A.G. Nassiopoulou and G. Kaltsas

THERMAL CONDUCTIVITY OF POROUS SILICON LAYERS PROBED BY MICRO-RAMAN SPECTROSCOPY

D. PAPADIMITRIOU AND P. TASSIS

National Technical University of Athens, Department of Physics, GR-15780 Athens, Greece.

L. TSOURA, C. TSAMIS AND A.G. NASSIOPOULOU

IMEL/NCRS "Demokritos", P.O.Box 60228, 15310 Aghia Paraskevi, Athens, Greece.

The thermal conductivity of as prepared and oxidized porous silicon layers was estimated by means of micro-Raman Spectroscopy. The shift in peak position of the Raman Stokes spectra from the same sample spot obtained at two different values of the laser beam power was used in this respect. This shift is due to local temperature increase by increasing the laser power. The sample temperature for a given laser power was estimated in two ways: a) from the temperature dependence of the Raman Stokes frequency, and b) from the intensity ratio of the Raman Stokes to Raman Antistokes peaks. Raman spectra from oxidized porous silicon layers exhibited lower frequencies and they were broader than those from the as-prepared layers due to the presence of smaller crystallites within the material. The thermal conductivity decreased by increasing the oxidation temperature. In oxidized samples, it was almost one order of magnitude smaller than in the as-prepared material. Depth profiling did not show any significant changes of thermal conductivity across the porous layer.

1 Introduction

Scanning probe microscopy is one of the techniques[1] used to map the thermal conductivity variations of surfaces on the nanometer scale and to estimate the thermal conductivity of conductors and thin insulating films deposited on top of conductors. Micro-Raman spectroscopy was also proposed[2] recently as a direct and non-contact method for thermal characterization of bulk and reduced dimensionality semiconductors. The thermal conductivity of porous silicon formed on crystalline silicon was estimated. In this work, micro-Raman spectroscopy is applied to measure the thermal conductivity of as prepared porous silicon layers and to compare it with that of porous layers oxidized at high temperatures. The thermal conductivity profile as a function of depth is also estimated by probing micro-Raman spectra on a sample cross-section.

2 Experimental results and discussion

Porous silicon layers, 30 μm thick, were grown on (100) crystalline silicon by anodization in an $HF:C_2H_4(OH)_2$-solution 3:2 at a current density of 80mA/cm^2. Some of the samples were oxidized at 300°C for 1h in an O_2-atmosphere and some of them were oxidized in two steps: at 300°C for 1 hour and at 800°C for another hour.

Raman scattering was excited by the 476.2 and 647 nm lines of a Kr⁺-laser. It was spectrally analyzed by a Jobin-Yvon micro-Raman spectrometer (T-64000) equipped with a CCD-detector. The laser beam was microscopically focused to a spot of 1 µm diameter. To avoid damage or structural changes of the porous layer due to local heating by the laser beam, the beam power did not exceed 1 mW. Raman measurements were performed on a cleaved face of the supported porous sample. Two spectra were recorded from the same sample spot at two different values of beam power. Usually, low- and high value were different by 0.6-0.8 mW. By measuring the Raman-Stokes frequency for the two different heating power values P_1 and P_2, $P_2 > P_1$, the corresponding temperatures T_1 and T_2 can be estimated, and the thermal conductivity k of the layer can be obtained from a simple linear relationship between $\Delta P = P_2 - P_1$ and $\Delta T = T_2 - T_1$ given in ref. /1/:

$$k = \frac{2\Delta P}{\pi \alpha \Delta T} \quad (1)$$

α is the heating source diameter.

For porous silicon, calibration of Raman frequency shift versus temperature was found in ref. /2/ to be the same as that of bulk silicon, e.g. 0.025 cm⁻¹/°C. The temperature dependence of Raman frequency of the bulk is well known from previous investigations[3].

The local sample temperature can be also estimated from the intensity ratio of Raman-Stokes to Raman-Antistokes spectra:

$$T = 1.439 \frac{\omega_o \, (cm^{-1})}{\ln(I_S/I_{AS})} \quad (2)$$

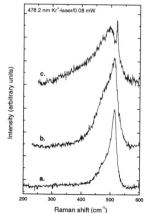

Raman spectra of porous silicon layers as prepared and oxidized in one (300°C) or two steps (300°C and 800°C) are shown in Fig. 1. The spectra can be fitted by a superposition of Lorentzians and Gaussians as demonstrated in Fig. 2. Different spectral

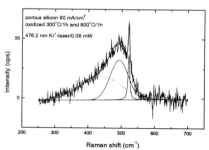

Fig. 1 Raman spectra of porous silicon layers: a) as prepared, b) oxidized for 1h at 300° C, and c) oxidized for 1h at 300° C followed by a second oxidation step at 800° C for 1h.

Fig. 2 Evaluation of Raman spectra: fitting with a combination of Lorentzian and Gaussians.

components are attributed to the presence within the porous layer of crystallites with different sizes. Large and small size crystallites are represented by high and low frequency Raman modes respectively (HF, LF), while the phonon mode at 521 cm^{-1} belongs to the crystalline Si substrate (c-Si). By increasing the oxidation temperature, crystallites of smaller size are formed, so that the spectra of oxidized layers, compared with the spectra of non-oxidized layers, are shifted to lower frequencies and are broader. The larger frequency shift from c-Si peak in the spectra of oxidized samples makes separation of porous-Si and c-Si bands possible (Figs. 1c and 2).

The local temperature increase was obtained from the Raman frequency shift with the increase of the power of the laser beam. Frequency shifts can be also observed in the case of porosity gradients across the porous layer or when stresses are present due to the lattice mismatch between the porous film and substrate[4]. Thus, only the frequency shift of spectra measured at the uppermost layer of the film can be directly connected with the local temperature increase in the sample. In all other cases, confirmation of the estimated temperature values by Raman-Stokes /Antistokes measurements is necessary.

Thermal conductivity values were calculated according to Eq. (1) for both Raman modes shown in Fig. 2. High and low frequency modes of Raman spectra represent, as already stated, different subsets of crystallites with different sizes within the porous layer: high frequency modes represent large size crystallites with bulk-like character, while modes of lower frequency are representative of smaller size crystallites. The distribution of these values across the porous layer is shown for as-prepared and oxidized samples in Figs. 3 and 4 respectively. Since the high

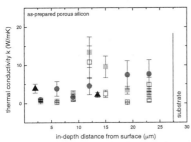

Fig. 3. Thermal conductivity of as prepared porous silicon layers estimated from the low frequency mode of Raman spectra excited by 647 nm Kr$^+$-laser: ■ $\Delta T(\Delta\omega$-Stokes), □ $\Delta T(\Delta\omega$-Antistokes), ● $\Delta T(I_S/I_{AS})$, and 476.2 nm Kr$^+$-laser: ▲ $\Delta T(\Delta\omega$-Stokes).

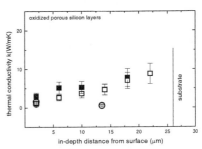

Fig. 4. Thermal conductivity of porous silicon layers oxidized at 300°C for 1h: ■ (HF), □ (LF), and at 300° C, 1h followed by a second step at 800° C, 1h: ● (HF), ○ (LF) by excitation of Raman scattering by 476.2 nm Kr$^+$-laser.

frequency mode in the spectra of as prepared samples cannot be separated from the c-Si mode (compare Fig. 1c and Fig. 2 with Fig. 1a), the thermal conductivity values in Fig. 3 were calculated from the shift (■, □, ▲), or the intensity ratio I_S/I_{AS} (●), of the low-frequency Raman mode of porous-Si. The thermal conductivity values estimated from the intensity ratio of the Raman-Stokes/Antistokes spectra were consistent with those obtained from the Raman frequency shift as a function of laser power. In Fig. 4, the thermal conductivity of samples subjected to one- or two-step oxidation is shown for both Raman modes. The difference in values obtained from the low and high frequency modes of Raman spectra is reduced after one-step oxidation (Fig. 4, ■, □) and diminishes after two step oxidation of the porous layer (Fig. 4, ●, ○) depending on reduced or diminishing influence of the c-Si substrate mode on the porous-Si modes. The observed increase of thermal conductivity near the interface of the porous layer with the bulk silicon is associated with the influence of the higher thermal conductivity of the bulk (thermal leakage to the bulk). In the average, the thermal conductivity of highly oxidized samples (k_2=1W/mK, Fig. 4 /circles) is about 4x smaller than that of pre-oxidized samples (k_1=4W/mK, Fig. 4 /squares) and almost one order of magnitude smaller than the maximum conductivity (k=10W/mK) observed in non-oxidized samples (Fig. 3). This is in excellent agreement with theoretical predictions[5] on the heat conduction of meso-porous silicon which for a given porosity depends on initial crystallite size and on degree of oxidation. The average conductivity of as prepared samples seems to remain constant across the porous layer, which implies that the porous films are almost free of structural inhomogeneities.

3 Conclusions

The thermal conductivity of as prepared and oxidized porous silicon layers was estimated by micro-Raman spectroscopy. It was found: **a)** to decrease by increasing the oxidation temperature and thus increasing the surrounding oxide and decreasing the silicon core of each crystallite composing the material, and **b)** to be almost one order of magnitude smaller in oxidized layers than in the non-oxidized ones.

Depth profiling did not show any significant changes of thermal conductivity across the porous layer, confirming the in-depth homogeneity of the films.

References

1. M. Nonnenmacher and H.K. Wickramasinghe, Appl. Phys. Lett. 61(2), 168 (1992).
2. S. Périchon, V. Lysenko, B. Remaki, D. Barbier, and B. Champagnon, J. Appl. Phys. 86(8), 4700 (1999).
3. R. Tsu and J.-G. Hernandez, Appl. Phys. Lett. 41(11), 1016 (1982).
4. D. Papadimitriou, J. Bitsakis, J.M. López-Villegas, J. Samitier and J.R. Morante, Thin Solid Films 349, 293 (1999).
5. V. Lysenko, L. Boarino, M. Bertola, B. Remaki, A. Dittmar, G. Amato and D. Barbier, Microelectronics Journal 30, 1141 (1999).

Silicon Integrated Technology and Integrated Circuit Design

A MCM-L BOARD FOR THE BASEBAND PROCESSOR OF A DUAL MODE WIRELESS TERMINAL

C. DROSOS, C. DRE, K. POTAMIANOS AND S. BLIONAS

Intracom S.A., 19.5Km. Markopoulou Ave.,Peania, 19002,
GREECE
E-mail: cdro@intracom.gr

An advanced System-On-A-Chip (SoC) design, capable to undertake all baseband signal processing, is presented is this paper. The development and design aspects of an innovative baseband processor, namely the ASPIS processor, for a dual mode DECT/DCS-1800 wireless terminal are covered in the first part of this paper. The architecture of the system and all the hardware design techniques, testing and software development methodologies are given in details. The second part of this paper concerns the fabrication of this SoC design and the technology used for the building of a MCM-L board that incorporates this baseband processor and its whole memory sub-system.

1 Introduction

In the last few years, there is a continuously growing demand for wireless terminals integrating sophisticated multi-service applications. The achievements of the modern digital design technology make possible the idea of a mobile terminal that can operate and support different communications protocols, covering in this way the lack of a universal standard. On the other hand, modern, multi-mode, mobile terminal devices require very complex system implementations. These devices are usually implemented using the System-On-A-Chip methodology, as they involve multi-processor solutions (DSP and microprocessor), and require advanced hardware-software co-development. The role of the DSP in such applications is to implement the computational intensive algorithms, while the microprocessor is responsible for the execution of the upper layers of the communication protocol stacks.

Today, the idea of a multi-mode terminal becomes more attractive and mature [1]. The ASPIS processor is the first multi-mode baseband processor available for DECT/GSM/DCS1800 wireless terminals into an one-chip solution. Other commercial available multi-mode terminals use more than one different chipsets in order to implement the multi-mode operation, resulting into no flexible and cost-inefficient designs. The ASPIS processor incorporates an ARM7TDMI core, a DSP core and several peripherals that interface the chip with the RF parts, the audio parts, the keypad, the display, and the other general I/O peripherals [2].

The design of this advanced System-On-Chip processor is given in chapter 2, while its fabrication issues, first as an embedded gate array chip and then as a die, that is hosted in a MCM-L board based on advanced fine line/gaps and microvia technologies, are presented in chapter 3. The development of the ASPIS was partially supported by the projects ASPIS #20287 and FLINT #23261, that were funded by European Union in context of ESPRIT IV.

2 ASPIS architecture

The general architecture of the ASPIS processor is based on a multi-processor system-on-chip architecture with an internal system bus. The ASPIS processor is capable to implement all the baseband signal processing required in a multi-mode terminal, the protocol stacks for the DECT and GSM/DCS-1800, the Man-Machine Interface (MMI) as well as the overall control of the operation of the complete dual mode DECT/GSM handset. A detailed view of the chip's internal architecture is presented in figure 1.

The ASPIS processor incorporates an ARM7TDMI core, a DSP core and several peripherals that interface the chip with the RF parts, the audio parts, the keypad, the display, and the other general I/O peripherals. The role of the ARM core is the execution of the protocol stacks, while the DSP handles all the heavy signal processing functions required by the DECT and the GSM standards, such as voice processing and radio signal equalisation.

During the hardware development of the ASPIS processor all the usual low power techniques, such as gated clocks, were used. Due to the internal bus structure of the system and the low power requirements a novel strategy was adopted for the integration of the diverse testing strategies that are used in the ASPIS processor [3]. The strategy adopted for the testing is a non-scan-based technique, exploiting the high-bandwidth and control features of the chip's internal bus. This technique uses the internal bus of the system as the medium to apply test vectors to the system's components. Scan-based testing is employed in the DSP core, modified to fit the bus-based testing technique. Also, the ARM7TDMI core features a JTAG port plus an auxiliary boundary-scan chain, used for die-to-PCB connectivity testing. A dedicated BIST controller attached to the internal bus is also employed for testing of the embedded SRAM arrays (augmented March-C+ algorithm).

Modern SoC methodologies require parallel development of the hardware and software parts. To meet this requirement an advanced co-development and co-simulation environment was created and used in order to assure that both hardware and software components would be developed according to the system's specifications. Two co-simulation environments have been developed: a low-level VHDL co-simulation environment and, an advanced high-level software model of the ASPIS processor to be used within the ARMulator instruction set simulator tool. The first co-simulation environment was being used for the validation of the ASPIS processor design, while the second, for the cross-development of the DECT/GSM software parts of the protocols for the target ASPIS platform in parallel with the hardware design.

3 MCM-L implementation

The ASPIS processor was designed entirely in VHDL and fabricated using an embedded gate array. The operating voltage is 3.3V and its initial package is a 208-lead plastic QFP with a size of 28x28x3.4 mm. The technology used was Mitel Semiconductor's cla200 at 0.35um. The developed random logic for the ASPIS peripherals equals to 200,000 gates, while the total gate count of the chip is 280,000 equivalent gates. The maximum power dissipation of the ASPIS processor is 250mW, while its maximum operating frequency is 27.648MHz.

In order to enhance the ASPIS-based wireless terminal, Intracom also implemented a demonstrator MCM-L board for the dual mode handset terminal. Specifically, the basic processor chip (ASPIS die) as well as two fast SRAMs were attached as COB (Chip On Board) components (ASPIS die minimum pad pitch 77μm with staggered pads) on the same module along with the FLASH which was provided as CSP (Chip Scale Package) in μBGA package. The final board (MCM-L) achieved more than 50% reduction in board size/area compared to the similar discrete component version exceeding by far the company expectations.

The final MCM-L design uses fine lines and clearances of 100μm width and microvias (photovia processing) of 100μm diameter, each drilled with 23 or 25 mils copper annular ring. The MCM-L external pad arrangement is that of a JEDEC BGA case (BG352, MO-192-BAR-2), while its external dimensions are 35mm x 35mm. The ball pitch under the MCM is 1.27mm with balls Sn/Pb eutectic made and size (diameter) 0.75mm nominal. On the MCM bottom side, each pad has 29 mils soldermask opening. Varnish recovers pad exists on top and bottom sides. The MCM board consists of six layers: one for VCC, one for GROUND and four routing layers.

The size of a conventional board hosting the ASPIS chip, two static RAM modules and one FLASH module 6.75cm x 4.3cm. The size of this board is not sufficient enough for use in a mobile terminal. In this case the advantages of the developed MCM-L become more than apparent. The size of the developed MCM-L board is only 3.5cm by 3.5cm, a significant reduction compared with the size of the standard board. A view of the MCM-L board for an ASPIS based terminal is given in figure 2.

4 Conclusion

The development process, the architecture and the features of a SoC design, the ASPIS processor, were presented in this paper. The ASPIS processor is the first available multi-mode baseband processor available for DECT-GSM/DCS1800 into an one-chip solution. Advanced hardware and software design techniques and methodologies were used during the development of the processor. In order to enhance the flexibility of the ASPIS based wireless terminal an MCM-L board was also developed with microvias (photovias) and fine line technology.

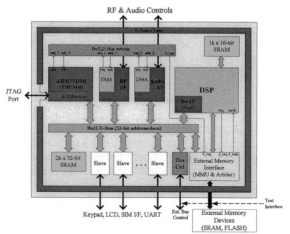

Figure 1. ASPIC SoC architecture.

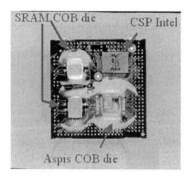

Figure 2. ASPIS MCM-L board.

References

1. Mobile Communications International 21 (Apr. 1995).
2. H. Katathanasis, C. Dre, D. Metafas, S. Blionas, "Designing a DSP for DECT and GSM/DCS-1800 Baseband Processing", Real-Time Magazine 96/3, 1996
3. F. Ieromnimon, C. Dre, et. al., "On the Integration of Diverse Testing Strategies in a Low-Power Processor", DATE 2000, March 27-30, Paris, France.
4. C. Drosos, M. Zayadine, et. al. "Embedded real-time communication protocol development using SDL for ARM microprocessor", Dedicated Systems Magazine, Jan 2001.

A CAD TOOL FOR AUTOMATIC GENERATION OF RNS & QRNS CONVERTERS

M. M. DASIGENIS, D. J. SOUDRIS, S. K. VASILOPOULOU AND A. T. THANAILAKIS

Department of Electrical and Computer Engineering
Democritus University of Thrace, GR-671 00 Xanthi, Greece

This paper presents a CAD tool, that can generate the structural description of converters for the Residue Arithmetic System (RNS). This description is based on a previously proposed architecture[1], which has been reported as the most efficient, known to date, in terms of area and delay. This structural description can be the input to a Design Environment, in order to take real silicon measurments of area and delay, or to fabricate the converter. The whole process can be done automaticly, in both PC and HP platforms, aiding the engineer to explore the different design possibilities.

1 Introduction

High Speed Digital Signal Processing (DSP) units increasingly demand efficient arithmetic algorithms, as well as improved VLSI architectures. The hardware implementation of an arithmetic algorithm is largely affected by the choice of specific numbering system. RNS and Quadratic RNS (QRNS)[2] is considered as a nonconventional numbering system whose major characteristics are: carry free, fault isolating and modular design[3]. In the scientific and technical literature there is a great number of publications about RNS processors[4,5,6].

A systematic design methodology for converters from Binary to RNS, to QRNS and vice versa, is introduced, based completely on FAs as building blocks, which outperform conventional look-up table based structures when very high throughput rates are required. In order to implement the modulo m operation, a recursive bit-level algorithm is used. The proposed methodology provides a systematic design framework for deriving array architectures, for any m. The resulting architectures can achieve high throughput rates. The concept of the design is based on a unified methodology, which means that the converters have similar stages and, thus, a low complexity and a small delay can be achieved, making designs simpler to fabricate.

The main contribution of the present work lies in the automatic implementation of RNS and QRNS converters through a Computer Aided Design (CAD) tool. The tool is implemented in C++ and its output is a synthesizable Very High Speed Integrated Circuit (VHSIC) Hardware Description Language (VHDL) code[7]. Furthermore, with the use of this tool, we were able to take measurements concerning chip area and path delay using the CADENCE Design Enviroment. The hardware compiler, presented here, can be very helpful to designers of RNS systems, considering that it is possible with this tool to explore all the different design possibilities, without reference to any abstract measurements from mathematical formulas. In addition, the usefulness of this CAD tool arises from the autogenerated extraction of the VHDL code, without the need for the designer to describe the required compiler.

2 The Proposed Methodology

Generally the unified methodology presented here consists of: (*i*) Preprocessing stage, (*ii*) Bit reduction stage, (*iii*) Multiplication stage, and (*iv*) Final mapping stage. More specifically, the first stage is used in the converters that require to calculate a sum before proceeding to the modulo calculation, i.e. in the QRNS converters. The second stage, consisting of many sub-stages, is a FA based design used for the bit reduction and the computation Y_r, which has the properties $\langle Y_r \rangle_m = Y_{out}$, where $\langle \rangle$ is the modulo operation. The third stage is used when we want to calculate the multiplication of a variable number with a constant, i.e. in the RNS to binary converters. The multiplication is achieved without using a ROM lookup table, but instead by continuously using shifts and sums. Finally the fourth stage, the final mapping stage, is used for mapping the output Y_r of the bit reduction stage to its residue modulo m value. This can be achieved based on the equation:

$$Y_{out} = \begin{cases} Y_r, & Y_r < m \\ Y_r - m, & Y_r \geq m \end{cases}. \tag{1}$$

Thus, a n-bit adder is used instead of an undesirable ROM in order to maintain the high throughput, and it is implemented using 1-bit adders.

2.1 Design of the FA-Based Architecture

A methodology for the systematic derivation of the array architectures, starting from the algorithmic level description, is developed. The detailed design of every converter's FA-based stage is obtained as a sequence of four steps: (*i*) Calculation of the number of input bits, (*ii*) Calculation of the number of FAs, (*iii*) Derivation of the Dependence Graph, and (*iv*) Design of the Final Mapping stage. Details concerning the FA-based architectures may be found in a previous publication[1].

2.2 Analysis Methodology of the CAD Tool

The CAD tool developed and presented in this paper, is a hardware compiler written in C++, and it takes as input: i) the required number of input bits l, ii) the moduli number m, and iii) the desirable type of converter. This tool can be used as an architectural exploration tool, both in PC and HP Workstation Environment. The compiler, which is programmed in accordance with the technical and mathematical foundations of residue number arithmetic, calculates all the stages that are required.

More specifically, the hardware compiler consists of a number of modules, Fig. 1. The core of the compiler is responsible for the right sequence of modules and their synchronization. The inputs to the first module of the CAD tool are: a) the type of converter, b) the moduli number m, and c) the number of input bits l. This module is named Stage Calculation Module (SCM) and it specifies the number of stages s and their type, where s is given by $s = p + b + f + c$, where p, b, f and c are the numbers of preprocessing stages, bit reduction stages, multiplication stages and final mapping stages, respectively, required for the implementation of the converter. Subsequentrly a loop from $k = 1$ up to $k = s$, where k is the current stage, is being executed. This means that gradually every stage of the converter is being compiled. Inside this loop are: i) the Stage Type Module (STM), which provides the corresponding equations of every type of converter, ii) the Bit Calculation Module (BCM), that calculates the number of output bits of every stage, iii) the FAs Calculation Module (FCM), where the calculation of the number of FAs included in these stages takes place, and iv) the Dependence Graph (DG) Extraction Module (DGEM), that gives the final form of the DG of every stage.

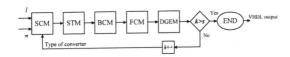

Figure 1. Analysis of the hardware compiler

When $k > s$, the loop execution terminates, and the CAD tool uses the VHDL generation module, to compile the VHDL file. The VHDL code is synthesizable and it can easily be compiled, elaborated and synthesized in the HDL-Synergy Cadence Design Environment 4.1, in order to extract the floorplanning, placement, routing, and eventually the layout of the specific converter. Finally, realistic measurements of chip area and path delay can be taken from the generated layout.

The VHDL generation module combines some parametrisable basic elements in VHDL form, like the clock component, the delay unit, the full adders and the FA arrays, in order to create more complex structures and it is the result of the following sequence of calculations. The DG is the input to a matrix, in the lattice points of which the FAs and the delay units are placed. In fact, the placement of the FAs is such that it includes the timing constraints. These constraints are extracted in such a way that the FA cell depenencies are not violated, according to a set of seven lemmas[1]. This provides us with a valid VHDL code, that describes the function of the converter. Subsequently, the set of lemmas mentioned above allows us to achieve maximum locality, and thus, produce an optimal placement of the FA cells in the matrix. The VHDL generation module takes as input this matrix and the interconnection information of its lattice points, and creates the respective port mapping clauses between them. This analysis leads to the conclusion that the output VHDL code is not only a valid one, but also optimal, due to the fact that it best gives the architectural description of the required converter. Finally, the VHDL code produced is easy to be read and analyzed, due to the insertion of automatically generated comments.

3 Conclusions and Future Work

In this paper, we attempt to bridge the gap between proposed architectures of RNS converters and their harware implementation. The CAD tool developed in this work can aid and automate the design of RNS systems, minimizing the time required for the designer to investigate possible alternative solutions. The ouput of the tool is a VHDL structural description, which is syntesizable in a Design Environment. This hardware compiler could be usefull to system designers, who evaluate the use of RNS in practical systems, or to researchers who investigate the RNS arithmetic.

References

1. D.J. Soudris, M.M. Dasigenis, A.T. Thanailakis, *A unified methodology for designing RNS & QRNS FA based converters*, IEEE Int. Symp. on Circuits and Systems (ISCAS), May 28-31, 2000, Geneva, vol. I, pp. 20-23.
2. N. S. Szabo and R.T. Tanaka, *Residue Arithmetic and its Applications to Computer Technology*, New York, McGraw Hall, 1979.
3. F. J. Taylor, *Residue Arithmetic: A turorial with examples*, IEEE Trans Computers, pp. 50-62, May 1984.
4. Stanistaw J. Piestrak, *A High-Speed Realization of a Residue to Binary Number System Converter*, IEEE Transactions on Circuits and Systems Part II, vol. 42, No. 10, October 1995, pp. 661-663.
5. T. Stouraitis, *Efficient convertors for residue and quadratic residue number systems*, IEE Proceedings-G, Vol. 139, No. 6, Dec. 1992, pp.626-634.
6. Stanislaw J. Piestrak, *Design of Residue Generators and Multioperand Modular Adders Using Carry-Save Adders*, IEEE Transactions on Computers, Vol. 423, No. 1, January 1994, pp.68-77.
7. *IEEE Standard VHDL Language Reference Manual*, IEEE Std 1076-1993, June 1994.

MOSFET MODEL BENCHMARKING USING A NOVEL CAD TOOL

NIKOLAOS A. NASTOS

National Technical University of Athens (NTUA) Electrical and Computer Engineering Department, Iroon Politechniou 9, Zografou 15773, Athens, GREECE

E-mail: nikos@elab.ntua.gr

YANNIS PAPANANOS

National Technical University of Athens (NTUA) Electrical and Computer Engineering Department, Iroon Politechniou 9, Zografou 15773, Athens, GREECE

E-mail: papan@elab.ntua.gr

In this paper a novel CAD tool is presented suitable for benchmarking MOSFET models. It is implemented under Cadence® environment and its purpose is to perform various tests on existing MOS models in order to investigate their accuracy and reliability.

1 Introduction

Since the early beginning of IC design, simulators were a valuable and useful part of the whole design process [1],[2]. They create and solve thousands of equations in order to describe and investigate the electrical performance of the circuit. This is implemented using models that describe the electrical characteristics of each device used in the design.

Unfortunately the characterization of each electrical element is not always precise. Specifically, the accuracy of MOSFET models for SPICE-like simulators is a real headache for IC designers especially to those who are concerned with the implementation of analogue or/and high frequency circuits [1],[2],[5]. This is evident in today's deep submicron technologies where a lot of phenomena difficult to describe take place.

The main reasons for this situation are the non-linearities that appear during the transistor operation, the existence of complicated equations that have to be simplified so as their solution consumes less computational time during simulation and the numerical errors that take place [2],[5]. All these factors combined with the fact that many different phenomena appear in the regions of MOSFET operation arise difficulties on the accurate characterization of the transistor.

From the above description one concludes that key to first time success is to check the validity of the models which will be used during the design. This can be achieved by applying benchmark tests on the model. These tests inform the IC designer about the limitations of the model under use so as to take them into account during the designing process.

In this paper, a novel CAD tool is presented which can estimate the accuracy and indicate the validity of the MOS models. It is designed to run under the Cadence environment and is written in the SKILL programming language provided by Cadence [3]. The tool is modular and additional benchmark tests can be easily adapted.

In the second section some examples of critical benchmark tests are given, while in the third section the mathematical functions performed, the tool's flow and some indicative tool menus and results are displayed.

2 Benchmark Tests: Description, Usage and Examples

There are two ways to check the validity of a MOSFET model [1], [2]. The first one is to measure a transistor and compare the results with those derived from simulation. This is a difficult and impractical way to examine it. The other way is to perform several benchmark tests that have been implemented and presented in bibliography [1], [2], [4]. These tests are simple circuits designed to focus on a certain MOSFET region of operation, quantity or model parameter neglecting or minimising the effect of the others.

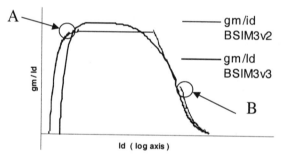

Figure 1: Gm/Id benchmark test as it was given by the tool presented

A typical and well-known example of benchmark tests is the one shown in Fig.1. In this figure, the g_m/I_d versus I_d is plotted for the same circuit using two different MOS models. These two models produce quantitatively and qualitatively different results even on a parameter of such significance for analogue designs. This plot reflects the need for accurate MOS modelling.

The presented tool categorizes the benchmark tests according to the examined parameter so as to simplify their selection. Thus, categories for checking the validity of DC and small-signal values exist. Other categories check model accuracy over frequency and finally, a set of tests is dedicated to noise performance evaluation.

3 Tool's Flow and Results Processing

3.1 Tool's Flow

The tool performs the following operations in order to benchmark a MOSFET model: At first the user specifies the model path, the biasing conditions he is interested in and which category(s) of tests he wants to examine. Next the tool automatically loads and executes the suitable simulation netlists and then it collects and checks the simulation data. Finally, windows with the result waveforms and a complete status report appear informing the user on the benchmarking outcome.

3.2 Mathematical Data Processing

The main feature of the tool presented is its capability to search and find quite successfully the regions where the model is inaccurate. There are several kinds of model inaccuracies, both qualitative and quantitative. In the first case, the main concern is that the extracted data does not exhibit any unusual behaviour (derivative discontinuities, totally wrong prediction for some phenomena etc) while in the second case we are interested in the exact value prediction. The tool searches for qualitative errors as most benchmark tests do.

At the present state of development, the tool searches for derivative discontinuities and the procedure to predict other kind of errors is currently under investigation. Derivative discontinuities are a significant and common error that can lead to disastrous results. This error appears in several old (SPICE level 1 and 2 models) and newer models as well (BSIM3). The presented tool uses the following algorithm to find such areas: It calculates the derivatives of each set of points and when a sudden change appears it assumes the existence of a discontinuity. This idea is mathematically expressed via the following equation where (x_i, y_i) are the coordinates of the examined point and W is a positive constant.

$$\left| \frac{y_i - y_{i-1}}{x_i - x_{i-1}} - \frac{y_{i-1} - y_{i-2}}{x_{i-1} - x_{i-2}} \right| > W$$

The W constant defines the "sudden slope change". The correct estimation of its value is very important for the success of the above algorithm. If W has a small value then the algorithm is might wrongly interpret a significant slope change (like the one of point A in fig. 1) as a discontinuity. On the contrary, if W is large then the tool may not indicate some minor discontinuities (point B, Fig.1) and regard them as continuous slope variations. In order to avoid these events two criteria are set: The first one is that the constant should have a value that is a portion of the mean slope. The other one is that W should depend on the waveform shape. In this case, W is empirically extracted after the execution of several simulations with different MOSFET dimensions and models.

3.3 Tool's Output

In this paragraph some of the tool's output windows are demonstrated. Firstly, in Fig2, a part of the detailed report and the window that controls the graphical output are presented. Next, in Fig3 some indicative waveform windows are displayed.

Figure 2: The report and the results control window

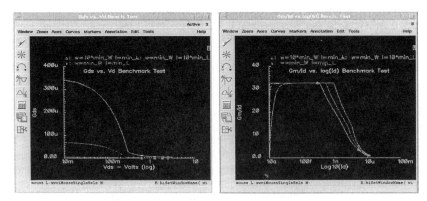

Figure 3: Output plots of some benchmark tests as the tool displays them

References:

1. Y. Tsividis and K. Suyama, "MOSFET Modelling for Analog Circuit CAD: Problems and Prospects". IEEE J. of Solid-State Circuits, vol. 29, March 1994.
2. Yannis Tsividis, "Operation and Modelling of The MOS Transistor". McGraw-Hill, Second Edition.
3. SKILL Language Reference, Cadence, SKILL ver. 04.50
4. IEEE Recommended Practices #P1485 on: "Test Procedures for Microelectronic MOSFET Circuit Simulator Model Validation" (Working Draft – completed 05/21/97), "http://ray.eeel.nist.gov/modval/database/contents/reports/micromosfet/standard.html"
5. C. Enz and Y. Cheng, "MOS Transistor Modelling for RF IC Design". IEEE J. of Solid-State Circuits, vol. 35, pp 186-202, February 2000.

AN HBT-BICMOS LASER DRIVER WITH INDEPENDENTLY ADJUSTABLE DC AND MODULATION CURRENTS FOR HIGH SPEED OPTICAL INTERCONNECTIONS

P. ROBOGIANNAKIS
Physics Department, Univ. of Crete, Heraklion, Greece
E-mail: pavrob@physics.uoc.gr

S. G. KATSAFOUROS, E. D. KYRIAKIS-BITZAROS, N. HARALABIDIS, G. HALKIAS
Institute of Microelectronics, NCSR "Demokritos", Athens 153 10, Greece
E-mail: halkias@imel.demokritos.gr

A SiGe HBT-BiCMOS Laser Diode Driver (LDD) capable of operating at multi-Gb/s data rates is reported. The design allows for independent adjustment of both DC and modulation current components. Simulation results have verified that the DC bias current of the Laser Diode can be varied between 0 and 12.5mA, whereas the modulation component ranges from 0 to 4mA in order to retain pulse symmetry and still produce a clear eye diagram at 20 Gb/s. The full-custom layout technique used for the implementation, minimizes space requirements as well as parasitic effects. The circuit has been designed using the AMS 0.8μm SiGe HBT-BiCMOS process.

1 Introduction

The high transmission capacity of optical fiber technology is in great need of respective integrated circuits capable of supporting multi-Gb/s rate networks. The technology used to implement such circuits up to now has relied upon GaAs or Si pure-bipolar technologies. The recently available choice of SiGe HBT-BiCMOS technology offers an equal performance potential for high-speed operation.

The field of application of the work presented here, lies in low-power high-speed optical interconnection systems. The required performance features are more efficiently realized through an optical interconnection scheme utilizing a Vertical Cavity Surface Emitting Laser (VCSEL). Such devices are reported to efficiently operate at high speeds with a threshold current close to 1mA.

Section 2 of this paper reports on design techniques supplemented by simulation results, while the layout and improved implementation aspects that assure design efficiency are presented in section 3.

2 Circuit Design and Simulation Results

The LDD, shown in Fig. 1, consists in principle of two cascaded differential pairs Q5-Q6 (DI) and Q1-Q2 (DO). The overall circuit performance depends on DO's high frequency characteristics and the quality of the signal driving it [1].

The input signal driving the Q5 base is a 0-3 V voltage swing. DI is biased at 3.6mA through current source Q8. The base of Q6 is biased at a 2V DC reference voltage (Vref). The latter is produced by the bias network consisting of devices D1-D4, the transistor Q9 and R5-R6 voltage divider. The Vref value has been chosen in order to assure the best noise margin possible. Capacitor C1 is utilized to eliminate the high frequency ripple on Vref. Due to its quite large value (~50pF) it is connected off-chip.

Resistors R3-R4, equal in value, set the differential output of DI. Q3-Q4 emitter follower buffers provide impedance matching and level shifting [2]. The modulation current component is set by the mirror pair Q10-Q12 through external bias voltage adjustment (Vcs). The DC current component is set by the mirror pair Q13-Q15 through adjustment of the Vdc bias voltage. The maximum DC and modulation current sum is limited by non-linear operation of the Q2 device. Vcs and Vdc can be varied independently and as a result the laser DC and modulation components can be adjusted individually according to the requirements of the application.

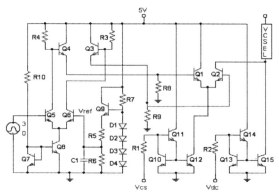

Figure 1. Circuit topology of the LDD.

In order to test the performance potential of our LDD we have performed simulations using a nonlinear equivalent circuit of a specific VCSEL. The chosen VCSEL has a threshold current of 2.5mA and a 3V maximum voltage drop [3]. Consequently, the external voltage bias for it should be carefully chosen to guarantee proper Q2 operation. The employed equivalent circuit of the VCSEL is based on a non-linear current-controlled voltage-source, as proposed in [3], in shunt with a parasitic capacitance of 0.3pF.

The simulations carried out using the CADENCE™ design framework showed that the LDD is capable to drive the specific VCSEL at a data rate of 20 Gb/s. Fig. 2 shows the transient response of the LDD when driven at 20 Gb/s and the modulation current is set at 4mA with a zero DC component (on-off operation). The expectation of a clear eye diagram is readily justified as the output current totally switches between 0 and 4mA maintaining distinct low/high levels. It is worthwhile noting that, from the 20-80% rise and fall time of the signal in Fig. 2, a data rate of 25Gb/s is attainable for the employed VCSEL. In the on-on mode of operation, Fig. 3 shows the possibility of maintaining a 4mA output pulse on top of a DC component ranging from 2.5mA to 12.5mA, at 20Gb/s.

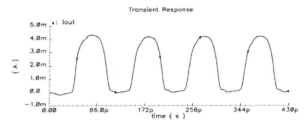

Figure 2. Transient response of the LDD at the Bit Rate of 20Gb/s.

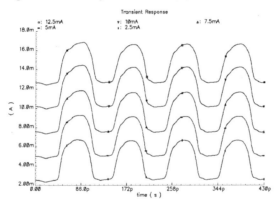

Figure 3. 20 Gb/s LDD response at various DC pre-biases (2.5mA-12.5mA) and 4mA modulation current.

3 Layout aspects

Special care has been taken for the design of buffers Q3 and Q4. Their outputs have been symmetrically connected and as close as possible to the DO inputs. This, along with wide metal interconnections, reduces the parasitic inductance of the buffers due to their low output impedance [4]. In general, a symmetrical concept has been followed to reduce electrical and thermal deviations. Careful routing is performed in order to avoid parasitic capacitance caused by first and second metallization crossings. Substrate contacts have been widely utilized to minimize resistive paths in the substrate that could lead to substrate coupling effects deleterious for the high-speed signal integrity. The layout of the designed circuit is shown in Fig. 4. The total area is (640 x 340) µm² including the pads.

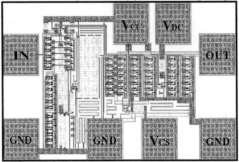

Figure 4. Layout of the designed HBT-BiCMOS Laser Diode Driver.

4 Conclusions

The work presented points out the potential of the SiGe HBT-BiCMOS technology in realizing ICs for multi-Gb/s optoelectronic links. Systematic design procedures resulted in an LDD capable of operating with a 0-12.5mA DC component and 4mA modulation current. Simulation results, based on AMS 0.8µm SiGe HBT-BiCMOS process, demonstrated the ability of the designed LDD to operate at data rates as high as 20Gb/s. To the best of our knowledge, this performance is the best ever reported in BiCMOS technology.

References

1. H. M. Rein, M. Moeler, "Design Considerations for Very-High-Speed Si-Bipolar IC's Operating up to 50 Gb/s", IEEE JSSC, Vol. 31, No 8, Aug. 1996, pp. 1076-1090.
2. H. M. Rein, R. Schmidt, P. Weger, T. Schmidt, T. Herzog, R. Lachner, "A Versatile Si-Bipolar Driver Circuit with High Output Voltage Swing for External and Direct Laser Modulation in 10 Gb/s Optical-Fiber Links", IEEE JSSC, Vol. 29, No 9, Sept. 1994, pp. 1014-1021.
3. Y. Ohiso, K. Tateno, Y. Kohama, A. Wakatsuki, H. Tsunetsugu, T. Kurokawa, "Flip-chip bonded 0.85-µm bottom emitting vertical cavity laser array on an AlGaAs substrate", IEEE Phot. Tech. Lett., vol. 8, Sept. 1996, pp. 1115-1117.
4. R. Schmid, T.F. Meister, M. Rest and H.-M. Rein, "SiGe Driver Circuit with High Output Amplitude Operating up to 23 Gb/s", IEEE JSSC, Vol. 34, No. 6, June 1999, pp. 886-891.

ON THE DESIGN OF A LOW POWER MODULATOR/DEMODULATOR FOR DECT/GSM

C. DROSOS, C. DRE AND S. BLIONAS
Intracom S.A., 19.5Km. Markopoulou Ave.,Peania, 19002,
GREECE
E-mail: cdro@intracom.gr

D. SOUDRIS
VLSI Design and Testing Center, Dept. of Elec. & Comp. Eng., Democritus University of Thrace, Xanthi, 67100,
GREECE

G. KALIVAS
Applied Electronics Lab, Dept. of Elec. & Comp. Eng., University of Patras, 26110 Patra,
GREECE

Recent advances in electronic technology and integration coupled with increasing needs for more services in portable communications favors the development of high performance dual-mode terminals. Here, we present the complete architecture and implementation of a novel GMSK/GFSK modulator/demodulator design. The main features of the modulator/demodulator, the methodologies that were followed and the design techniques used are described. The whole architecture of the modulator/demodulator was described in VHDL and then synthesized and implemented in Xilinx environment.

1 Introduction

In the last few years, there is a continuously growing demand for wireless terminals, integrating sophisticated multi-service applications. The achievements of the modern digital design technology make possible the idea of a mobile terminal that can operate and support different communications protocols, such as DECT and GSM/DCS-1800. Besides their functional characteristics these terminals should have small size and low-power consumption, so low-power design of integrated circuits becomes an interesting and challenging problem.

The purpose of this paper is to describe the architecture and the innovations that were followed during the design and implementation phase of a low power modem, the LPGD, for use in multi-mode DECT/GSM terminals. The receiver's modem is the basic building block of any wireless terminal. Today's heterodyne receivers have reached a high technical standard, but they still suffer from high production costs because of many expensive RF and IF components and mechanical integration problems. An alternative solution lies in the form of the direct conversion (homodyne) receivers. The direct conversion concept reduces the number of the RF circuits, transfers the signal processing to the baseband and avoids mirror frequency problems [1]. Furthermore this receiver architecture is insensitive against carrier and local oscillator frequency offset and drift. For the above reasons, the direct conversion architecture was selected for the implementation of the DECT receiver inside the LPGD modem. The development of the modem was supported by the LPGD #25256 project, funded by European Union in context of ESPRIT IV.

2 LPGM modem

LPGD is the only available dual-mode GMSK/GFSK low power modem, supporting both DECT and GSM standards. It supports a direct conversion terminal architecture and it incorporates a DECT differential detector based on a novel, low complexity I-Q algorithm, suitable for low power, digital implementation. The direct conversion architecture is based on a zero IF solution and allows digital processing, detection and implementation in the baseband part of the multi-mode receiver. The LPGD receiver solution is based on In phase (I) and Quadrature (Q) channel demodulation. It also features a GMSK digital modulator suitable for both DECT and GSM operation and suitable FIR filters for the receive path.

The main building blocks of the implemented Modem, as well as the interface links and signals with a reference dual-mode terminal are presented in **Figure 1**.

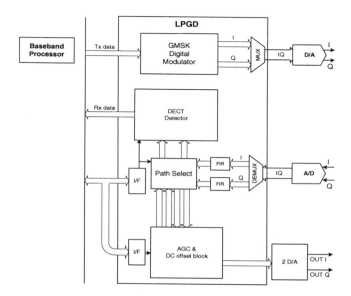

Figure 1. LPGD Modem with interfaces.

The architecture of the digital GMSK modulator was designed to satisfy both GSM and DECT standards. The modulator receives data from the processor and produces two output streams, I and Q, which are eventually the input of a D/A converter. The outcome of the modulator is two GMSK modulated digital streams, Iout and Qout of 10 bits each. The modulator uses internally an optimized look-up-table in order to implement the GSMK algorithm [2].

The Path Select block of the LPGD circuit functions as a multiplexer, controlled by the Receiver. The operation of this multiplexer is to drive signal inputs of the DECT Detector and the AGC block with valid data. So it can drive the Detector with data from either the output of the AGC block or the FIR filters. The AGC block is fed with signals from the FIR or the output of the Detector.

The operation of the AGC & DC Offset block is essential for the correct operation of the selected DECT detection algorithm. This block is responsible for the digital elimination of the DC offset in the I and Q samples. It also adjusts the amplitude of these samples, so that the quality of the signal fed to the detector to be as good as possible, regardless of the characteristics of the received signal.

Filtering in most digital communication systems is accomplished by a combination of analog and digital low-pass filters. Digital low-pass filters are designed as Finite Impulse Response (FIR) filters. In this way there is more flexibility in the design and on the other hand it is easier to obtain some very desirable characteristics like phase linearity and high accuracy in the response.

The operation of the DECT Detector/Receiver block is to accept as its input a quantized IQ stream consisting of a pair (I, Q) of 6-bit vectors in size-magnitude format. The processing of the above stream by the internal blocks of the Detector yields the bit stream of the data section contained in a DECT slot.

3 LPGD verification and testing

Special care was given to the verification and testing of the LPGD implementation, and specifically of the DECT receiver block. For these purpose, a multi-mode terminal testbench was created according to the direct conversion architecture. The LPGD modem was the device under test of this terminal.

The performance of the output of the digital modulator developed for the LPGD modem was tested and analyzed in terms of spectral compliance. Figure 2 displays the spectrum of the generated I and Q signal from the modulator in DECT mode. The maximum power levels as specified by ETSI for the adjacent channels are also shown. It is clear that the filtered output signal satisfies the maximum power levels. The performance in GSM mode is similar and within the GSM protocol's specifications.

A number of bit error rate (BER) measurements have been performed in order to evaluate the performance of the LPGD DECT Receiver block and its conformance with the specifications. Figure 3 depicts the performance of the detector for gaussian noise channel. The reference BER of 10^{-3} is achieved for a signal to noise ratio (SNR) of 10.5dB. These measurements were performed with the FIR filters in the LPGD receiver path. Same measurements with the FIR filters disabled showed a degradation of about 4dB. The performance of the algorithm for a frequency difference of 25kHz between the transmitter's and the receiver's reference frequencies is degraded about 3dB. Figure 4 displays the performance of the detector when an adjacent DECT channel is present. The detector achieves the reference BER with a SNR of 19.5dB, even if the of the interferer channel is 13dB more than the usable signal as specified in the DECT standards.

4 Conclusion

A low-power DECT/GSM modem is a required block for any dual-mode wireless terminal. A novel direct conversion architecture was selected and implemented, due to its advantages, for the DECT detector block.

The LPGD modem was designed entirely using VHDL and implemented in a Xilinx series FPGA. The modem design is of 39,000 equivalent gates, its maximum power dissipation is 8,204 mW/MHz and its maximum operating frequency is 9.216MHz. Special care was given to the verification and testing of the design in order to prove the correct operation of the LPGD modem. The overall performance of the LPGD modem and its direct conversion architecture proved to be quite satisfactory and promising and the developed architecture met its initial specifications and targets.

References

1. E. Metaxakis, A. Tzimas, G. Kalivas, "A Low Complexity Baseband Receiver for Direct Conversion DECT-based Portable Communication Systems", IEEE 1998 Int' Conf. on Universal Personal Communications.
2. D. Soudris, M. Perakis, et al, "A Low Power Design of a DCS1800-GSM/DECT Modulator/Demodulator", IEEE International Symposium on Circuits and Systems (ISCAS'00), Geneva, Switzerland, May 28-31.

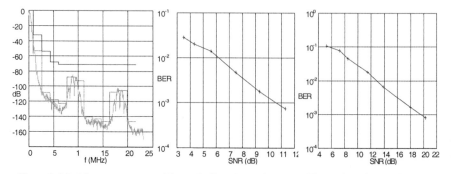

Figure 2. Modulator spectrum.   Figure 3. Gaussian noise.   Figure 4. Adjacent channel.

BLUETOOTH ENCRYPTION / DECRYPTION ALGORITHM ARCHITECTURE AND IMPLEMENTATION

PARASKEVAS KITSOS AND ODYSSEAS KOUFOPAVLOU

VLSI Design Lab, Electrical and Computer Engineering Department, University of Patras, GREECE
E-mail: {pkitsos, odysseas}@ee.upatras.gr

In this paper a new VLSI implementation of the Bluetooth Encryption algorithm, for efficient usage in portable Bluetooth telecommunication systems, is presented. The algorithm modulo operation of polynomials was implemented by using Linear Feedback Shift Registers (LFSR). The measured power dissipation at 32 Mbps is only about 1,12 mW. The whole Bluetooth Encryption / Decryption algorithm was captured by using VHDL language. For the synthesis and simulation commercial available tools were used, but for the power measurements a custom design tool developed in our University was used.

1 Introduction

Bluetooth [1] uses a secret key encryption technique that offers key lengths of 0, 40, 64 or 128 bits by a pair of devices. The encryption of the payloads is carried out with a stream cipher called E_0 that is re-synchronized for every payload. The stream cipher system E_0 consists of three parts. One part performs the initialization (generation of the payload key), the second part generates the key stream bits and the third part performs the encryption and decryption. The key stream bits are generated by a method derived from the summation stream cipher generator attributable to Massey and Rueppel [2].

2 The E_0 Algorithm and System Architecture

The hardware implementations of E_0 algorithm content two different components [1]. The first is the initialization part that uses 48 bits of Bluetooth address, 26 bits of the master real time clock and the encryption key K_c. Within this part the encryption key K_c is modified into another key denoted K'_c. The second entity, which generates the key stream use four linear feedback shift registers (LFSR's) whose output is combined by a finite state machine (called the summation combiner) with 16 state. The output of this state machine is the key stream sequence. The overall principle is shown in Figure 1.

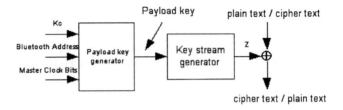

Figure 1: Stream ciphering for Bluetooth with E_0.

The key stream generator (Figure 2) uses four linear feedback shift registers (LFSR's) with lengths $L_1=25$, $L_2=31$, $L_3=33$, $L_4=39$, with feedback polynomials ($f_i(t)$) as specified in Table 1 below.

Table 1: The four feedback polynomials

i	L_i	Feedback $f_i(t)$	Weight
1	25	$t^{25}+t^{20}+t^{12}+t^8+1$	5
2	31	$t^{31}+t^{24}+t^{16}+t^{12}+1$	5
3	33	$t^{33}+t^{28}+t^{24}+t^4+1$	5
4	39	$t^{39}+t^{36}+t^{28}+t^4+1$	5

The total length of the register is 128. The Hamming weight of all the feedback polynomials is chosen to be five for reasonable trade-off between reducing the number of required XOR gates in the hardware realization and obtaining good statistical properties of the generated sequences [3].

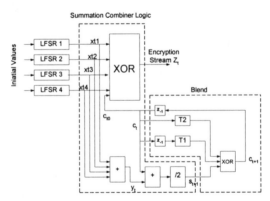

Figure 2: The key stream generator.

Let x_t^i denote the tth symbol of LFSR$_i$. From the four-tuple $x_t^1,...,x_t^4$ derived the value y_t as

$$y_t = \sum_{i=1}^{4} x_t^i \quad (2)$$

where the sum is over the integer. Thus y_t can take the values 0,1,2,3 or 4. The output of the summation generator is given by the following equations

$$z_t = x_t^1 \oplus x_t^2 \oplus x_t^3 \oplus x_t^4 \oplus c_t^0 \in \{0,1\} \quad (3)$$

$$s_{t+1} = (s_{t+1}^1, s_{t+1}^0) = \left[\frac{y_t + c_t}{2}\right] \in \{0,1,2,3\} \quad (4)$$

$$c_{t+1} = (c_{t+1}^1 + c_{t+1}^0) = s_{t+1} \oplus T_1[c_t] \oplus T_2[c_{t-1}] \quad (5)$$

where $T_1[c_t]$ and $T_2[c_{t-1}]$ are two different linear bijections over Galois Field Functions GF(4) [4]. We can write the elements of GF(4) as binary vectors. Since the mapping is linear, we can realize them using XOR gates, as

$$T_1 : (x_1, x_0) \mapsto (x_1, x_0) \quad (6)$$

$$T_2 : (x_1, x_0) \mapsto (x_0, x_1 \oplus x_0) \quad (7)$$

3 Hardware Implementation

The whole Bluetooth Encryption / Decryption algorithm was captured by using VHDL language. For the synthesis and simulation commercial available tools were used. All the measurements are made using a 0.7 μm CMOS standard-cell library. Power measurements have been acquired using toggle count during logic-level simulation under real-time delay model, capacitance estimates provided by the tools, and 365 randomly generated input vectors. To implement the measurements flow (synthesis-simulation-power estimation) we have used a in-house plug-in to Mentor Graphics framework [5], which automatically generates scripts, invokes the tools and parses their reports. The critical path of our architecture is 0,574 nsec, the Bit-rate is 32 Mbps while the maximum clock frequency is 1,7 GHz. In table 2 the architecture simulation and synthesis results are shown.

Table 2: The characteristic of our Architecture

E_0 Algorithm Measurement	Value
Power Dissipation	1.12 mWatt
Estimation Area	557,87 mils^2
Number of Transistors	8069
Number of Nets	759
Library Cells	383

4 Conclusions

A low power and very high-speed architecture of the Bluetooth Encryption / Decryption algorithm is described. All the measurements are made using a 0.7 μm CMOS standard-cell library. The critical path of our architecture is 0,574 nsec and the throughput is 32 Mbps. The operation clock frequency is up to 1,7 GHz. For this reason, this architecture can be applied in the next generation low power Bluetooth digital cellular and personal communication services applications.

5 Acknowledgements

We thank Nikos Sklavos for his valuable help.

References

[1] *"Specification of the Bluetooth System"*, Specification Volume 1, December 1st 1999, pp. 159-168.
[2] J. Massey and R. Rueppel, *"Linear Ciphers and Random Sequence Generators With Multiple Clocks"*, In T. Beth and N. Cot and I. Ingemarsson, editors, Advances in Cryptology: Proceedings of EUROCRYPT 84, volume 209 of Lecture Notes in Computer Science, pages 74-87. Springer-Verlag, 1985, 9-11 April 1984.
[3] J. Massey, *"Shift-Register Synthesis and BCH Decoding"*, IEEE Transactions on Information Theory, Vol. IT-15, NO 1, January 1969.
[4] Bruce Schneier, *"Applied Cryptography"*, 1996, Wiley editions.
[5] N. Zervas, S. Theoharis, D. Soudris, C. Goutis, and A. Thanailakis, *"Generalized Low Power Design Flow"*, LPGD Project (ESPRIT 25256), Deliverable Report LPGD/WP2/UP/D1.3R1, pp. 11-15, Jan

PASSIVE ELEMENT DESIGN ISSUES FOR FULLY INTEGRATED RF VCOs

ARISTIDES KYRANAS AND YANNIS PAPANANOS
Microelectronic Circuit Design Group
National Technical University of Athens
9 Iroon Politechniou, Zografou 15773, Greece

In this paper the issue of designing fully integrated RF VCOs is addressed. The emphasis is given in passive device integration in silicon technologies. On-chip inductors and variable capacitors are presented along with design examples at 5 and 6GHz. An overview of different resonator topologies is made and simple models of passive devices are used to compare them.

1 INTRODUCTION

The need for ubiquitous communications demands low power portable devices with small form factors. One of the key issues that need to be addressed is the level of integration of the various subsystems, especially the RF circuitry. Higher level of integration can reduce PCB area and increase the reliability of the system. Recent research activities have demonstrated the potential for increasing the level of integration of RF subsystems into the same IC [1]-[2].

One of the major building blocks in an RF transceiver is the Local Oscillator (LO) signal generator. A voltage-controlled oscillator (VCO) is used in a PLL to generate the LO signal and drive the up- and down-conversion mixers. For high frequency LO generation, harmonic oscillators are almost exclusively used due to their superior performance in terms of phase noise, harmonic content and current consumption. They are formed by the parallel connection of a passive resonator and an active circuit that compensates for the resonator losses. One of the most important parameters for a VCO circuit is phase noise, which determines the spectral purity of the output signal and depends heavily on resonator quality factor. The integration of high-Q resonators can be problematic, since on-chip passive components have lower quality factors compared to their discrete equivalents.

In recent research works, designs of integrated RF VCOs have been presented [3]-[6]. In this paper emphasis is given in passive device integration potential in silicon technologies. Passive component design examples are also presented.

2 PASSIVE COMPONENTS FOR INTEGRATED VCOs
2.1 INTEGRATED INDUCTORS

Integrated inductors are implemented as planar spiral structures in one or more metal layers. The main design parameters for a spiral inductor are its inductance L and quality factor Q. Three distinct regions of operation exist [7]:

- Low frequencies where the inductor has relatively constant inductance. This is the useful frequency region where the spiral can be used as an inductor.

- Mid frequencies where a transition from inductive to capacitive behaviour occurs. The inductance changes with respect to frequency significantly. At the self-resonance frequency f_{SR} the spiral behaves as a tuned LC tank. F_{SR} depends on spiral parasitic capacitances to ground and is sensitive to process spreads.
- At higher frequencies the spiral behaves as a capacitor and cannot be used.

The quality factor Q is a measure of the power dissipated as heat. The main loss mechanisms are substrate and metal track resistances. The substrate behaves like a lossy dielectric due to the induced substrate currents from the inductor magnetic field, increasing inductor losses. The ohmic resistance of the spiral metal tracks contributes to the inductor series resistance. Wide metal tracks reduce series resistance, but at the same time increase parasitic capacitance to ground and reduce self-resonance frequency f_{SR}. Care should be taken so that f_{SR} is not at the frequency band of operation, making the inductor unusable.

The CAD tool presented in [8]-[9] can be used to simulate the spiral structure. Inductance and quality factor are extracted from the spiral geometry. The series connection of an inductor L and a resistor R_{SL} can model the behaviour of the spiral in a narrow band of interest. R_{SL} can be computed as $R_{SL} = \frac{2\pi f L}{Q}$.

The design of inductors that operate at high frequencies is more challenging because small inductances are needed. The spiral should have small geometrical characteristics and interconnections play a significant role to overall inductance. **Figure 1** illustrates an example of an integrated inductor design at 5GHz. It is designed in AMS's 0.8um HBT BiCMOS process. The simulated inductance is 0.8nH with a quality factor of 6 at 5GHz. Figure 1 shows simulations of L and Q with and without interconnections, along with measurement results. Excellent agreement exists between inductance measurements and simulation, however the discrepancy in the quality factor is being investigated.

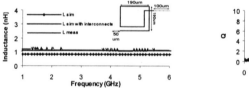
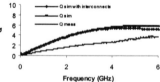

Figure 1: Small inductance performance

2.2 VARIABLE CAPACITORS

Variable capacitors are used to tune the resonance frequency. The most common implementation in silicon technologies is the reverse biased p/n junction. Figure 2 shows a simplified electrical equivalent for the variable capacitor along with implementation examples in bipolar/BiCMOS and pure CMOS technologies.

Junction capacitance C_j is given as $C_j = C_0 \left(\frac{A}{A+V_R} \right)^m$ where V_R is the reverse bias voltage, C_0 the capacitance at zero bias, and A, m constants that depend on the technology and diode geometry.

Figure 2: Variable capacitor equivalent circuit and Bipolar/BiCMOS and CMOS implementations

At high frequencies the effect of R_s dominates and the quality factor is computed as $Q \approx \frac{1}{\omega R_s C_j}$. For a given frequency, Q depends on reverse bias voltage. As the reverse bias voltage increases, both junction capacitance C_j and series resistance R_s decrease increasing Q. In bipolar and BiCMOS processes the base-emitter junction has the advantage of low emitter resistance, but the drawback of low breakdown voltage, in the order of 2V for a 0.8um HBT BiCMOS process, and thus reduced tuning range. The base-collector junction can withstand higher reverse bias voltages, as its breakdown voltage is in the order of 15V and can provide increased tuning range. However, the collector resistance is large, increasing losses.

In a CMOS process the variable capacitor can be implemented as shown in Figure 2. To reduce diode losses, many unit structures can be connected in parallel and the strapping method be used in the layout.

Figure 3 shows measurement results for a floating CMOS varactor. The structure is designed in AMS's 0.8um HBT BiCMOS process in an interdigitized form of p+ and n+ regions. The capacitance is computed from S_{11} and S_{22} parameters as the imaginary part of the input impedance evaluated at 6GHz. Port1 is connected to the anode and Port2 to the cathode of the diode. Measurement results clearly demonstrate the influence of the parasitic diode to ground.

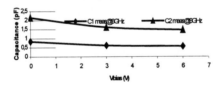

Figure 3: Varactor measurement and simulation results

3 RESONATOR DESIGN

The resonator is formed by the series or parallel connection of an inductor and a variable capacitor. In this paper we consider the parallel connection. The resonator Q is computed as $Q_{\tan k} \approx \sqrt{\frac{L}{C}} \frac{1}{R_{sL}+R_{sC}}$. It can be increased if we either use high-Q passive elements, or a high inductance value. Fabrication technology poses a limitation to the highest Q of passive devices. By increasing L and in order to keep oscillation frequency constant, the tuning capacitance should be decreased, decreasing also the tuning range. For the parallel resonator considered in this work, there are three basic different configurations, as shown **Figure 4**.

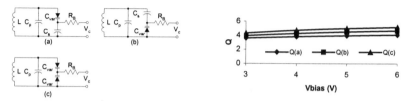

Figure 4: Parallel resonator topologies **Figure 5**: Parallel resonator simulation results

From the analysis in [10], the configuration of **Figure 4(c)** is advantageous for integration, as its resonance frequency is more stable with respect to the oscillator signal and the effect of the biasing resistor is constant throughout the whole tuning range. **Figure 5** shows the simulated quality factor for the three resonator configurations. The inductor is the one presented in **Figure 1**. The varactors are the ones presented in **Figure 3**, the fixed capacitor is 5pF and the biasing resistor 1K. The Q for each resonator is calculated as $Q = \pi f_0 \tau_d$ where τ_d is the group delay of the resonator and f_0 the frequency where group delay is maximized.

REFERENCES

[1] P. R. Gray, R. G. Meyer, "Future Directions in Silicon ICs for RF Personal Communications", Proc. 1995 CICC.
[2] M. A. Copeland et. al., "5-GHz SiGe HBT Monolithic Radio Transceiver with Tunable Filtering", IEEE Transactions on MTT, Vol. 48, pp. 170-181, Feb. 2000.
[3] M. Soyer et. al., "A 3-V 4-GHz nMOS Voltage-Controlled Oscillator with Integrated Resonator", IEEE JSSC, pp. 2042-2045, Dec. 1996.
[4] M. Soyuer et. al., "An 11-GHz 3-V SiGe Voltage Controlled Oscillator with Integrated Resonator", IEEE JSSC, pp. 1451-1454, Sep. 1997.
[5] P. Kinget, "A Fully Integrated 2.7V 0.35um CMOS VCO for 5GHz Wireless Applications", Proc. 1998 ISSCC, pp.226-227.
[6] A. Kyranas, Y. Papananos, "A 5GHz Fully Integrated VCO in a SiGe Bipolar Technology", Proc. 2000 ISCAS.
[7] Y. Koutsoyannopoulos, Y. Papananos, "Systematic Analysis and Modelling of Integrated Inductors and Transformers in RF IC Design", IEEE TCAS-II, Vol. 47, pp. 699-713, Aug. 2000.
[8] Y. Koutsoyannopoulos et. al., "A Generic CAD Model for Arbitrary Shaped and Multi-Layer Integrated Inductors on Silicon Substrates", Proc. 1997 ESSCIRC.
[9] Y. Koutsoyannopoulos, Y. Papananos, "SISP: A CAD Tool for Simulating the Performance of Polygonal and Multi-Layer Integrated Inductors on Silicon Substrates", Proc. 1997 ICVC, pp. 244-246, Oct. 1997.
[10] A. Kyranas, Y. Papananos, "Design Issues Towards the Integration of Passive Components in Silicon RF VCOs", Proc. 1998 ICECS.

POWER AMPLIFIER LINEARISATION TECHNIQUES: AN OVERVIEW

NIKOS NASKAS AND YANNIS PAPANANOS

Microelectronic Circuit Design Group, National Technical University of Athens
9 Iroon Politechniou, Zografou 15773, Greece

The design of a linear and power efficient PA is a large and extensively active research area for a wide range of applications in mobile and terrestrial communication. A number of linearisers are available for commercial use until now, while others are being in a varying degree of development. In this paper techniques that linearise non-linear high efficiency power amplifiers are being briefly presented.

1 Introduction

An important feature of a modern telecommunication system is bandwidth efficiency. Bandwidth efficiency is directly related not only to the type of modulation used but also to the transmitter linearity. A bandwidth efficient system uses generally a non-constant envelope modulation scheme. Forming a variable envelope signal to a non-linear transmitter would probably be catastrophic because of bandwidth regrowth that increases adjacent channel interference (ACI) and intermodulation distortion (IMD) above acceptable limits. In such case the design of a linear transmitter is essential. This requires a linear power amplifier (PA), which has usually low power efficiency. In a mobile communication system the efficiency of the PA is one of the most important parameters, as it determines power consumption that results in battery size, talk time etc.

Some of the well-published linearisers that combine good efficiency with high linearity are: *Feedforward, Cartesian loop, Predistortion, Envelope Elimination and Restoration (EE&R)*.

2 Feedforward

Feedforward technique is an open loop architecture that applies the signal correction directly to the output of the non-linear PA [1,2]. Figure 1 depicts the feedforward in its simplest form. The RF signal is split into two paths, with one going to the PA (main amplifier) and the other to the error amplifier. Portion of the distorted signal in the upper path is subtracted from the time-delayed original signal generating an error signal. This signal after appropriate linear amplification (error amplifier) cancels the distortion in the main amplifier path. The resulting output is a linear amplified version of the original input signal.

Improvements in distortion performance of 30dB can be achieved using the feedforward in its basic configuration, although better improvement can be achieved by using the feedforward in combination with another lineariser (e.g. RF predistortion) or by using a dual loop feedforward.

As an open loop system the feedforward is unconditionally (or almost unconditionally) stable something that makes it attractive in broadband applications. However, its poor adaptability and power efficiency (primary determined by the error amplifier) has limit its use only in high power basestation transmitters.

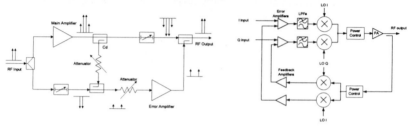

Figure 1. A generic feedforward system.   Figure 2. Cartesian loop block diagram.

3 Cartesian Loop

An excellent technique with low complexity, high linearity and good power efficiency is Cartesian Loop. In contrast to feedforward, it is an inherently adaptive technique, but because of its "closed loop nature" it has restrictions in signal bandwidth it can linearise [3].

Figure 2 demonstrates a simplified block diagram of the Cartesian loop lineariser. Cartesian loop using the in-phase and quadrate components of the complex signal can remove both amplitude to amplitude (AM/AM) and amplitude to phase (AM/PM) distortion. The translation of the signal to the desired RF carrier frequency can be direct through a quadrature modulator or after some intermediate stage. The RF signal after passing trough the non-linear amplifier is sampled by the directional coupler. The coupled signal, after the appropriate attenuation, is down converted and subtracted from the forward baseband I and Q signals. By this mechanism apart to the non-linearities of the RF power amplifier, the non-linearities of the whole upconversion chain are linearised as well.

Similar to the Cartesian Loop is the Polar Loop [4]. The difference between them is that the baseband signal is now presented in its equivalent polar from, Amplitude/Phase instead of I/Q. The Polar Loop has not to offer something better than Cartesian Loop and because of its increased complexity its use is restricted.

4 Predistortion

Predistortion technique like the feedforward technique is not an inherently adaptive system. It is generally an open loop system that can be used both in narrowband and broadband applications [5]. In Figure 3 the basic concept behind predistortion is illustrated. The original signal is passed through a predistortion component that introduces a non-linearity (complementary to the PA non-linearity), which when cascaded to the PA, results in a linear amplification.

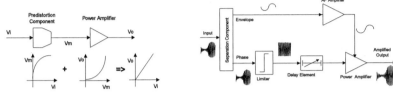

Figure 3. Predistortion basics.   **Figure 4**. The EE&R technique.

Depending on the frequency in the upconversion chain, the predistortion solution could take one of the following forms:
- Baseband predistortion - in this case the predistortion characteristic is stored in a look up table memory (LUT), which with the aid of a DSP predistorts the baseband signal before upconversion.
- IF predistortion - the predistortion is performed in a convenient intermediate frequency. This architecture allows the same predistortion circuitry to be used to linearise a number of different carrier frequencies.
- RF predistortion and RF postdistortion - the predistorter operates at the final carrier frequency, while the predistortion is performed before (predistortion) or after (postdistortion) the PA.

Adaptive predistortion is a variant of the basic predistortion that is used when increased adaptability is required [6]. The baseband signal is presented in a quadrature form in a similar topology to the Cartesian Loop. In such case, the predistortion coefficients are being adaptive while the linearity is preserved at high levels.

5 Envelope Elimination and Restoration

EE&R was originally proposed by L. Kahn in 1952 and thenceforth it has been extensively used in high power TV and Radio broadcasting [7,8]. The basic block diagram of this technique is being presented in Figure 4.

The key issue in the principle of operation of the EE&R is that the RF signal is amplitude modulated through the power supply of the PA. The complex signal is split into two components, the amplitude information and the phase information. The splitting can be done in baseband part using a DSP method or in the RF part using an envelope detector. The phase information is driven to a limiter so as to have no amplitude variation and then is been amplified in a non-linear high efficiency PA. A switched audio frequency amplifier (AF) effectively amplifies the envelope information. The signal after the AF amplifier, directly amplitude modulates the PA, restoring the envelope information to the transmitted signal. As the AF and PA can be 100% power efficient, this technique has a theoretical potential of 100% power efficiency.

6 Summary

Feedforward is probably the most suitable technique for wideband and high linearity applications. However, the poor adaptability and power efficiency in addition to full integration restrictions have made feedforward technique intractable for mobile communication systems.

The Cartesian loop is an excellent technique with relatively low complexity and easy integration to a single ASIC. It offers high linearity and good power efficiency but its use is restricted to narrowband applications such as a single 25KHz carrier or several 5KHz carriers.

Adaptive predistortion is a potentially growing method that offers good linearity and system flexibility. However, it encounters practical difficulties because of its slow adaptation time and increased complexity.

Finally, EE&R, as it can achieve high power efficiency and good linearity in a relatively easy integration, could have promising practical implementations.

References

1. J. Dixon, "A Solid-State Amplifier with Feedforward Correction for Linear Single-Sideband Applications," IEEE ICC, June 1986, pp. 728-732.
2. J. Cavers, "Adaption Behavior of a Feedforward Amplifier Linearizer," IEEE Tran. on Vehicular Tech., Vol. 44, No.1, Feb.1995, pp. 31-39.
3. A. Bateman, D.M. Haines and R.J. Wilkinson, "Direct conversion linear transceiver design," IEE 5th International Conference on Mobile Radio and Personal Communications, Warwick, UK, December 1989, pp. 53-56.
4. V. Petrovic and W. Gosling, "Polar Loop Transmitter" Electronic Letters, 1979, Vol. 15, pp. 1706-1712.
5. L. Sundstrom, "Chip for Linearisation of RF Power Amplifiers Using Digital Predistortion", Electronics Letters, July 1994, 30, (14), pp. 1123-1124.
6. J. Cavers, "Amplifier Linearisation Using a Digital Predistorter with Fast Adaptation and Low Memory Requirements," IEEE Tran. on Vehicular Tech., Vol. 39, No. 4, Nov. 1990, pp.374-382.
7. L. Kahn, "Single-Sideband Transmission by Envelope Elimination and Restoration", Proc. I.R.E., July 1952, pp. 803-807.
8. D. Su, W. McFarland, "An IC for Linearizing RF Power Amplifiers Using Envelope Elimination and Restoration", IEEE Journal of Solid-State Circuits, Vol. 33, No. 12, December 1998.

LOW-POWER IMPLEMENTATION OF AN ENCRYPTION/DECRYPTION SYSTEM WITH ASYNCHRONOUS TECHNIQUES

NIKOS SKLAVOS and ODYSSEAS KOUFOPAVLOU

VLSI Design Laboratory, Electrical and Computer Engineering Department
University of Patras, Patras, GREECE
{nsklavos, Odysseas}@ee.upatras.gr

> An asynchronous VLSI implementation of the IDEA encryption/decryption algorithm is presented in this paper. In order to evaluate the asynchronous design a synchronous version of the algorithm was also designed. VHDL hardware description language was used in order to describe the algorithm. By using Synopsys commercial available tools the VHDL code was synthesized. After placing and routing both designs were fabricated with 0.6 μm CMOS technology. The two chips were tested and evaluated comparing with the software implementation of the IDEA algorithm. Although our synchronous design has low power dissipation, the asynchronous one has significantly very low power consumption. The asynchronous implementation achieves 20-40% total power savings comparing with the synchronous one. These integrated circuits can be applied as very fast and low power encryption/decryption devices in high speed networking protocols such as ATM (Asynchronous Transfer Mode) and WEP (Wireless Encryption Protocol).

1 Introduction

In general terms, there are two types of key-based encryption algorithms: Symmetric and public key ones. Symmetric algorithms, which are the ones of interest here, are these algorithms in which the encryption key can be calculated from the decryption key and vice versa while in most of them both (encryption and decryption) keys are identical. Symmetric algorithms can be further divided in two other categories: The first category includes the ones that operate on the plaintext a single bit at a time and they are called stream algorithms or stream ciphers. The second category includes the algorithms which operate on the plaintext in groups of bits (usually 64, since this length is considered long enough to preclude analysis and small length to be workable). These algorithms are called block algorithms or block ciphers [1]. The latter category will be the one of interest in this document.

2 Idea Algorithm – Design Principles

The IDEA is a block oriented encryption algorithm that operates on a 64-bit plaintext and uses a 128-bit length key. The algorithm has been developed and published by Lai and Massey [2]. The design philosophy of this algorithm is based on the concept "of mixing operations from different algebraic groups". These three algebraic operations are the following:

- Bit-by bit XOR (denoted as $\oplus$)

- Addition of integers modulo 2^{16} (denoted as $\otimes$) with inputs and outputs treated as unsigned 16-bit integers
- Multiplication of integers modulo 2^{16} (denoted as $\square$) with inputs and outputs treated as unsigned 16-bit integers.

The 64-bit data block of the input plaintext is divided into four 16-bit sub-blocks: X1, X2, X3 and X4. These four sub-blocks become the necessary input to the first round of the algorithm. There are eight rounds in the algorithm in total. Each iteration (round) takes four 16-bit sub-blocks as inputs and produces four 16-bit output blocks. In each round, the four sub-blocks are XORed, added and multiplied with one another and with six 16bit sub-keys. The subblock keys are denoted as $Z_1^{(r)},\ldots,Z_6^{(r)}$, for every (r) round. Between rounds, the second and third sub-blocks are swapped. Finally, the four sub-blocks are combined with four subkeys, $Z_1^{(9)},\ldots,Z_4^{(9)}$, in an output transformation, producing 16-bit output blocks which are concatenated to form the 64-bit ciphertext. Figure 1 gives an overview of this process.

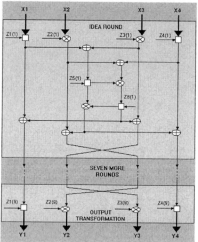

Figure 1. IDEA Algorithm Architecture

3 System Architecture

Two different designs were fabricated in this work in order to implement the IDEA algorithm in hardware. These two designs are similar in their functionality but different clocking techniques were used in each of them. The first design is synchronous and the second asynchronous. The architecture consists of the five rounds (4 normal IDEA rounds and the output transformation) and a feedback loop. The IDEA rounds all use encryption sub-keys generated from the user-defined 128-bit key. The schematic diagram of each of the two versions, synchronous and asynchronous are showed in figures 2 and 3.

327

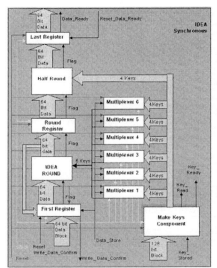 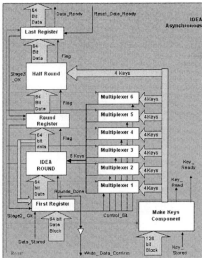

Figure 2. IDEA Synchronous Version **Figure 3.** IDEA Asynchronous Version

4 Asynchronous VLSI Design

In the asynchronous version, special control units have been designed in order to replace control operations based on clock, which the synchronous implementation uses. The main demand of the asynchronous control units is to avoid their non-deterministic behavior. In order to achieve this, these circuits must be hazard-free under circuit delay model. Certain forms of MIC behavior can be tolerated. The most general signal concurrency must be controlled by arbitration in order to avoid circuit hazards. C Element is a component that can be used for arbitration [3 - 5]. The C Element, showed in figure 4, is a commonly used asynchronous circuit component. The use of this element prevents another pending request from passing through the arbiter until after the active request cycle has cleared.

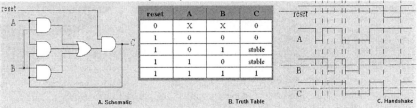

Figure 4. A two inputs C element

5 Chip Characteristics

The characteristics of the two chips are shown below in Table I

Code	Parameter	Value (Syn/Asyn)	Unit
N I/O	Number of I/O pins	40 / 40	pins
Package	ASIC Package	DIL40 / DIL40	Package type
X	Length	8920 / 8431	um
Y	Height	5990 / 5932	um
Area	Chip effective area	53,43 / 50,01	mm^2
Ngates	Equivalent gates	47555 / 44708	2 input NAND
Ntran	Transistor Count	190223 / 178835	transistors
VDD	Power Supply	4.5-5.5	Volt
F	Operation frequency	up to 8	MHz
Supply Current	Standby Mode	7.51 / 4.25	(mA)
	Operating Mode	10.40 / 8.40	
Power Supply	Standby Mode	37.55 / 21.25	(mW)
	Operating Mode	52.00 / 42.00	

Table I. Chip Characteristics of the IDEA Synchronous and Asynchronous Chip

The two designs were fabricated with 0.6 μm CMOS technology. Both implementations operate with 5 MHz frequency. This value produces a period clock of 200 ns. The four rounds and the last transformation of IDEA's implementation need 5 (4+1) rounds * 200 ns =1000 ns / data block. In other words the ASIC produces 64 bits every 1000 ns, so the throughput is 64 Mbits/sec. The testing procedure gives us results of the power consumption. While there are three different scenarios in the power measurement operation the conclusion is still the same for all of them. Asynchronous implementation has the lowest power consumption for all the operation modes of the ASIC. The difference of the values in percentage units is 20-40 %. This proves that the asynchronous version is more useful and performs better in terms of power consumption than the synchronous.

References

1. "Applied Cryptography", Bruce Schneier, 1996, Wiley editions
2. Xuejia Lai and James L. Massey " A proposal for a new Block Encryption Standard" proceedings of Eurocrypt'90, Aarhus, Denmark, May 21-24 1990, pp. 389-404.
3. S. Hauck, "Asynchronous Design Methodologies: An Overview", Proceedings of the IEEE, vol. 83, pp. 69-93, January 1995.
4. Al. Davis, and Steven M. Nowick, "Asynchronous Circuit Design: Movitation, Background, and Methods", Asynchronous Digital Circuit Design, pp 1-49, Workshops in Computing, Springer Verlag 1995.
5. J.V. Woods, P. Day, S.B. Furber, J.D. Garside, N.C. Paver, and S.Temple, "AMULET1: An Asynchronous ARM Microprocessor", IEEE Transactions on Computers, vol. 46, no. 4, pp. 385-398, April 1997.

A 0.25μm CMOS FAST CURRENT AMPLIFIER WITH LEAKAGE CURRENT COMPENSATION FOR SOLID STATE DETECTOR APPLICATIONS

Emmanuel Zervakis

Institute of Microelectronics, N.C.S.R. "Demokritos", Athens 15310, Greece

E-mail: ezervak@imel.demokritos.gr

Nikos Haralabidis

Institute of Microelectronics, N.C.S.R. "Demokritos", Athens 15310, Greece

E-mail: N.Haralabidis@imel.demokritos.gr

A fast current amplifier has been designed for use in solid state detectors systems. It provides a differential output and the baseline can be externally adjusted. The dynamic range is high up to 400mips and the gain is switchable. The total rms noise in the case of a 45pF detector capacitance is 2550e$^-$. Additional circuitry is included in order to compensate for the leakage current of the detector up to 5μA. The design has been implemented in a standard 0.25μm 1P/6M CMOS process.

1 Introduction

Multichannel VLSI readout electronics are extensively used in High Energy Physics experiments, biomedical applications and instrumentation setups. Depending on the critical target specifications in each case, different front-end architectures are employed. Fast signal evolution and pile-up problems in the case of high counting rates, suggest the use of a transimpedance amplifier.

The specifications for this design request to cover a dynamic range of 0-400mips. In order to improve the signal to noise ratio in small input charges the amplifier has two gain settings. At the high gain region the input charges range from 1-160fCb. For large input charges from 160 to 1600fCb we switch to the low gain region. The baseline of both differential outputs can be externally adjusted. Finally a self-adjustable circuit is included for compensating the leakage current of the detector due to aging.

2 Circuit Description

2.1 Preamplifier topology

The complete schematic of the preamplifier is shown in Fig.1 [1]. The current source in parallel with the capacitor at the input is the equivalent representation of the detector. When a particle crosses the detector a current pulse is produced which is directed through the feedback to the source of M5. The M5 is driven by a fast core amplifier that forces M5 to sink the current and discharge the input node. The current of the branch is multiplied by the current mirror M7-M8 and produces the output voltage on RL1. In order to have differential output we use two more current mirrors, M7-M9 and M10-M11 and the opposite voltage swing is produced on RL2.

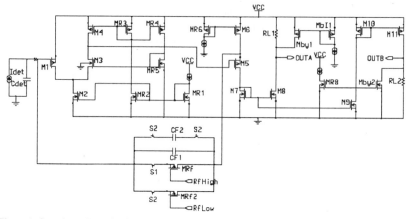

Figure 1. Complete schematic of the preamplifier.

The core amplifier is a folded cascode, consisting of M1-4, with a PMOS as an input device for reducing the overall noise. The feedback consists of the capacitor C_{F1} connected in parallel with a resistance, which is implemented by a MOS transistor in the linear region. The two additional MOS in parallel with the output resistors, are used to increase the dynamic range of the preamplifier. Most of the dc current of the branch is directed to these transistors allowing small voltage drop across the resistor. These transistors are biased independently so we can externally adjust the baseline. In order to change operation from high gain mode to low gain we use a set of switches. The switches are implemented by MOS devices and controlled through one external pin.

2.2 Switchable gain operation

In high gain operation only one capacitor is connected to the feedback, as for the resistive element we use a suitable sized MOS. The dynamic range is up to 40mips whereas the gain of the preamplifier is high and is expected to be around 10mV/fCb for a peaking time around 20ns.

To switch to the low gain mode we connect another capacitor in the feedback in parallel with the first one and we also change the resistive element. Finally we adjust appropriately the biasing of the core amplifier in order to achieve the desirable gain. The dynamic range is from 40 to 400mips which corresponds to large values of input charges, up to 1600fCb, whereas the gain is 1mV/fCb.

2.3 Leakage current compensation

After extended time of operation the detector is expected to degrade and present a leakage current of few µA's. This dc current affects the operation of the front-end stage by shifting its dc operating point. In order to compensate for the leakage current, we add a self-adjustable circuit in the preamplifier as it is shown in Fig.2 [2]. It consists of a differential amplifier, which is a two stage Miller Operational Transconductance Amplifier(OTA), the integrating capacitor and the two NMOS connected to the input of the preamplifier. The source of M5 is sensitive to leakage current and in presence of that is shifted to lower dc values. This dc value is compared to an externally set threshold. The output voltage controls the NMOS, which sinks the leakage current of the detector preventing it from flowing through the feedback network. The load capacitance is used to set the bandwidth of the loop very close to dc, in order not to be sensitive to fast signals during operation.

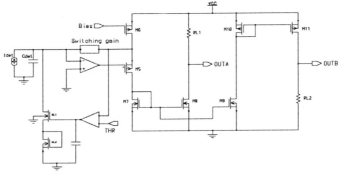

Figure 2. Self-adjustable leakage current compensation circuit.

3 Simulation Results

The amplifier has been designed in a 0.25µm CMOS process. Fig.3 shows the transient response of the differential output of the preamplifier for both modes of operation. For the high gain mode the input charge is 80fCb, while for the low gain is 500fCb. The high gain is 10mV/fCb, the low 1mV/fCb and the peaking time for both cases is 20ns. The recovery time for the high gain mode of operation is 600ns while for the low gain is much smaller, 200ns.

The preamplifier has been optimized to operate with a 45pF detector capacitance. The Equivalent Noise Charge(ENC) vs the detector capacitance is shown in Fig.4. The calculated noise relationship is $660e^-+43e^-/pF$ resulting in 2550 e^- rms for the 45pF input capacitance.

Finally, in the presence of leakage current both the gain and the peaking time are not affected and the baseline remains stable ensuring proper operation even after detector performance deterioration.

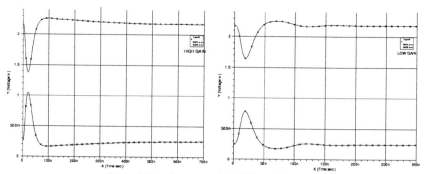

Figure 3. Transient responses for high and low gain modes.

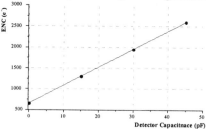

Figure 4. ENC vs. detector capacitance.

4 Conclusions

We have designed a fast current amplifier which provides differential output, and the baseline can be externally adjusted. The dynamic range is 400mips using switcable gain. For small input charges, up to 160fCb, the gain is 10mV/fCb, while for large input charges, 160 to 1600fCb, is 1mV/fCb. For both modes of operation the amplifier recovers quickly to the baseline so pile-up problems are avoided. The design has been optimized for a 45pF detector capacitance with a total rms noise of 2550e⁻. A self-adjustable circuit compensates for the leakage current up to 5 µA. The design has been implemented in a standard 0.25 CMOS process.

References

1. N.Haralabidis and K.Misiakos, "An Ultra Fast Current Amplifier with Active Feedback Loop", IEEE Transactions on Nuclear Science, vol. 44, No. 3, Jun. 1997, pp. 370-373.
2. F. Krummenancher, "Pixel detectors with local intelligence: an IC designer point of view", Nuclear and Methods in Physics Research, A305, 1991, pp.527-532.

DESIGN AND SIMULATION OF ON-CHIP BANDPASS FILTERS

A.T. Kollias, E. D. Tsamis and J. N. Avaritsiotis

Department of Electrical and Computer Engineering National Technical University of Athens, 9 Heroon Polytechneiou Street 15773 Zografou-Athens

This work establishes design procedures and directions in the analysis and design of a miniature bandpass filter based on TFRs. Taking advantage of the BVD's simplicity a linear two-port network theory has been applied. Material's electromechanical coupling factor k_{eff} and quality factor Q define the filter bandwidth (BW) and insertion losses (IL) in the bandpass zone respectively. Consequently the knowledge of Q, k_{eff} and the thickness of piezoelectric material is enough for simulating the frequency response. A case study consisting of four TFRs in order to implement an on chip bandpass filter with central frequency 1.8 GHz is given as an example.

1. Introduction

The most attractive solution for the minimization of radio communication systems is the use of Thin Film Resonators (TFRs) that can be integrated with other analog-digital subsystems. The TFR combines both the advantage of small dimensions and the high quality factors that can be achieved.

Several electrical circuits have been proposed for the modeling of TFRs. The Mason model provides coupling between mechanical and electrical sections of the TFR and it is suitable for describing the effect of multiple piezoelectric and non-piezoelectric layers [1]. Other researchers have used a direct analogous of Mason model in order to study the central frequency shift of a ladder filter for small variations of the multiple layers [2]. Finally the Butterworth-Van Dyke model offers simplicity and can be used near a resonance frequency. The validity of BVD model has been checked in previous work, where the electromechanical coupling factor of the piezoelectric material was obtained [3].

Simulated and experimental frequency responses of stacked crystal filters and ladder networks have been reported in the past [4]. In this work filters in the form of a ladder network are studied where two port theory can be applied. Experimental data for piezoelectric Zinc Oxide, required for the simulation, were taken from previous work [3].

2. Design and simulation

The two port network, which shown in figure 1, is the building element of a ladder filter. A number of same basic elements can be cascaded to form a ladder filter.

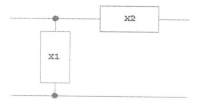

Figure 1. Basic element (X1: shunt, X2: series resonators)

The series, shunt resonators impedance A, B could be simulated either with the BVD model [5] or the Mason model but this does not affect the overall analysis of the network. The T matrix of the basic element has the following form

$$T = \begin{bmatrix} 1 & A \\ \dfrac{1}{B} & 1+\dfrac{A}{B} \end{bmatrix}$$

Consequently the T^k of a k-element ladder filter has the form

$$T^k = \begin{bmatrix} t_{11}^k & t_{12}^k \\ t_{21}^k & t_{22}^k \end{bmatrix} = \begin{bmatrix} t_{11}^{k-1} & t_{12}^{k-1} \\ t_{21}^{k-1} & t_{22}^{k-1} \end{bmatrix} \times \begin{bmatrix} 1 & A \\ \dfrac{1}{B} & 1+\dfrac{A}{B} \end{bmatrix} = \begin{bmatrix} t_{11}^{k-1}+\left(\dfrac{1}{B}\right)t_{12}^{k-1} & At_{11}^{k-1}+\left(1+\dfrac{A}{B}\right)t_{12}^{k-1} \\ t_{21}^{k-1}+\left(\dfrac{1}{B}\right)t_{22}^{k-1} & At_{21}^{k-1}+\left(1+\dfrac{A}{B}\right)t_{22}^{k-1} \end{bmatrix}$$

From the above recurrence types it is easy to obtain the form of $t_{11}^k, t_{12}^k, t_{21}^k, t_{22}^k$

$$t_{11}^k = \sum_{i=1}^{k} \mu_i^{k-1}\left(\dfrac{A^{i-1}}{B^{i-1}}\right), \; t_{12}^k = \sum_{i=1}^{k} \lambda_i^k \left(\dfrac{A^i}{B^{i-1}}\right)$$

$$t_{21}^k = \sum_{i=1}^{k} \lambda_i^k \left(\dfrac{A^{i-1}}{B^i}\right), \; t_{22}^k = \mu_1^k + \sum_{i=1}^{k} \mu_{i+1}^k \left(\dfrac{A^i}{B^i}\right)$$

where $\mu_i^k, \lambda_i^k \; i = 1,...,k$ are number sequences which can be computed from the recursive types $\mu_i^k = \lambda_{i-1}^{k-1} + \mu_i^{k-1} + \mu_{i-1}^{k-1}, \; \lambda_{i-1}^k = \lambda_{i-1}^{k-1} + \mu_{i-1}^{k-1}$
$\mu_1^k = 1, \lambda_1^k = k$

Furthermore the S_{21} parameter can be computed from the following equation

$$S_{21} = \dfrac{2}{t_{11}^k + t_{22}^k + \left(\dfrac{t_{12}^k}{Zload}\right) + \left(Zload \cdot t_{12}^k\right)}$$

where Z_{load} is the reference impedance which is usually the characteristic impedance of the transmission line in which the network is embedded.

3. Experimental

Overmoded resonators were constructed on silicon substrates using rf reactive sputtered zinc oxide thin films as a piezoelectric layer. Assuming that the shunt and series resonators hypothetically constructed at the optimum deposition conditions,

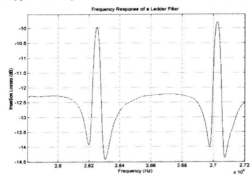

Figure 2. Simulated insertion losses of a two elements ladder filter based on experimental data

with the appropriate thickness of the ZnO layers, the resulted frequency response of the ladder filter would have the form of the curve in figure 2. The simulated filter presents great insertion losses at the passband due to the low quality factor of the resonators. Exploiting the experimental data of ZnO thin films, as deposited in our sputtering system in specific optimum conditions, we examined the improvement in filter's performance with the use of an acoustic Bragg reflector.

The reflector's stack consists of six layers as shown in table 1. The addition of two layers in reflector's stack compress the variation in phase difference, leading to improved performance as shown in figure 3. In both cases the filter consists of four elements.

Table 1 Reflector's layers (all numbers refers to microns)

6 layers	SiO_2	AlNOx	SiO_2	AlNOx	SiO_2	AlNOx		
	0.99	2.27	0.99	2.27	0.99	2.27		
8 layers	SiO_2	AlNOx	SiO_2	AlNOx	SiO_2	AlNOx	SiO_2	AlNOx
	0.99	2.27	0.99	2.27	0.99	2.27	0.99	2.27

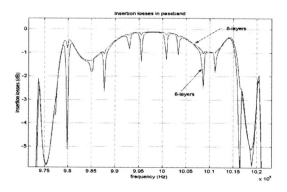

Figure 3. Simulated insertion losses in passband of a four element ladder filter with 6 layers and 8 layers reflector

4. Conclusions

In conclusion this work presents the analysis and design of ladder filters based on TFRs. The simulation shows that the quality factor of TFRs plays a dominant role in the frequency response of the filter. The use of a Bragg reflector drastically increases the quality factor but the reflector's leakage affects filter's characteristics.

5. Acknowledgement

The authors acknowledge the support of the Institute of Communications and Computer Systems of National Technical University of Athens.

References

[1] Lakin, K.M.; Kline, G.R.; McCarron, K.T., High-Q microwave acoustic resonators and filters, IEEE Transactions on Microwave Theory and Techniques, Vol.41, Issue12, Dec.1993, pp. 2139 –2146
[2] Olutade B. L, Hunt W. D, Proceedings of the 1997 IEEE International Frequency Control Symposium pp. 737-742
[3] Kollias A. T, Tsamis E. D, Avaritsiotis J. N, Active and Passive Electronic Components,vol. 23, pp. 82, 2000
[4] Lakin K. M., Kline G. R., McCarron K. T., Proceedings of the 1992 IEEE International Ultrasonics Symposium pp. 471-476
[5] Joel F. Rosenbaum, "Bulk Acoustic Theory and Devices", 1988, Arthech House Inc.

THE DESIGN OF A RIPPLE CARRY ADIABATIC ADDER

V. PAVLIDIS, D. SOUDRIS, AND A. THANAILAKIS

VLSI Design and Testing Center, Department. of Electrical & Computer Engineering, Laboratory of Electrical and Electronic Materials Technology,

Democritus University of Thrace, 67100 Xanthi, Greece

The novel design of a ripple carry adiabatic adder based on pass-transistor logic is introduced. The architectural design of the adiabatic adder and a formula for delays, are presented. The performance of the ripple carry adiabatic adder, in this work, against the performance of its CMOS counterpart, is discussed. More specifically, the adder (conventional, CMOS or adiabatic) was simulated by PowerMill tool for power dissipation, latency and energy efficiency. In addition, a first estimation of area was done by the transistor count. Both, the conventional and adiabatic adders were simulated at 3.3V and 5V, for a broad range of frequencies, from 5MHz to 50MHz. Experimental results indicate that the adiabatic adder outperforms the corresponding conventional adder in terms of power consumption, and it exhibits a lower hardware complexity.

1 Introduction

Reducing power dissipation in digital systems has, nowadays, become of great importance because of the restrictions that mainly portability imposes. Considering the significance of addition as an undoubtedly basic function, in any data processing system, we may conceive the benefits we receive from minimizing the power consumed by the arithmetic units [1].

Energy recovery is another low power approach, which is based on the principle of adiabatic switching, [1],[2],[4]. Theoretically, energy recovery circuits can achieve an arbitrarily small dissipation by increasing the charging time T of a node from 0V to a voltage *V*. Energy recovery requires the usage of power clock supplies which generate ramped voltage waveforms. By preserving small potential differences across devices a minimal amount of the delivered power is consumed as heat, whereas during the falling edge of the ramp the charge flows back to the power source.

In this paper, a novel 1-bit full adder (FA), based on pass-transistor logic, with a lower hardware complexity than an existing one [1] is proposed. The architectural design of the adder, as well as a new formula for delays, are presented. For comparison, a conventional CMOS adder with the same features, i.e. supply voltage, wordlength, and frequency, was also designed. Power consumption, transistor count, worst case delay, and energy saving were measured.

2 The proposed energy-recovery adder

2.1 General features

The low power adder designed in this work is based on the nMOS pass–transistor logic due to latter's intrinsic simplicity in constructing adiabatic gates. Because the pass–transistor adiabatic gates need the complements of the signals, we had to use inverters (internally) in datapaths. Another aspect in designing with pass–transistor logic is the voltage drop due to the voltage threshold, V_{th}. To overcome the voltage degradation, we took the advantage of bootstrapping effect, especially in output nodes, without using extra amplifiers. A detailed presentation of bootstrapping can be found in [5].

Our adiabatic adder has been designed and simulated for 8-bit and 16-bit word length. An identical conventional CMOS adder has also been implemented for comparison. The adder performs a complete operation within three phases. Firstly, the inputs are latched, then the sum and carry are evaluated, and finally the sum is driven to the output. Both, the adiabatic and conventional adders operate with two-phase non-overlapping clocking scheme. To accommodate the delay, as well as the driving capabilities of the longest paths, we used transistors with a W/L ratio among the values 2/1, 4/1, and 8/1. Assuming that t' is the delay of the inverters, t is the delay of each adiabatic gate, and n is the number of input bits, an equation estimating, theoretically, the delay of the proposed adder, has been derived. However, this formula is not so accurate, because it does not take into account the delay coming due to the interconnections and the variable size of transistors.

2.2 Energy-Recovery Ripple Carry Adder Design

Here, we introduce a new 1-bit FA with a reduced number of nMOS transistors. Writing down the logical functions of the sum S and carry out C_{out}, of two 1-bit operands A and B with a carry input C_{in}, we have:

$$S = A \oplus B \oplus C_{in} \tag{1a}$$
$$C = (A \oplus B)C_{in} + AB \tag{1b}$$

Rearranging eq. (1a) and (1b), we obtain:

$$S = (A \oplus B)\overline{C_{in}} + \overline{(A \oplus B)}C_{in} \tag{2a}$$
$$C = (A \oplus B)C_{in} + AB = (A \oplus B)C_{in} + \overline{(A \oplus B)}A \tag{2b}$$

Noticing the two common logical terms, namely $(A \oplus B)$ and $\overline{(A \oplus B)}$, the circuit shown in Figure 1a, consisting of nine nMOS transistors, can implement eqs. (2a) and (2b) (i.e. 1-bit FA). The voltage waveforms of sum node and carry output node are shown in Figure 1b, for all the individual values of A, B and C_{in}. We can pinpoint the correct voltage value of the sum and the voltage drop of the carry out.

We attacked the problem by adding two inverters, achieving a satisfactory voltage level of carry out and receiving its complementary signal, which is necessary for the next stage. Thus, we have a basic FA cell of 13 nMOS and 2 pMOS transistors producing S, C_{out}, and $\overline{C_{out}}$.

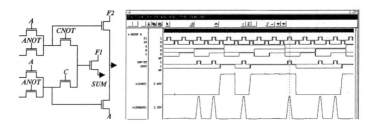

Figure 1. a) The circuit of the novel adiabatic 1-bit FA, b) The output waveforms of adiabatic 1-bit FA

The adiabatic 8-bit and 16-bit carry ripple adders were designed by simply replicating the 1-bit adiabatic full adder unit. This repetition results in a delay of

$$t_{delay,crp} = t(n+1) + 2t'n \qquad (3)$$

Eq. (3) includes: i) a delay of $2t+2t'$ is produced by the first addition, and ii) a delay of $t+2t'$ for the propagation of the carry across the next $(n-1)$ FA.

3 Experimental Results

In this section, we present PowerMill simulation results for 8-bit and 16-bit adders, we developed in CMOS and pass-transistor logic. The adders were designed with Cadence in a 0.35µm AMS CMOS technology and were simulated using the ACE (Analog Circuit Engine) option of PowerMill. The comparative study of the proposed adiabatic adder and the conventional one is performed in terms of the total power consumption, the delay, the transistor count, and of the energy saving.

Two sets of simulations were done at 3.3V and 5V, changing the operating frequency from 5MHz to 50MHz, with an incremental step of 5MHz. Figure 2 shows the power consumption of adiabatic and CMOS adder for various parameters. Measurements show clearly that although simple clocked conventional inverters were used in the adiabatic circuit, the power consumption by the adiabatic adder is significant less that the conventional one.

Table 1 shows the delays of the adiabatic adder for 3.3V & 5V at the frequency of 25 MHz. Beyond power dissipation, the transistor count of adiabatic adder is reduced over 60%. Finally, from the aspect of energy efficiency the proposed adder achieves energy saving of about 52% of the delivered power. This percentage is altered according to frequency by approximately 10%.

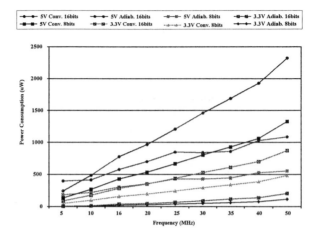

Figure 2. Power cosumption of CMOS and adiabatic carry ripple adder for 3.3 & 5 volts, 8 and 16-bits.

Table 1. The worst case delay and the transistor count of conventional and adiabatic adder.

Adder Type	# bits	Voltage Supply		Transistor count		% reduction
		3.3 volts	5 volts	Adiabatic Adder	CMOS Adder	
Carry ripple adder	8	6.16 ns	3.63 ns	216	574	62 %
	16	13.24 ns	7.12 ns	438	1134	62 %

4 Conclusions

The design and experimental results of a new adiabatic adder based on pass transistor logic was presented. The new design was simulated and measured in terms of the power consumption, delay, and energy recovery, using PowerMill, a commercial CAD tool, for different supply voltages, bit wordlengths and frequency of operation. The experimental results showed that the adiabatic adder dissipates less power and exhibits a lower hardware complexity than the conventional one.

References

1. J. Rabaey and M. Pedram, "*Low Power Design Methodologies,*" KAP, 1996.
2. W.C. Athas, L. Svensson, J.G. Koller, N. Tzartzanis, and E.Y.C. Chou, "*Low-Power Digital Systems Based on Adiabatic-Switching Principles,*" in IEEE Trans. on VLSI Systems, vol. 2, no.4, Dec. 1994.
3. N. Tzartzanis, W. Athas, "Design and Analysis of a Low-Power Energy-Recovery Adder," Proc. of 5th Great Lakes Symp. on VLSI Design, Mar. 16-18, 1995, pp. 66-69.
4. S.G. Younis and T.F. Knight, Jr., "*A asymptotically zero energy split-level charge recovery logic,*" Proc. of IEEE Int. Workshop on Low Power Design, 1994, USA.
5. J.P. Uyemura, "CMOS Logic Circuit Design," Kluwer Academic Publishers, 1999.

MAXIMUM POWER ESTIMATION IN CMOS VLSI CIRCUITS

N.E. EVMORFOPOULOS AND J.N. AVARITSIOTIS

*National Technical University, Department of Electrical and Computer Engineering, Athens
157 73, GREECE
E-mail: abari@cs.ntua.gr*

G.I. STAMOULIS

*INTEL Corp., 2200 Mission College Blvd., Santa Clara,
CA 95052,USA
E-mail: georgios.i.stamoulis@intel.com*

The problem of maximum power estimation in CMOS VLSI circuits is addressed. Estimation of a chip's maximum power requirements, as they relate to electromigration and IR-drop failures in its supply bus is becoming important as we move into the deep-submicron where guardbanding that has been used up to now is no longer acceptable. An approach of statistical nature, based on recent advances in the field of extreme value theory is proposed. Experimental results establish our claims and they demonstrate the overall efficiency of the proposed approach in comparison to previously employed techniques.

1 Introduction

The issues regarding reliability of CMOS digital circuits has drawn considerable attention are even more important as we move into deep-submicron where the safeguarding measures that have been used up to now (plain fabrication of reliable transistors and wires) are no longer acceptable. This has given the spark for a substantial research activity in the field and has led to the development of various relevant techniques, which typically fall into the broad categories of simulative and non-simulative approaches [1-7]. The objective here is to obtain a reasonable estimate of the circuit's maximum power through efficient statistical analysis of a relatively small sample of randomly drawn input vectors from the entire population

The term "maximum power" refers to the worst-case instantaneous power (or equivalently current) that is drawn from the supply bus. In order to render the estimation problem precise, we have to consider the sources of power dissipation in digital CMOS circuits. These are [8] , P_{sc} and P_{leak} , and correspond to switching, short-circuit and leakage power respectively.

2 Procedure Description

The basis of the theory applied are discussed in [9]. The procedure begins with the generation of n vector pairs in random way or under certain constraints. This is equivalent to random sampling out of the unconstrained or constrained population of input vector pairs. The circuit due for maximum power analysis is then entered in a transistor-level or gate-level simulator such as SPICE or PowerMill. Each generated vector pair is fed as input for transient analysis to the simulator and the peak power during transition time is recorded. Special care has to be taken for the clock period to be longer than the maximum delay along any path from the inputs to the outputs of the circuit, so that there is enough time for the transition effects to spread to all internal nodes.

3 Experimental Results and Discussion

The proposed approach was applied on the ISCAS85 benchmark circuits and the results for the average, maximum and minimum estimated values, among 10 runs of the procedure [9] for each circuit, are shown in Table 1. Note that the current instead of the power figures were recorded, assuming a constant voltage supply throughout the circuit. The value of the test statistic appearing in column 2, the maximum current estimate in column 3 and the confidence intervals for confidence levels 95%, 99% and 99.99% in columns 4 to 6. Finally, column 7 gives our best estimate for the maximum current of each circuit, after very long simulation comprising of 1,000,000 units.

First of all it is evident that only one of the circuits examined (c7552) belongs to the type-II extreme model, while the rest (9 out of 10) are of type-I. For the sake of comparison, the average estimated value among the same 10 runs for c7552 treated as a type-I circuit was 47.096mA, with the 90%, 95%, 99% and 99.99% confidence intervals being ±1.570mA, ±1.870mA, ±2.458mA and ±3.713mA respectively.

Table 1. Results for the average estimated values of maximum current among 10 runs.

Circuit	Value of test statistic	Estimated maximum current (mA)	Confidence Interval (mA)			Observed maximum current (mA)
			95%	99%	99.99%	
c432	4.046	9.407	±0.618	±0.812	±1.227	9.053
c499	4.430	21.042	±1.267	±1.666	±2.516	17.778
c880	3.514	14.595	±1.010	±1.327	±2.005	12.254
c1355	4.889	17.777	±1.120	±1.472	±2.224	15.175
c1908	3.854	17.570	±1.031	±1.354	±2.046	14.760
c2670	4.536	21.395	±1.113	±1.462	±2.209	20.034
c3540	4.263	30.928	±2.041	±2.682	±4.051	25.774
c5315	4.591	47.358	±2.506	±3.293	±4.974	41.511
c6288	3.475	54.103	±2.869	±3.770	±5.694	46.570
c7552	9.194	41.866	±2.202	±2.894	±4.371	41.323

4 Conclusion

The problem of maximum power estimation in CMOS VLSI circuits has been investigated in this paper. We stated that a sound approach has to combine simulation which guarantees the accuracy needed for deep-submicron IC design and statistics so as to overcome the pattern dependence obstacle. We identified extreme value theory as the pertinent field of statistics for the problem at hand, and presented the theoretical background behind it. We subsequently formulated the problem of maximum power estimation and provided the means to address it within the framework of extreme value theory. The proposed approach was applied to the problem of global maximum power estimation in the power supply bus.

5 Acknowledgements

We thank Intel Corporation for partially supporting this work by an equipment grant.

References

[1] M. Pedram, "Power minimization in IC design: principles and applications", *ACM Trans. Design Automation of Electronic Systems*, vol. 1, no. 1, pp. 3-56, 1996.

[2] H. Kriplani, F. Najm and I. Hajj, "Pattern independent maximum current estimation in power and ground buses of CMOS VLSI circuits: algorithms, signal correlations and their resolution", *IEEE Trans. Computer-Aided Design*, vol. 14, no. 8, pp. 998-1012, 1995.

[3] C.-Y. Wang and K. Roy, "Maximum power estimation for CMOS circuits using deterministic and statistical approaches", *IEEE Trans. VLSI Systems*, vol. 6, no. 1, pp. 134-140, 1998.

[4] C.-Y. Wang, T.-L. Chou and K. Roy, "Maximum power estimation for CMOS circuits under arbitrary delay model", *IEEE Int. Symp. Circuits and Systems*, 1996.

[5] C.-Y. Wang and K. Roy, "COSMOS: a continuous optimization approach for maximum power estimation of CMOS circuits", *IEEE/ACM Int. Conf. Computer-Aided Design*, 1997.

[6] A. Krstic and K.-T. Cheng, "Vector generation for maximum instantaneous current through supply lines for CMOS circuits", *ACM/IEEE Design Automation Conf.*, 1997.

[7] Y.-M. Jiang, K.-T. Cheng and A. Krstic, "Estimation of maximum power and instantaneous current using a genetic algorithm", *IEEE Custom Integrated Circuits Conf.*, 1997.

[8] N. Weste and K. Eshragian, *Principles of CMOS VLSI Design: A Systems Perspective* 2nd ed., Addison-Wesley, 1993.

[9] N.Evmorfopoulos and J.N.Avaritsiotis, "A new statistical method for maximum power estimation in CMOS VLSI circuits",. *Active and Passive Electronic Components,* 22(3)2000, pp. 214-233.

DESIGNING A MICROWAVE VCO 945MHZ WITH COMPUTER AIDED DESIGN SIMULATION

S. PANAGIOTOPOULOS AND A. KOUPHOYANNIDIS

Research Center of INTRACOM S.A., Karaoli and Smirnis, 67100, Xanthi, Greece
E-mail: span@intracom.gr, akouph@intracom.gr

Voltage Controlled Oscillators (VCOs) are key components in a wide range of high-frequency applications. When operating in a phase-locked loop (PLL), the VCO provides a stable local oscillator (LO) for frequency conversion in superheterodyne receivers. The design and realization of a VCO with particular specifications is often a difficult and time consuming procedure for engineers with small experience. This article describes the design of VCOs by means of some basic calculations and CAD software for microwave circuit analysis. With a design of a VCO for DECT tranceiver operating at 945MHz, it is very clear that it is important to account for the following: the effects of paracitics generated by the active device, the circuit layout and component paracitics.

1 Introduction

The usage of the negative impedance model is common for high frequency Oscillators. Having the right point of view, we can treat a feedback oscillator as a negative resistance generator. In our example we dealt with a Clapp Oscillator described in fig.1a.

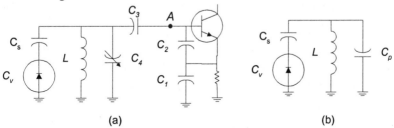

Figure 1. a)Tuning network of a VCO, b)Equivalent tuning circuit

The bipolar transistor and the capacitive feedback from C_1, C_2 (fig.1a), generates a negative resistance, according to [1]. If $X_{C1} << h_{ie}$, the input impedance is approximately equal to

$$Z_{in} \cong \frac{-g_m}{\omega^2 C_1 C_2} + \frac{1}{j\omega \left(\frac{C_1 C_2}{C_1 + C_2}\right)} \qquad (1.1)$$

So, the input impedance of the circuit is a negative resistor R in series with a capacitor C_{in}

$$R = -g_m / (\omega^2 C_1 C_2) \qquad C_{in} = C_1 C_2 / (C_1 + C_2) \qquad (1.2)$$

If we connect an inductor L (with its series resistance R_s) as shown in fig.1a, it is clear that the condition for sustained oscillation is

$$R_s = g_m / (\omega^2 C_1 C_2) \qquad F_o = 1 / 2\pi \sqrt{LC_{in}} \qquad (1.3)$$

where F_o is the oscillation frequency.

It is obvious from the above, that if we manage to have negative resistance at the input of network (point A), then with the addition of the appropriate tuning network, parallel with C_{in}, the circuit will oscillate at the desired frequency.

In the case of VCO design, we must take into account the tuning range and the center frequency of the VCO. In the next paragraph we describe a method to achieve the tuning range and the center frequency with good accuracy.

2 The Proposed Method

The tuning network that we propose for use is shown in fig.1a. The circuit right to point A creates the negative resistance and the circuit left to point A gives the ability for tuning in a desired frequency range.

The evaluation of the elements can be done with the help of the equivalent circuit of Fig.1b. This circuit is the equivalent circuit before the base of the transistor. The capacitor C_p represents the total equivalent capacitance across the inductor. The capacitor C_s blocks the thermal noise made by the varicap. This capacitor with the varactor C_v makes an equivalent variable capacitance C_D' with desired limits of minimum $C_{D\min}'$ and maximum $C_{D\max}'$ capacitance.

$$C_D' = \frac{C_v}{\left(1 + C_v / C_s\right)} \qquad (2.1)$$

The oscillation frequency is given from the equation (2.2) and it is obvious that $f_{\max} = f(C_p, C_{D\min}')$ and $f_{\min} = f(C_p, C_{D\max}')$.

$$f = 1 / \left(2\pi \sqrt{L(C_p + C_D')}\right) \qquad (2.2)$$

If R is the ratio of maximum to minimum operating frequency we can find the value of C_p (2.3).

$$R = f_{max}/f_{min} \Rightarrow C_p = \left(C'_{Dmax} - R^2 C'_{Dmin}\right)/\left(R^2 - 1\right) \qquad (2.3)$$

From the (2.2) and all the other elements known we can calculate the inductance L, where

$$L = 1/\left(4\pi^2 \left(C'_{Dmin} + C_p\right) f_{max}^2\right) \qquad (2.4)$$

3 Implementation of a VCO 945MHz with the above method

Our objective was a VCO operating at half the operating frequency of a DECT transponder, particularly at f_c=(1889.568)/2=944.784MHz, having a tuning bandwidth of 31.64MHz. As a varicap we chose the SMV1233-011 with $C_v(1V)$=3.6pF and $C_v(3V)$=2.2pF. For C_s we choose C_s=2.2pF. With the help of equations (2.1) to (2.4) we found C_p=2.733pF and L=7.16nF.

Working with Microwave office 2000 (a CAD software for microwave circuit analysis) and keeping in mind that components in high frequency design are not perfect, we tried to use models that included the parasitics elements instead of ideal component. During the design we followed the next steps.

First we designed the transistor bias and the feedback network, and as a transistor we used the linear model of BFR93 for specific bias and current I_C. Next we tried values of C_1, C_2 and C_3 to achieve negative input resistance. We found that if C_1=C_2=3.3pF and C_3=2.2pF, we have negative input resistance. After that we added the C_4=1.8pF, having in mind that C_p should be 2.733pF and L=6.8nF (this inductance is near to the ideal value). For these values we found that the oscillation frequency was at 934MHz. The parasitic elements, which we had not taken into account at our theoretical analysis –for example the parasitic capacitance of transistor–, were the reason for this deviation. The answer to this was to include these parasitic elements at the tuning network. This could be realized indirectly with the change of the value of the capacitance C_4. With C_4=1.5pF we had the desired oscillation frequency and a frequency range which was in the limits of our tolerance.

The final measurements could be done having the port 1 (which measures the S_{11} and S_{12} parameters) at the grounded edge of inductor L (first we removed the ground). The reason is that the tuning network as a whole can be seen as a series LC circuit.

The $Re(Z_{in})$=-3.45 Ohm, and this means that the unloaded Q_u of the inductor must be $Q_L \geq 11.7$

When we tested the circuit at our laboratory, the results we had were very close to our predictions.

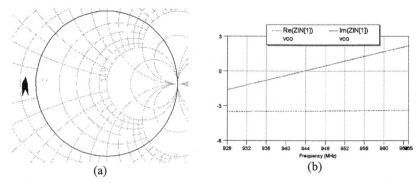

Figure 2. (a) S_{11} parameters at input (b) $Re(Z_{in})$ and $Im(Z_{in})$

4 Conclusion

The steps we followed and we propose as a designing method are:
1. Evaluate the proper L and C_p elements of the equivalent circuit of fig.1b, for the desired central frequency and frequency range.
2. Add the appropriate capacitive divider at the input of the transistor, in order to have negative resistance at the input.
3. Evaluate the rest elements of the tuning network with the help of the relations (2.1) - (2.4)
4. Adjust the capacitance C_4 in order to tune on the desire central frequency.

It must be noticed that the better models for our components we used, the better accuracy at our results we had. This method makes the design of a VCO just a typical routine. We tested it up to 2GHz and it works perfectly.

References

1. G.D.Vendelin, A.M.Pavio and U.L.Rohde, Microwave Circuit Design Using Linear and Nonlinear Techniques, A. Wiley-Interscience Publication.
2. Randal W. Rhea, Oscillator Design And Computer Simulation, 1995 by Noble Publishing Corporation, Atlanta.
3. David M. Pozar, "Microwave Engineering", 1990 by Addison Wesley Publishing Company.
4. Robert E. Collin, "Foundation for Microwave Engineering", 1992 second edition Mc Graw Hill.

POWER DISSIPATION CONSIDERATIONS IN LOW-VOLTAGE CMOS CIRCUITS

ALKIS A. HATZOPOULOS

Aristotle University of Thessaloniki, Dept. of Electrical & Computer Engineering,
54006 Thessaloniki, GREECE. Tel.: (+3031) 996305, FAX: (+3031) 996396
alkis@vergina.eng.auth.gr

The dynamic power dissipation dependence on the input rise/fall time in low-voltage CMOS circuits is studied and it is efficiently approximated by an additional term in the typical formula. Theoretical calculations using the proposed formula give results very close to those from the simulations.

1. Introduction

Power dissipation (PD) in sub-micrometer CMOS technology is one of the most important performance parameters in VLSI design. Simple but accurate PD models can help the designers to effectively estimate it and thereon to reduce it. Usually three terms of PD are identified in CMOS circuits [1, 2, 3]: the static part, the short-circuit part and the dynamic component P_d. The dynamic component is typically calculated by $P_d = \alpha C_L V_{DD}^2 f$, where α is the percent activity of the gate in a complex circuit and f is the clock frequency. This component is considered independent of the input slope. Therefore, the total PD P_{total} should be monotonically minimized when faster inputs are used. However, simulation results using typical bsim3v3 transistor models have shown that the P_d (and thus also P_{total}) is increased when very fast inputs are applied. This performance is studied in the following and it is efficiently modeled as a factor in the typical formula of P_d. Simulation results for the dynamic PD under various operating conditions are compared to the P_d calculated by the new proposed formula showing its sufficient accuracy.

2. Problem description

A typical CMOS inverter was simulated by using the Analog Artist simulator (CADENCE software) with 0.6 μm AMS technology parameters, under various load and supply voltage conditions. A clock input of 100MHz was applied with input rise/fall time varying from 0.3ns to 3ns. The plots of the total PD of the inverter P_{total} and the dynamic component $P_{d\_simulation}$ as calculated by the simulator (using the bsim3v3 MOS models) versus the input rise/fall time for V_{DD}= 2V, 2.5V and 3V and for C_{Load} = 50fF and 100fF are shown in fig. 1 (a to f). The value of the P_d as given by the typical formula $P_{d\_typ} = C_{Load} V_{DD}^2 f$ is also shown. It can be noticed in fig. 1d, for example, that the $P_{d\_simulation}$ is larger than P_{d_typ}= 62.5 μW (C_{Load}= 100fF and V_{DD}= 2.5V) and that it depends on the input rise/fall time. Similar observations can be made on the rest of the plots. This dependence affects the total PD P_{total}, especially in low voltage operation (V_{DD}= 2V, fig. 1a, 1b).

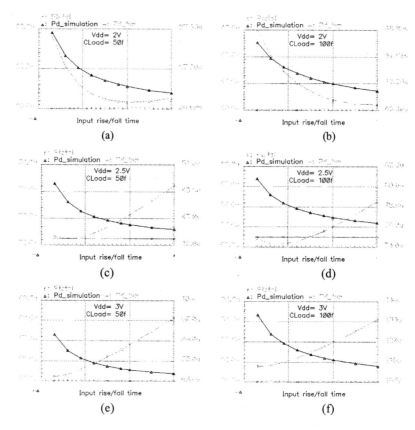

Figure 1. Plots of the power dissipated P_{total}, $P_{d\_simulation}$, $P_{d\_new\_theory}$ and P_{d_typ} (in μW) versus the input rise/fall time τ (in ns) for three supply voltages and two load capacitance values, as indicated on the plots.

3. Theoretical analysis

To model the dependence of P_d on the input rise/fall time τ we may consider the fact that in the sub-micrometer CMOS technology a contact resistance and a diffusion resistance give rise to parasitic source and drain resistance. Also, hot carrier degradation is another cause of parasitic drain resistance [4]. Therefore, the PD dependence on the input slope in a simple RC circuit should be analyzed first.

Let $V_{in}(t)$ be an input signal of a simple series RC circuit (L type with grounded capacitor) and $V_c(t)$ be the output voltage across the capacitor C. This voltage is:

$$V_c(t) = e^{-t/RC}\left[c + \frac{1}{RC}\int V_{in}(t)e^{t/RC}dt\right] \quad (1)$$

where c is the constant of integration. For an input ramp from 0 V to V_{DD} with slope s, i.e. $V_{in}(t)=st$, and considering the initial condition for $t=0 \rightarrow V_c(0)= 0$ V, eq. (1) will become:

$$V_C(t) = s(t - RC + RCe^{-t/RC}) \qquad 0 < t \leq \tau \qquad (2)$$

This equation is valid for $0 < t \leq \tau$, where $\tau = V_{DD}/s$ is the rise time of the input ramp. For $t > \tau$ ($V_{in}(t \geq \tau)=V_{DD}$) the output will follow the equation:

$$V_C(t) = sRC(1-e^{-\tau/RC})(1-e^{-t/RC}) \qquad t > \tau \qquad (3)$$

The current of the capacitor $i_c(t)$, using eq. (2) and (3) for $0 < t \leq \tau$ and $t > \tau$ correspondingly, will be:

$$i_C(t) = C\frac{dV_C(t)}{dt} = Cs(1-e^{-t/RC}) \qquad 0 < t \leq \tau \qquad (4)$$

$$i_C(t) = Cs(1-e^{-\tau/RC})e^{-t/RC} \qquad t > \tau \qquad (5)$$

Considering a periodical pulse with period T and equal rise and fall times τ, the PD on the resistor R is:

$$P_R = \frac{1}{T}\int_0^T i_C^2(t)Rdt = \frac{2R}{T}\int_0^\tau i_C^2(t)dt + \frac{2R}{T}\int_\tau^{T/2} i_C^2(t)dt \qquad (6)$$

By using eq. (4) and (5) in eq. (6) and after some mathematical manipulations, the power P_R dissipated on R will be given by:

$$P_R = \frac{2RC}{\tau}CV_{DD}^2 f - \left(\frac{2RC}{\tau}\right)^2 CV_{DD}^2 f\left[\frac{3}{4}+\frac{1}{4}e^{-2\tau/RC} - e^{-\tau/RC} +\frac{1}{4}(1-e^{-\tau/RC})(e^{-T/RC} - e^{-2\tau/RC})\right] \qquad (7)$$

By setting $A = 2RC/\tau$ and provided that for the values of R, C and τ in the case under consideration it is $RC < \tau \Rightarrow A < 1$, the exponential terms in eq. (7) are very close to zero and, thus, it becomes:

$$P_R = ACV_{DD}^2 f - (3/4) A^2 CV_{DD}^2 f \qquad (8)$$

For the practical values of A in our simulations (i.e. $A << 1$) only the first term of (8) may sufficiently approximate P_R, and it will be used so in the following.

In order to approximate the dynamic PD of a CMOS inverter, an effective resistance R_{eff} is assigned to each of the MOS transistors and the typical formula is enhanced by a parameter $A_{inv}=2R_{eff}C_{Load}/\tau_{rf}$, according to the first term of eq. (8), giving a new formula:

$$P_{d\_new} = \left(1+\frac{2R_{eff}C_{Load}}{\tau_{rf}}\right)C_{Load}V_{DD}^2 f \qquad (9)$$

The rise/fall time of the output pulse τ_{rf} can be found from the output delay time t_d as [1]:

$$\tau_{rf} = 2t_d = 2(\tau_{d\_step} + k_{in}\tau_{input}) \qquad (10)$$

where τ_{d_step} is the delay time for a step input pulse and k_{in} may be calculated by [5]:

$$k_{in} = (1+2|V_t|/V_{DD})/6 \qquad (11)$$

The value of τ_{d_step} is proportional to the value of the load C_{Load}:

$$\tau_{d\_step} = k_s C_{Load} \qquad (12)$$

where k_s depends on the device characteristics and supply voltage [1].

The R_{eff} has been observed to depend on the operating conditions and the device characteristics. In our implementation its value has been practically assigned.

4. Simulation results

A CMOS inverter with W_p= 9 µm and W_n= 3 µm was simulated and the results are given in fig. 1 (a to f), as described in section 2. In each plot the theoretical curve $P_{d\_new\_theory}$, approximating the dynamic PD dependence on τ according to eq. (9), has been included. The values of the parameters k_{in}, k_s and R_{eff} used for V_{DD}=2V, 2.5V and 3V are given in table 1. For the k_{in} a single approximate value has been used in each case, although slightly different values emerge from eq. (11) for V_{tp}= −0.78V and V_{tn}= 0.84V. For the k_s a single approximate value has also been used in each case, although different values should be assigned to the p and the n device [1]; this difference has been largely compensated by using transistor widths inversely proportional to their carrier mobility.

In all cases the maximum relative error between the simulated PD $P_{d\_simulation}$ and the theoretical approximation is less than 3.8%, which is very sufficient.

Table 1. Parameter values for the three supply voltages.

	V_{DD}= 2V	V_{DD}= 2.5V	V_{DD}= 3V
k_{in}	0.30	0.275	0.26
k_s	$4 \cdot 10^3$	$3 \cdot 10^3$	$2.1 \cdot 10^3$
R_{eff} (Ω)	300	170	100

5. Conclusions and further work

A simple analytical expression approximating the dynamic power dissipation dependence on the input rise/fall time in low-voltage CMOS circuits has been proposed. The new formula has been applied on a CMOS inverter under various operating conditions giving in all cases quit accurate results, compared to those emerged from a standard circuit simulator. It must be noted that this dependence mostly affects the total PD P_{total} in low voltage operation, as it is shown in fig. 1a, 1b for V_{DD}= 2V.

The evaluation of the R_{eff} parameter, based on MOS device characteristics and operating conditions, should be further investigated aiming to a straightforward analytical expression for R_{eff}. The application of the new expression in cases of large complex gates should also be studied.

References
1. Weste N., Eshraghian K., *Principles of CMOS VLSI design*. Addison-Wesley Publ. Co., 1993
2. Turgis S., Auvergne D., "A novel macromodel for power estimation in CMOS structures", *IEEE Tr. CAD of Integrated Circuits & systems*, vol. 17, no 11, Nov. 1998, pp 1090-1098.
3. Bisdounis L., Nikolaidis S., Koufopavlou O., "Propagation delay and short-circuit power dissipation modeling of the CMOS inverter", *IEEE Tr. CAS–I: Fundamental Theory and Applications*, vol. 45, no 3, March 1998, pp 259-270.
4. Sakurai T., Newton R., "Alpha-Power law MOSFET model and its applications to CMOS inverter delay and other formulas", *IEEE J. of Solid-State Circuits*, Vol. 25, no 2, April 1990
5. Hedenstierna N., Jeppson K., "CMOS circuit speed and buffer optimization", *IEEE Tr. on CAD*, vol. CAD-6, no. 2, March 1987, pp. 270-281.

Microelectronics Networks / Technology Transfer and Exploitation

EURACCESS: A EUROPEAN PLATFORM FOR ACCESS TO CMOS PROCESSING

Presented by

C.L. CLAEYS

IMEC, Kapeldreef 75, B-3001 Leuven, Belgium
also at E.E. Depart., KU Leuven, Belgium
E-mail: Cor.claeys@imec.be

Abstract

Euraccess is a European Network of suppliers of advanced technological steps in silicon processing.

1 Main objectives

The long term competitiveness and/or even possibly leading role of Europe in some specific areas of microelectronics is to a large extent depending on the advanced research activities, which are forming a nucleus for the generation of new and innovative material and device concepts.

In many cases, however, the researchers are lacking "access" to the appropriate advanced and well equipped facilities to process, fabricate, measure and/or characterise the practical implementation of new ideas. Such factor is dramatically hindering take-up of new European technologies.

The aim of this activity is to take the most useful profit of a network of platforms enabling academic and industrial institutions to execute advanced research projects in order to experimentally validate new research ideas and concepts and to study the industrial feasibility.

This network was established within the EURACCESS project referenced IST 10407.

Therefore the network will use and contribute to improve the structure providing access to the **EUR**opean **A**dvanced **C**entres for **C**oordinating and **E**nabling **S**upport in long term **S**ilicon research, further referred to as **EURACCESS**.

The timing of such initiative is well fitted in regard of the International Technology Road Show (ITRS), the technical domain adressed concerning 70nm and less technologies.

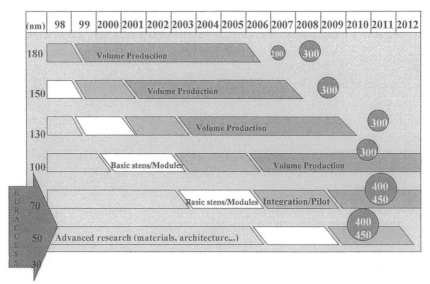

The main objectives of the Network are as follows:
1) To offer to academic and industrial institutions within Europe a platform to support the execution of research projects related to advanced materials, technologies and devices. This implies that <u>the partners, further referred as "suppliers" would make part of their infrastructure, facilities and technical know-how basis available</u> for European customers, referred as "users". Although the main focus would be oriented towards silicon based 50 nm and less structures and devices, so-called nano-electronics applications, more exotic material and device structures are not excluded. This infrastructure will be fully open for participation to other labs acting as suppliers.

Such infrastructure will include a <u>wafer logistic</u> between "suppliers" and "users" as well as <u>a management structure</u> involving both industrial end-users and technical experts.

A light office taking into account all the administrative issues will be set-up at the coordinator node.

2) To establish service activities and develop access for :
- execution of advanced processing steps, process modules and process flows.
- support to execution of joint research and development projects.
- access to state of the art structural and electrical material and device characterisation techniques, most of the advanced ones being available at the Technology Nodes.

If required for some specific projects, later on, the network could be open to:
- support with modeling and simulation activities, offering direct access to a large variety of commercially available process, device and circuit simulators, exploring their limitations and collaborating on the development of improved models
- access to state of the art design tools for device and circuit layout could be offered for the realization of novel device concepts. The design service would be strongly customized in order to take into account the often non-standard requirements to execute the advanced research projects.

2 Graphical representation of the project components

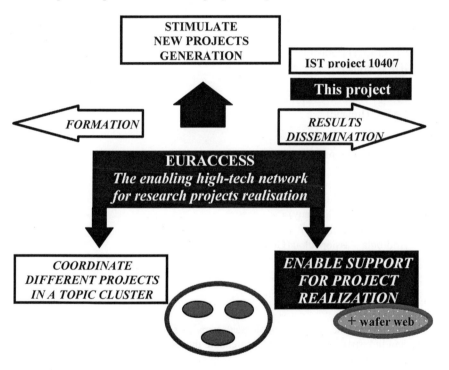

3 Main suppliers of technology

- CEA / LETI (France)
- IHP (Germany)
- IMEC (Belgium)
- IMEL / NCSR Demokritos (Greece)
- NMRC (Ireland)

4 Contact persons

The five contact persons are as follows:
- Jean-Charles Guibert (France)
- Bend Tillack (Germany)
- Cor Claeys (Belgium)
- Androula Nassiopoulou (Greece)
- Gareth Redmond (Ireland)

5 Conclusion

Euraccess offers advanced technological steps in silicon process, to both research labs and industrial partners in order to enable the execution of projects on advanced materials, devices and processes by people who does not have in-house such facilities.

MMN: GREEK NETWORK ON MICROELECTRONICS, MICROSYSTEMS AND NANOTECHNOLOGY

Presented by

A. G. NASSIOPOULOU

IMEL/ NCSR "Demokritos", 153 10 Aghia Paraskevi, Attiki, Greece
E-mail: A.Nassiopoulou@imel.demokritos.gr

1 Objectives

Gather together all scientists and engineers working in Greece in Microelectronics, Microsystems and Nanotechnology fields. Create linkages between research organisations and industry, promote technology transfer, promote educational activities in the above fields.

2 Activities

- Organization of the Conference MMN (Microelectronics, Microsystems, Nanotechnology) every 2 years.
- Organization of Seminars, Summer schools etc.
- Edition of the Journal of Micro- and Nanotechnologies.

3 Participating organizations

15 organizations participate at present in the network, including research centers, universities and SMEs.

4 Contacts

Contact person: Director NASSIOPOULOU Androula G.
Organization: Institute of Microelectronics, NCSR "Demokritos"
Address: 153 10 Aghia Paraskevi, Attiki, Greece, P.O. Box 60228
Telephone: 003016542783
Fax: 003016511723
Email: A.Nassiopoulou@imel.demokritos.gr
Website: http://imel.demokritos.gr

TECHNOLOGY TRANSFER IN ACTION

VASSILIOS TSAKALOS

Help-Forward Network, 5 Xenofontos str., 105 57 Athens, Greece

E-mail: vtsak@help-forward.gr

1. Introduction

HELP FORWARD (HELlenic Project FOR Wider Application of Research and Development) was established on the 1st of February of 1991, as a joint initiative of the Federation of Greek Industries (FGI) and the Foundation of Research and Technology-HELLAS (FORTH). The main aim was to bridge the gap between industrial manufacturing / production and scientific developments in technological research by the utilisation of relevant Community programs. The initial financing for the project came from FGI and FORTH with voluntary grants from industries, research institutes, universities as well as members of the FGI and other industrial associations.

Developing HELP FORWARD (Community Programmes) to its full capacity formed one of the projects which was financed by the General Secretariat for Research and Technology (GSRT) through the STRIDE-HELLAS program which was launched on the 1st of February 1992. Nodes were set-up in nine cities: Xanthi, Kavala, Thessaloniki, Ioannina, Larisa, Volos, Patras, Athens and Heraklion. Each node was created in collaboration with regional industrial associations, chambers of commerce and industry and the science and technology parks, supported by members of both the research and industrial domains. In its final form as a network, HELP FORWARD continued to inform, provide consulting services and perform partner searches for participation in Community programmes and initiatives.

In 1995, HELP FORWARD secures a two-year grant from the European Commission through the INNOVATION program and becomes one of the 53 Innovation relay centres (IRC) of the European Union. The operation of HELP FORWARD is under the hospices of three partners, FORTH, FGI and FING (Federation of Industries of Northern Greece) who closely collaborate with 5 regional industrial associations, 5 regional chambers of commerce and industry, 3 scientific and technological parks and 2 sectoral contract research companies. During this period, financial support from GSRT continued. The main aim of the HELP FORWARD network moved towards the promotion of technology transfer agreements (to and from Greece and to and from enterprises and research centres). Exploitation of EU programmes and initiatives formed an additional means of enhancing and developing collaborations.

HELP FORWARD (PRAXI in Greek), continued its operation under the hospices of the original partners and the original source of financing after the renewal of its contract in 1997 for a further 3-year-period, with a growing emphasis to supporting technology transfer projects. The need to support Greek enterprises in their effort to play a leading role in the Balkans led HELP FORWARD to contest and win EU funding to assist the creation and operation of equivalent centres in Bulgaria (1997) and Romania (1998).

In 1999, HELP-FORWARD supported the creation of an Innovation Relay Centre in Cyprus in the framework of an official agreement signed with the Cyprus Institute of Technology in July 1998.

With the commencement of the 5th Framework Programme, HELP-FORWARD was assigned by the General Secretariat for Research and Technology (GSRT) the task of the National Contact Point in Greece for the 5th European Framework Programme of Research and Technological Development.

In May 2000, HELP-FORWARD signed an official agreement with the Foundation for Research Promotion of Cyprus, for training the members of the Foundation for Research Promotion of Cyprus, training the Cypriot NCPs and to assist the planing and the running of the Foundations' activities with respect with the European projects.

The operation of HELP-FORWARD continues for the period 2000-2002 based partially to the funding of the INNOVATION Programme after H-F's succesfull application to the programme in 1999.

HELP FORWARD key dates

1991	Foundation of HELP FORWARD
1992	Financing from STRIDE-HELLAS program
1995	Financing from INNOVATION program
1997	Operation of HELP FORWARD as an Innovation Relay Centre
	Activities in the Balkan region
	Twinning agreement with the Bulgarian FEMIRC
1998	Twinning agreement with the Rumanian FEMIRC
1999	Support for the creation of IRC Cyprus
2000	Continuation of operation of HELP-FORWARD as an IRC

2. Structure

HELP FORWARD network is managed by a steering committee comprising of two members from FORTH, one of which being the president of the network, two members from FGI, one of which being the General Manager if FGI and one member from FING. This technical committee is responsible for planning the strategic development and direction of the network.

The scientific staff (core unit) of HELP FORWARD consists of the network coordinator and technology consultants based at the regional nodes. The scientific staff normally takes part in the steering committee meetings, puts forward new actions and initiatives for approval and presents the actions taken in order to materialise the network operation plan.

The action and operation plan is also submitted to the European Commission and is subjected to periodical inspection from the coordination unit of the Innovation Relay Centres. The national action and operation plan also includes initiatives on a national level.

In order to carry out the tasks concerning supporting enterprises and organisations in technology transfer projects, HELP FORWARD utilises, apart from its skilled and experienced consultants, the expertise from the collaborating research centres, scientists and members of industrial associations and enterprises (the human network of experts) who offer their services on a project basis.

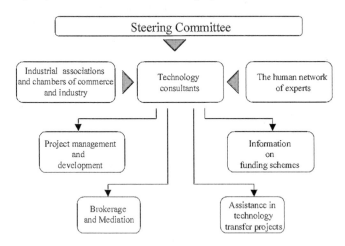

Structure and operation of HELP FORWARD

In order to achieve the aims set, HELP FORWARD depends on the effective operation of the European network it belongs in and of its national network operation. These networks are essential not only for covering effectively the whole of the Greek and European geographical region and a wide range of technological sectors but for the creative collaboration between the research and industrial world.

3. The Network

3.1 The Greek network

HELP FORWARD is not only a geographical network which covers the whole of Greece but is also a network of many different sectors of economical and research activities.

Consultants of the HELP FORWARD network are based in Athens, Thessaloniki, Patras, Heraklion and Larisa based in FGI, FING, FORTH – ICE/HT (Institute of Chemical Engineering and High Temperature Chemical Processes), FORTH – IESL (Institute of Electronic Structure and Laser) and the Science and Technology Park of Crete and ATI (Association of Thessalian Industries) respectively.

HELP FORWARD also has official collaboration agreements with the scientific and technological parks of Thessaloniki, Patras and Crete and several sectoral companies. Finally, HELP FORWARD utilises the expertise and experience of the researchers of the 7 institutes of FORTH.

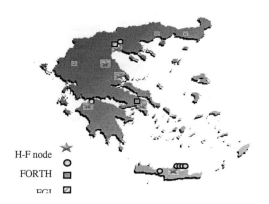

Figure 1: The HELP FORWARD network in Greece

3.2 The European network

HELP FORWARD network is part of a European network consisting of 68 Innovation Relay Centres (IRCs) which operate in 30 countries of Europe under the INNOVATION program. The aim of the IRCs is to provide support in transnational technology transfer projects, to promote innovation and dissemination of research results and to assist with technological development. The main beneficiaries of the IRC services are small medium sized enterprises (SMEs) and research centres. All IRCs are created from consortiums formed by associations, enterprises and research centres and often form geographical networks in their regions. HELP FORWARD collaborates with the other Innovation Relay Centres on a daily basis through Internet and written correspondence, common participation in meetings, conferences and technology transfer days. The consultants of HELP FORWARD collaborate with the consultants of the other IRCs on searching for suitable providers / users of technology or partners for common research and development to match the needs of Greek enterprises or research centres.

Figure 2: The HELP FORWARD network in Europe

3.3 South East European network

The need for supporting economical reforms in Central and Eastern Europe led to an initiative of the European Commission to back the creation of 10 Fellow Members to the Innovation relay Centres (FEMIRCs) in these countries. Their aim was to improve the competitiveness of local enterprises and to better utilise programs of the European Commission. Each FEMIRC was created and operates with the support and consultancy of an IRC through a formal agreement in the form of twinning.

The need to support Greek enterprises in their effort to play a leading role in the Balkans coupled with the recognition of the wider role which Greece could play in the sector of research, technological development and economical collaboration in the region led HELP FORWARD to contest and succeed in gaining financial support from the EU to help set up the FEMIRCs of Bulgaria and Romania. HELP FORWARD supports the operation of these two centres as well as undertakes common actions such as organising conferences and business meetings. The good working relationship HELP FORWARD has developed with these two centres, allow the immediate and effective service to whichever enterprise or organisation wishes to collaborate with Bulgarian or Romanian enterprises and organisations.

HELP FORWARD has also developed a close working relationship with the Cyprus Institute of Technology, which has created and operates a technology transfer centre with the consulting support of HELP FORWARD.

Figure 3: The HELP FORWARD South East European network.

4   Services

The central aim of the HELP FORWARD network is to support technology transfer from Greece to the European region and vice versa. HELP FORWARD consultants offer support services in each step of the long and often complicated chain of actions required for the completion and signing of a technology transfer agreement.

4.1 Promotion of Greek proposals for technological and business collaboration in Europe

➢ Audit and evaluation of technological and other relevant needs of enterprises. Search for and evaluation of technology and know-how in Greek enterprises or organisations for exporting to Europe.

Skilled consultants of HELP FORWARD visit enterprises and laboratories or have meetings with their members at the HELP FORWARD offices.

During the 1996 – 1999 period, 400 such meetings took place, 280 of which at the premises of the member enterprises, organisations and research laboratories of HELP FORWARD.

➢ Search for reliable providers, receivers or partners from Greece and abroad for technology transfer agreements or common technology or product development.

HELP FORWARD network utilises the other 68 members of the European Innovation Centre network (each being part of a network in their country) and also other contacts made with European organisations aiming to find the most appropriate collaborator. Furthermore, HELP FORWARD takes part in technology transfer days, organised by HELP FORWARD itself or by other centres abroad whose consultants are in contact with potential partners for Greek organisations.

During the past four - year period, 120 offers, technology or partner searches have been promoted from Greece in Europe and 400 replies and further contacts have been noted. HELP FORWARD has organised 5 technology transfer days[*] and consultants of HELP FORWARD have taken part in 11 technology transfer days abroad representing the interests of a total of 200 Greek enterprises and laboratories.

4.2 Promotion of European proposals for technological and business collaboration in Greece

➢ Promoting offers / technology requests or partner searches for undertaking

---

[*] Technology transfer days is one of the most effective tools for promoting transnational collaboration amongst the Innovation Relay Centre (IRC) network. These events are organised by one or more IRCs and normally take place during a large fair or a technical conference. They give the opportunity to small medium sized enterprises and laboratories who lack financial means, time or available staff to promote their new technologies / innovative products or to search for solutions to major technological problems through utilisation of the services and the mediation of their local Innovation Relay Centre. With the cost of participation shared between interested enterprises and organisations, the skilled consultants of IRCs represent their national portfolio in selected sectors.

common research and development or business actions originating from Europe to Greek enterprises and organisations.

These searches for collaboration come from the other 68 IRCs, from community databases, from fairs or conferences or from direct contacts of HELP FORWARD consultants with enterprises and organisations abroad. HELP FORWARD consultants first evaluate the searches and the ones considered to be of interest for Greece are translated and channeled through to Greek enterprises and organisations. This is done either by targeting specific enterprises and organisations or through the monthly newsletter of HELP FORWARD, the web pages of HELP FORWARD on the internet and in specialised publications in newspapers.

During the 9 years of its operation, HELP FORWARD has evaluated approximately 3500 searches for collaboration from abroad from which 1500 were partner searches for undertaking common research and development projects, 1000 were offers and 400 were requests of technology and 600 were proposals for common business action. A total of 1800 of these searches were channeled through in Greece and interest was expressed for almost all. Two hundreds of these led to further contacts or / and negotiations.

4.3 Consultancy support on searching for financial means to fund technology transfer or developments

➤ Financing possibilities from national and Community funds.

HELP FORWARD consultants offer consultancy to its members enterprises and organisations on available financial packages most suited to their situation.

In the 9 years of operation of HELP FORWARD, 1200 members of the network took advantage of this service.

➤ Consultancy support on drafting detailed proposals requesting funds for technology transfer or technology development projects.

The experience and special skills of HELP FORWARD consultants is precious to all enterprises and organisations not familiar with certain subjects such as matching their objectives to the call for proposals for a program, the advised budget for a project, eligible costs and other.

Since 1991, 1000 enterprises and organisations submitted proposals for funding in Greece or the European Union with a major or minor contribution of HELP FORWARD from which 350 were approved for funding.

4.4 Consultancy support in technology transfer agreement negotiations

Active support to Greek enterprises or organisations during negotiations with potential technology suppliers or receivers through to the final signing of the contract. Help to members seeking legal advisers or consulting firms specialising in subjects such as market research, business plans or legal support.

Some of the services the Innovation Relay Centre HELP FORWARD offers to its members during negotiations are consulting services, mediation in communicating with foreign partners where needed, checking the credibility of potential partners, co-funding travel in cases of advanced negotiations, technical translations etc.

During the period of 1996 – 1999, 40 enterprises and organisations took advantage of the complete package of services offered by HELP FORWARD at this level. Eight of these enterprises / organisations successfully completed a technology transfer agreement whereas a further 10 are in progress.

5 Members

Member - subscribers of HELP FORWARD network are enterprises, organisations and research laboratories. The prerequisite for subscription of an enterprise or organisation in the archive of members of HELP FORWARD is the submission of a completed form containing features and facts of the enterprise / organisation, which best describe their activities and research interests.

Members of the HELP FORWARD network receive the monthly newsletter and are given the opportunity to utilise the services of HELP FORWARD such as help in technology transfer matters, promotion of technology offers and requests and search for research and business partners.

The recipients of the newsletter are close to being 2300 with members of enterprises making up 70%, members of research and academic organisations 23% and members of public sector companies and organisations making up the rest of 7% of the subscribers to HELP FORWARD.

The active members, e.g. members who have used the services of HELP FORWARD at least once during 1999 are more than 400.

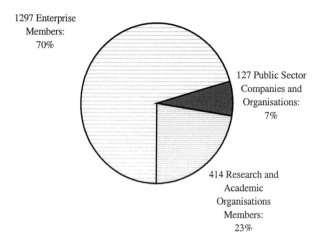

Figure 4: HELP FORWARD newsletter Reader distribution (1998 data)

As far as the activities of the members of the HELP FORWARD network, 42% are in the manufacturing sector, 24% are in the service sector, 2% in trade whereas 32% have mixed activities (manufacturing – trade). This is clearly shown in figure 5.

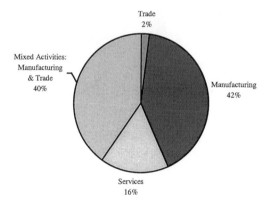

Figure 5: Distribution of enterprise member by activity (1998 data)

The distribution of the 1200 enterprise members of HELP FORWARD according to their sector of activity is shown in figure 6.

Of the 1200 enterprise - members of HELP FORWARD, the strongest participation is from the food and drinks – agricultural products sector (food and drinks, 11% – agricultural products, 5%), the consulting firms (18%), the information technology and telecommunication sectors (a combined 13%) and the plastics and chemicals sectors (7%).

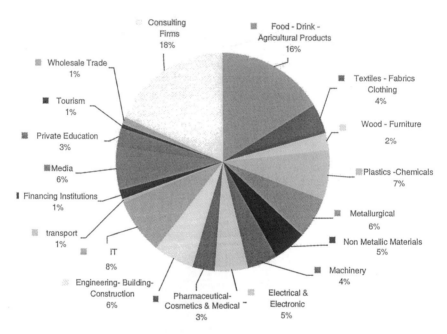

Figure 6: Distribution of enterprise – members by sector (1998 data).

The geographical distribution of the enterprise and organisation members of the HELP FORWARD network in Greece is shown schematically in figure 7. Fifty percent of the members of HELP FORWARD are located in the Attica and Thessaloniki prefectures. Given the population distribution in Greece, it is worth noting the participation of enterprises and organisations from Thessaly, Crete, Peloponnese, Sterea Ellas and Epirus.

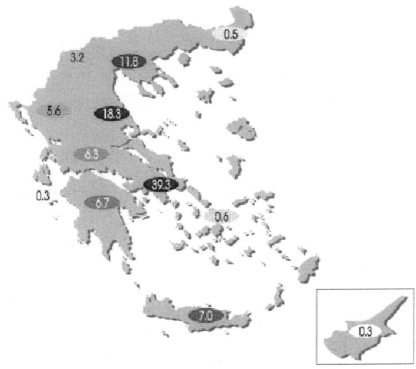

Figure 7: Geographical distribution in Greece of enterprise – members (1998 data)

6 Presence

HELP FORWARD network utilises the most efficient means of communicating with its members as well as with the wider business and research world. The most important ones are briefly described below.

6.1 Monthly newsletter

This is edited 11 times a year and is sent to the 2300 members - subscribers of HELP FORWARD. It contains information on new calls for proposals for programme participation, detailed calendar of proposal submission deadlines, short description of technology offers, requests and partner searches from the rest of Europe (in Greek), information on conferences and technology transfer days and a detailed presentation of a chosen technology offer under the heading "Technology of the month".

In its present form, the HELP FORWARD network newsletter circulates non-stop since October 1996 with continuous improvements and additions of new columns.

6.2 Permanent collaboration with newspapers and magazines

HELP FORWARD collaborates with the national newspaper "Kathimerini" ("Daily" - readership of 25,000) in the weekly publication of the deadline calendar for proposal submissions. It also publishes extra information reports on new calls for proposals. Selected searches for technological and business collaborations are also presented in the largest county newspaper of the Larisa region (readership of 23,000), "Eleftheria" (Freedom). "Technology of the month" is extensively presented with full colour photographs in the monthly engineering magazine "Techniki Epitheorisi" (Engineering Review) with a readership of 25,000. Further to these permanent collaborations, reports and announcements of HELP FORWARD are often hosted in local and national newspapers and magazines.

6.3 Internet web site (http://www.help-forward.gr)

HELP FORWARD network maintains a web site on the network since 1996. It contains information concerning programs and searches for collaboration and additionally it offers the following possibilities to the visitor / user:

- On-line subscription to the HELP FORWARD network member database
- On-line submission of searches for collaboration in Europe.
- Presentation of Greek searches for collaboration (in English), aiming at promoting them to the visitors of the web site from abroad.
- Useful links to European databases.
- Urgent announcements for conference and technology transfer days participation or for extra funding possibilities.

The material contained within the web site is updated daily. Since its creation in 1996, 25000 visits to the web site were noted from Greece and abroad (20% from abroad).

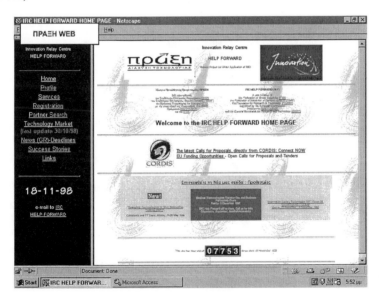

Printout of the web site of HELP FORWARD

6.4 Organising information days & taking part in conferences and fairs

Since 1991, HELP FORWARD has organised about 130 information days (figure 8) to brief on calls for proposals of national and Community programs around Greece. It often takes part (with own stand) in conferences and sectoral fairs (about twice a year) whereas its consultants visit the most important fairs around the country.

Participation in such events allows the direct contact of HELP FORWARD with the business (mainly) and research world of Greece and informs them of its activities and its role in achieving collaboration agreements and technology searches abroad

Figure 8: Information days & events of HELP FORWARD in Greece (1991-98).

SUPPORTING EUROPEAN RESEARCH ALLIANCES IN MICROELECTRONICS THE MINATECH PROJECT

D. A. TSAMTSAKIS

Help-Forward Network Morihovou Sq. 1,
546 25 Thessaloniki, Greece
E-mail: dimosthenis@help-forward.gr

Minatech is an Economic and Technological Intelligence (ETI) project of the Innovation and SME, EU programme. It aims at producing updated, both national and European (with comparison with US and Japan), state of the art and medium term evolution trends of micro and nano technologies and of applications and markets of these technologies in 4 selected areas: information technology and communications (ITC); health and biotech; automotive and instrumentation. Moreover, it aims at stimulating, selected SMEs and more generally all SMEs, through dissemination seminars, workshop, technological audits, technical and marketing information on web sites, on newsletters, on technical journals, to participate to EU initiatives for R&D (Exploratory Awards, CRAFT and RTD projects).

1 Introduction

The evaluation of the peculiar features of the market open to micro and nano technologies is part of the objectives of this project. It appears however that the fraction of this market accessible to SME's consists in several niches with a risk of fast obsolescence. From the point of view of the European dimension, each of these niches in the whole EU territory may be the target segment for one or more companies which have to compete in short time against aggressive foreign (USA, Japan,...) companies. If this does not happen in the next years, it will leave to European companies the role of followers for a long time in markets which be constantly dominated by the leaders.
More than that, micro and nano devices are so important-and more will be in the future- for the products in which they will be embedded, that their timely disposal for users/applications adapted design and production will be absolutely strategic, like microchips were for computers. This explains why it is important to address the realm of these technologies with a European strategy: the size of these niches in each country is too small for a successful growth of a high-tech company, even if stimulated by some national initiative.

The European Commission's aim is to maintain Europe's competitiveness in the global market place, improve the quality of life of the European citizen and address unemployment. This is done by encouraging and supporting R&D activities. Under a four-year initiative called 'Framework Programme 5', a budget of 14.96 billion euro has been allocated to fund research and development projects in a variety of topics ranging from biological sciences to manufacturing processes, IT solutions and environmental concerns.

In order to facilitate the uptake of European funding there is a network of National Contact Points across Europe and a variety of intermediary organisations. These two types of organisations can also apply for support to promote the research funding opportunities to a certain sector.

MINATECH is such an action and aims to highlight these opportunities to Small and Medium Sized Enterprises (SMEs) in the Micro and Nano technologies sector but is taking it a step further with initial technological investigations. These investigations, via literature surveys and interviews of experts, involve;

- Establishing the state of the art and medium term evolution trends of micro and nano technologies
- Establishing the applications and markets of micro and nano technologies in four selected areas:
    - Information Technology and Communications
    - Health and Biotech
    - Automotive and Instrumentation

A transnational approach is being taken with a total of ten partners representing eight countries (Italy, Czech Republic, Denmark, Germany, Greece, The Netherlands, Spain and UK). The results of the investigation will therefore produce individual national reports and combined European reports. The information generated will be used in the technological audits of appropriate SMEs. The overall aim of this 33-month project is to generate European funded micro and nano projects.

2 Activities Description

The main project's activities are:

- Evaluation of European state of art and evolution trends of micro and nano technologies, also versus US and Japan (1 European and 8 national reports):
    - survey of available patents and technical literature (country road map included);
    - direct interviews with leading scientists and researchers of the 8 countries involved in the project, taking in to account participants to past and present EC (and Eureka) initiatives (Nexus, Europractice, Eurimus, etc.).
- Evaluation of market trends in the fields of micro and nano technologies applications in ITC, health and biotech, automotive, instrumentation areas (1 European and 8 national reports):
    - analysis of available literature and surveys;

- direct interviews with experts of organisations that have relevant know how (manufacturers, users, industrial associations, etc.).
- Assessment of SMEs' needs (1 European and 8 national reports), through audits to SMEs selected among actual and possible producers and end users of micro and nano devices. The selection will be made by screening the answers to a questionnaire sent to an expanded mailing list. The audits will be carried out by experts who will receive a common targeted training.
- Dissemination of information about the opportunities offered to SMEs by the 5th Framework Programme (5FP). This will be done by seminars, dedicated workshops, notice in web sites, in newsletters (of the project partners) and in technical journals.
- Gathering Expression of Interest (EOI) of SMEs by each partner and their immediate cross dissemination among the partners themselves in order to establish transnational groups aiming at various European initiatives (Exploratory Awards, CRAFT and RTD projects). In case of EOIs not matched inside the partner group, they will be diffused through CORDIS.
- Setting up of at least one Brokerage Event, directed specifically to the four application sectors (information technology and communications, health and biotech, automotive and instrumentation). The aim is again that of assembling transnational groups even if EOIs are not previously formalised.
- Assistance to SMEs in contacting possible partners (also through Cordis network) to carry out research projects, and in pre-evaluating and preparing project proposals.
- As technology transfer possibilities will emerge, at least one Technology Transfer Day will be organised.

3 Assistance to SMEs

According to the aim of Minatech to stimulate, and assist, SMEs to participate in 5FP initiatives for R&D in the field of micro and nano technologies and their applications, the project is organised to give SMEs (in complete confidentiality):
- information on micro and nano technologies and their applications/markets;
- information on EC financing of R&D projects in the 5FP;
- technological evaluation (no cost audits) through visits by trained experts;
- assistance in partners finding and in R&D projects definition and formulation;
- specific workshops and information dissemination.

SMEs interested in knowing more about micro and nano technologies and their applications, about 5FP and its financing opportunities for R&D projects and about Minatech and its assistance services, can contact the Minatech partner of their respective country, so entering in the Minatech assistance network (8 countries, 10 partners).

Each Minatech partner is able to give SMEs full assistance and support in his country.

4 The Micro & Nano Technologies And The Components

The following micro & nano technologies and components will be considered:
1. Technologies
- Microelectronics (not standard)
- Microoptics
- Micromechanics
- Microfluidics
- Micro / Nano technologies in chemical field
- Micro / Nano technologies in materials field
- Micro / Nano technologies in medical field
- Micro / Nano technologies in biotech field
- Designing technologies (for micro / nano)
- Assembling, packaging technologies (for micro / nano)
- Measuring technologies (for micro / nano)

2. Components, Devices, etc.
- Microsensors
- Microactuators
- Micromachines
- Microsystems
- Micro electro-mechanical systems

5 The Micro & Nano Applications And Services

The following micro and nano applications and services will be considered:
1. Application areas
Information & Telecommunications (not standard)
- Peripheral subsystems of information systems (displays, smart cards, special recognition subsystems, etc.);
- Electro-optics for telecomm systems;
- Cellular phone and video-peripheral telephony;

Automotive
- Cars (navigation systems; logistic systems; engine control; tyres control; petrol consumption control; exhaust control; passengers safety and comfort, etc.)
- Heavy transportation means for goods and tourism (navigation systems; logistic systems; engine control; tyre control; petrol consumption control; load control; passenger safety and comfort, etc.)

Health and biotech
- Implantable systems;
- Diagnostic systems (autodiagnosis; transportable systems; non-invasive diagnosis, etc.);
- Non-invasive surgery;
- Pharmaceutical applications (skin disposable, etc.)
- Biotechnological applications (chip lab, etc.)
- Bio-functional devices;

Instrumentation
- Ambient conditions control;
- Industrial automation;
- House control and domotics;
- Agro-industrial production control (safety, products ageing, etc.);
- Metrology.

2. Services
- Micro / Nano design and simulation
- Micro / Nano foundries
- Micro / Nano machining, assembling and packaging
- Micro / Nano measuring and characterising

6 Participating Countries and Organisations

GREECE	IRC HELP FORWARD
UNITED KINGDOM	BETA TECHNOLOGY
GERMANY	VDI/VDE
DENMARK	EUROCENTER
SPAIN	FUNDACION TEKNIKER
ITALY	AIRI,CNR,APRE
THE NETHERLANDS	SENTER/EG-LIAISON
CZECH REPUBLIC	TECHNOLOGY CENTRE

SIMULATIONS OF MOLECULAR ELECTRONICS

SOKRATES T. PANTELIDES
Department of Physics and Astronomy, Nashville, TN 37235
Oak Ridge National Laboratory, Oak Ridge, TN 37861
E-mail: pantelides@vanderbilt.edu

MASSIMILIANO DI VENTRA
Department of Physics and Astronomy, Nashville, TN 37235
Department of Physics, Virginia Polytechnic Institute and State University,
Blacksburg, VA24061

NORTON D. LANG
IBM T. J. Watson Research Center, Yorktown Heights, NY 10598

The paper gives an overview of recent work by the authors in first-principles, parameter-free calculations of electronic transport in molecules in the context of experimental measurements of current-voltage (I-V) characteristics of several molecules by Reed et al. The results show that the shape of I-V characteristics is determined by the electronic structure of the molecule in the presence of the external voltage whereas the absolute magnitude of the current is determined by the chemistry of individual atoms at the contacts. A three-terminal device has been simulated, showing gain. Finally, recent data that show large negative differential resistance and a peak that shifts substantially as a function of temperature have been accounted for in terms of rotations of ligands attached to the main molecule, a phenomenon that is not present in semiconductor nanostructures.

1. Introduction

Silicon-based microelectronics is reaching the level of miniaturization where quantum phenomena such as tunneling cannot be avoided and the control of doping in ultrasmall regions becomes problematical. Though it is likely that silicon-based technology will simply move to a different paradigm and continue taking advantage of the existing vast infrastructure and manufacturing capabilities, novel and alternative approaches may give new insights and ultimately may usher a new era in nanoelectronics. Molecules as individual active devices are obvious candidates for the ultimate ultrasmall components in nanoelectronics. Though the idea has been around for more than two decades,[1] only recently measurements of current-voltage characteristics of individual molecules have been feasible.

Methods for the calculation of current in small structures placed between two metal electrodes have been developed over the years, but actual implementations have been scarce. For molecules, semiempirical methods have been used to study the dependence of current on various aspects of the problem,[2] but quantitative predictions for direct comparison with data are not possible because values of parameters under current conditions cannot be determined independently.

In the 1980's, one of us (NDL) developed a practical method to calculate transport in the context of imaging atoms with scanning tunneling microscopy.[3] The method has all the ingredients needed to compute current-voltage (I-V) characteristics of single molecules. We recently developed a suitable Hellmann-Feynman theorem for the calculation of current-induced forces on atoms that allows us to study the effect of current on relaxations and ultimately the breakdown of molecules.[4] With these tools, we have carried out extensive studies of transport in molecules whose core is a single benzene ring. Such molecules have been synthesized and measured by Reed and coworkers.[5,6] In this paper we summarize the most important results of the recent work. The method of calculation and more details of the results can be found in the original papers.[7-9]

2. Transport in a single Benzene Ring

Fig. 1 shows schematics of a single benzene ring bridging the gap between macroscopic gold electrodes. A sulfur atom at each end joins the benzene ring to the electrodes. The experimental I-V characteristic is shown in the top panel. The middle panel shows the theoretical results. We will discuss the third panel shortly.

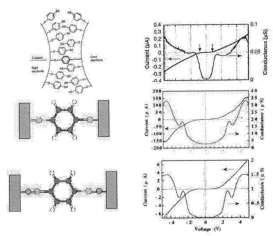

Figure 1. Schematics of benzene ring configurations between electrodes and I-V characteristics. Top row: Experiment (Ref. 5). Middle row: Theory with S atoms only between benzene ring and macroscopic electrodes (Ref. 7). Lower row: Theory with Au atoms inserted between the S atoms and the macroscopic electrodes (Ref. 7).

It is clear from the figure that the theory reproduces the shape of the I-V curve quite well, but the absolute magnitude of the current is off by more than two orders of magnitude. We address each of these issues separately.

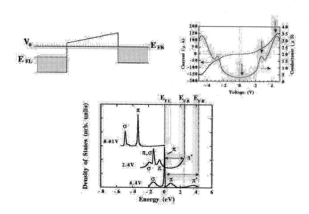

Figure 2. Schematic of the setup showing the left and right quasi Fermi levels; the I-V characteristic as in Fig. 1, and the densities of states discussed in the text.

In Fig. 2 we show the density of states of the molecule for three different voltages: 0.01 V, 2.4 V and 4.4 V and mark out the energy window between the left and right quasi Fermi levels. States within this window contribute to transport. We see that there is virtually no density of states in the small window at small voltages, in agreement with the slow initial rise of the I-V curve. At 2.4 V, the π^* states of the molecule enter the transport window and give rise to the first peak in the spectrum. At 4.4 V, the π states of the molecule enter the window while the π^* continue to participate, giving rise to the second peak in the spectrum. The peak at 2.4 V is somewhat more pronounced in the theoretical curve. The observed smoothing is likely to be caused by interactions between the electrons and vibrational modes.

In order to explore the mechanism that controls the absolute magnitude of the current we performed calculations by inserting an extra gold atom between the sulfur atom and the macroscopic electrode at each end of the molecule as shown in the lower part of Fig. 1. There was a dramatic decrease in the current, bringing its value much closer to the experimental value (lower panel in Fig. 1). The decrease is attributed to the fact that gold atoms have only one s electron available for transport and s electrons do not couple with the π electrons of the molecule. The gold atoms act as a quantum mechanical constriction. To test the idea we performed calculations by replacing the gold atoms with aluminum atoms. The latter have p electrons that should couple well with the π electrons of the molecule. Indeed, the current jumped to its initial value. An additional test was carried out with three gold atoms instead of a single gold atom. The current was again at its full value because the three s orbitals on the three gold atoms can form the appropriate linear combinations to produce sufficient coupling.

It is clear from the above that molecules determine the shape of the I-V characteristic, but the nature of individual atoms at the molecule-electrode contact determines the absolute magnitude of the current. The results illustrate the power of theory to contribute to device design , especially "contact engineering".

3. A three-terminal device

As an example of a three-terminal device we considered the same benzene ring as above but put a third "terminal" in the form of a capacitive gate: an external electric field across the molecule produces polarization that affects the current through the molecule. Figure 3 shows that at a very small (0.01 V) source-drain voltage, the current is a strong function of the gate voltage. In particular, we demonstrate substantial gain at the resonant peak. Further discussion of the results can be found in the original paper. Three terminal molecular devices have not yet been realized.

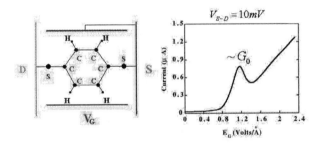

Figure 3. Schematic of the three-terminal device discussed in the text and the theoretical I-V characteristic as a function of the gate voltage for a small source-drain voltage.

4. A benzene ring with a ligand

Chen et al.[6] recently reported I-V characteristics of molecules consisting of chains of three benzene rings with ligands attached at various places. The most interesting result is large negative differential resistance evinced by a relatively sharp spike in the I-V characteristic. The spike is found to broaden and shift on the voltage axis with increasing temperature. The shift, by about 1 V, is very unusual. In semiconductor nanostructures resonant peaks have been found to broaden (by standard electron-phonon interactions) but they never shift appreciably.

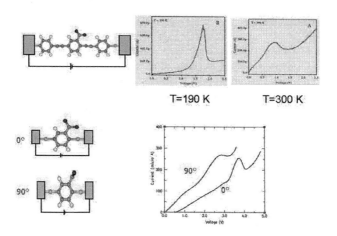

Figure 4. Experimental data by Chen et al. (Ref. 6) at two temperatures and theoretical I-V curves for two different orientations of the ligand. See text for more details.

Calculations for three-benzene-ring molecules are not practical, but we pursued the question by calculations for a single benzene ring with an NO_2 ligand replacing one of the hydrogen atoms. We found that the energy levels of the ligand move substantially with increasing voltage and push the π levels into the active window. Thus the main peak in the current arises primarily from π electrons instead of π^* electrons. We then explored the effect of rotating the ligand. We found that a rotation by 90° shifts the peak to lower voltage by almost 1 V, in agreement with the observations (Fig. 4). The interpretation is that higher temperatures excite the rotational modes of the ligand. Calculations of the total energy of the molecule as a function of ligand rotation show that the effective rotational quantum of energy is only 3 meV. Thus, even at the relatively low temperatures of the experiment, a large number of quanta are excited, making the ligand effectively a classical rotor that spends most of its time at the extrema of the amplitude. Ligand rotation is of course a unique phenomenon of the molecular world, explaining why large voltage shifts are not observed in semiconductor nanostructures.

5. Conclusions

The calculations summarized above show that theory has now advanced to the point where quantitative predictions can be made about transport in single molecules. Such calculations are expected to play a major role in the evolution of molecular electronics, the way that simple drift-diffusion calculations of current in semiconductor structures have played in the evolution of silicon-based microelectronics.

6. Acknowledgement

This work was supported in part by a grant from DARPA/ONR, by the National Science Foundation and by the William A. and Nancy F. McMinn Endowment at Vanderbilt University.

References

1. A. Aviram and M. A. Ratner, *"Molecular Rectifiers"*, Chem. Phys. Lett. 29, 277 (1974). See also *Molecular Electronics*, edited by A. Aviram and M. A. Ratner (New York Academy of Sciences, New York, 1998).
2. 2. S. Datta, W. Tian, S. Hong, R. Reifenberger, J. I. Henderson, and C. P. Kubiak, Phys. Rev. Lett. 79, 2530 (1997); S. N. Yaliraki, A. E. Roitberg, C. Gonzales, V. Mujica, and M. A. Ratner, J. Chem. Phys. 111, 6997 (1999); J. K. Gimzewski and C. Joachim, Science 283, 1683 (1999).
3. See N. D. Lang, Phys. Rev. B 52, 5335 (1995); ibid. 49, 2067 (1994).
4. M. Di Ventra and S. T. Pantelides, Phys. Rev. B 61, 16207 (2000).
5. M. A. Reed, C. Zhou, C. J. Muller, T. P. Burgin, and J. M. Tour, Science 278, 252 (1997).
6. J. Chen, M. A. Reed, A. M. Rawlett, and J. M. Tour, Science 286, 1550 (1999).
7. M. Di Ventra, S. T. Pantelides, and N. D. Lang, Phys. Rev. Lett. 84, 979 (2000).
8. M. Di Ventra, S. T. Pantelides, and N. D. Lang, Appl. Phys. Lett. 76, 3448 (2000).
9. M. Di Ventra, S.-G. Kim, S. T Pantelides and N. D. Lang, Phys. Rev. Lett. 86, 288 (2001).

AUTHOR INDEX

Amimer K., 171,197
Andreopoulos A., 271
Andriotis A.N., 25
Androulidaki M., 171,197, 201
Antonopoulos J.G., 183
Aperathitis E., 179,187,193
Apostolopoulos G., 143
Argitis P., 103,127
Arnaud d'Avitaya F., 33,73,77
Avaritsiotis J.N., 238,259,275,333,341

Balestra F., 153,213
Bassani F., 73,77
Beltsios K., 29, 69, 267,271
Berashevich J.A., 33
Birbas A.N., 209
Blionas S., 293,309
Blondy P., 234
Borisenko V.E., 33,41
Boudouvis A.G., 139
Boukos N., 143
Brida S., 254
Brini J., 167
Butturi M.A., 254

Calamiotou M., 171
Carotta M.C., 254
Cashmore J., 127
Cefalas A.C., 127
Cengher D., 179,193
Cengher M., 135,179,193

Chatzandroulis S.,250
Chatzigiannaki A., 217
Christides C., 13
Cimalla V., 135
Claverie A., 115
Clayes C.L., 91,355
Constandinidis G., 135,197,201
Constantoudis V., 123
Cornet A.,197

Danilyuk A.L., 33
Dasigenis M.M., 297
Davazoglou D., 107,131
De Moor P., 246
Deligeorgis G., 179,193,234
Diakoumakos C.D., 103,127
Dimakis E., 201
Dimitrakis P., 213,279
Dimitriadis C.A. ,167
Dimitropoulos P.D., 238
Dimotikali D., 103
Dimoulas A., 175
Di Ventra M., 380
Dontas I., 111
Douvas A., 103
Drakakis G., 135
Dre C., 293,309
Drosos C., 293, 309

Eickhoff M., 227
Evmorfopoulos N.E., 341

Farmakis F.V., 167
Fikos G., 217,221
Foukaraki V., 187
Fragoulis D.K., 259
Froudakis G., 25

Garoufalis C.S., 81,85
Gastal M., 246
Gautier J., 69
Georgakilas A., 135,171,179,193,197,201
Georgoulas N., 119
Giacomozzi F., 254
Giles L.F., 115
Girginoudi D., 119
Gleizes A., 131
Glezos N., 147
Gogolides E., 123,127,139
Goustouridis D., 250
Grousson R., 21
Grunewald P., 127
Guarnieri V., 254

Halkias G., 179
Halkias G., 305
Haralabidis N., 305,329
Hatzopoulos A.A., 349
Hatzopoulos Z., 179,187,193,197
Hionis G., 45
Hoang A., 3
Holliger P., 69

Ioannou – Sougleridis V., 73,77
Iordanescu S., 234

Jaguiro P.V., 41
Jomaah J., 213
Jourdan F., 69

Kakabakos S. E., 103,263
Kalivas G., 309
Kaltsas G., 242,283
Kamarinos G., 167
Kanellopoulos N., 267,271
Kapetanakis E., 29,69
Karakostas Th., 183,201
Karavolas V.C., 49
Katsafouros S.G., 305
Katsaros F., 271
Katsikini M., 183
Kavadias S., 246
Kavouras P., 183
Kayambaki M., 187
Kehagias Th., 201
Kennou S., 111
Kholod A.N., 33
Kitsos P., 313
Kokkorakis G.C., 17
Kokkoris G., 139
Koliopoulou S., 250
Kollias A.T., 333
Komninou Ph., 183,201
Konstantinidis G., 234
Kornilios N., 234
Kostopoulos A., 135,201
Kotsani M., 119
Koufopavlou O., 313,325
Kouphoyannidis A., 345

Kouvatsos D.N., 37,65,107
Kyranas A., 317
Kyriakis-Bitzaros E., 305

Ladas S., 111
Lagadas M., 234
Lang N.D., 380
Lazarouk S.K., 41
Leshok A.A., 41

Maes H.E., 91
Margesin B., 254
Martinelli G., 254
Mastichiadis C., 103, 263
Menon M., 25
Mikroulis S., 135,201
Misiakos K., 103,263
Mitsinakis A., 119
Mittler F.,3
Modinos A., 17
Muller A., 234

Naskas N., 321
Nassiopoulou A.G., 7,61,65,73,77, 242,283, 287,359
Nastos N., 301
Normand P., 29,69

Omri M., 115
Ouisse T., 37,65,77

Pagonis D., 283
Paloura E.C., 183
Panagiotopoulos S., 345
Panayotatos P., 187

Pantelides S.T., 380
Papadimitriou D., 287
Papaefthimiou V., 111
Papaioannou G.J., 213,279
Papananos Y., 301,317,321
Papandreopoulos P., 143
Patsis G.P., 123
Pavelescu E.M.,201
Pavlidis V., 337
Pecz B., 171
Peiro F., 197
Petrini I., 234
Photopoulos P., 37
Pignatel G., 254
Pilatos G., 271
Pisliakov A.V., 254
Potamianos K., 293

Robogiannakis P., 305
Rosilio C., 3
Ruzinsky M., 187

Saly V., 187
Sarantopoulou E., 127
Sarrabayrouse G., 217
Sfendourakis M., 179
Siokou A., 111
Sirotny P., 187
Siskos S., 217,221
Skarlatos D., 115
Sklavos N., 325
Soncini G., 254
Soterakou E., 267
Soudris D.J., 297,309,337
Stamoulis G.I., 341

Stefancich M., 254
Stoemenos J., 29, 115,167,227

Tassis P., 287
Tegou E., 128
Thanailakis A.T., 119,297,337
Theonas V., 279
Toth L., 171
Travlos A., 69,143,175
Triberis G.P., 21,45,49,53
Trohidou K.N., 147
Tsagaraki K., 171,197,201
Tsakalos V., 360
Tsakas E.F., 209
Tsamis C., 115,205,283,287
Tsamis E.D., 275,333
Tsamtsakis D., 375
Tsangaris G.,267
Tsaousidou M., 53
Tserepi A., 127
Tsetseri, M. 21
Tsoukalas D., 29,69,115,205,250
Tsoura L., 287

Vamvakas V., 107
Van den Berg J., 29
Van Hoof C., 246
Vasilache D., 234
Vasiliev A.A., 254
Vasilopoulou S.K., 297
Velessiotis D., 147
Vellianitis G., 175
Vidal S., 131
Vincenzi D., 254
Vinet F., 3
Voliotis V., 21

Xanthakis J.P., 17,85

Zappe S., 227
Zdetsis A.D., 81,85
Zen M., 254
Zervakis E., 329
Zervos M., 135,197
Zianni X., 61,147
Zuburtikudis I., 57